University of Michigan Publications

CONTRIBUTIONS FROM THE MUSEUM OF PALEONTOLOGY UNIVERSITY OF MICHIGAN

(*Continuation of Contributions from the Museum of Geology*)

VOLUME III

CONTAINING PAPERS PUBLISHED BETWEEN NOVEMBER 10, 1928, AND APRIL 9, 1932

CONTRIBUTIONS

FROM THE

MUSEUM OF PALEONTOLOGY
UNIVERSITY OF MICHIGAN

EDITED BY

EUGENE S. McCARTNEY

ANN ARBOR

UNIVERSITY OF MICHIGAN PRESS

1932

The occasional papers comprising this volume were collected and bound in December, 1932

Printed and bound by CPI Group (UK) Ltd, Croydon, CR0 4YY

CONTENTS

With dates of publication of the several papers

ILLUSTRATIONS

PLATES

FIGURES IN THE TEXT

MAP

Vol. III, No. 1, pp. 1–14 (1 pl.) November 10, 1928

INDICATIONS OF A COTYLOSAUR AND OF A NEW FORM OF FISH FROM THE TRIASSIC BEDS OF TEXAS, WITH REMARKS ON THE SHINARUMP CONGLOMERATE

By E. C. Case

THE expedition from the Museum of Paleontology of the University of Michigan to the Upper Triassic, Dockum, beds of western Texas in 1927 recovered the anterior portion of the lower jaw of a small Cotylosaur belonging in the Family *Procolophonidae,* which includes the genera *Telerpeton, Koïloskiosaurus, Procolophon, Sclerosaurus* and, perhaps, *Thelegnathus.*

This specimen, the first member of the *Cotylosauria* known to occur in the Triassic of North America, was discovered by Mr. W. H. Buettner near Walker's Tank in Crosby County, Texas. It bears the catalogue number 2338 and is the type of the genus and species, *Trilophosaurus buettneri.*

It is possible that Gilmore's *Hypsognathus fenneri,* recently described,[1] is the complement of this member of the Triassic fauna on the eastern side of the continent.

The characters of the specimen are best appreciated from a study of the illustrations, Plate I, Figures 1–5. The teeth are complete and perfect but the bone has suffered somewhat from solution. The anterior end of the bone is broken away, but only a small part of the extremity is lost, which did not, in all probability, carry a tooth. The first tooth is represented by the base only; the tooth was small, cylindrical, and evidently conical.

[1] Gilmore, Chas. G., "A New Fossil Reptile from the Triassic of New Jersey," *Proc. U. S. National Museum,* Vol. 73, Art. 7, 1928.

The second tooth is also represented by the base only, but this, with the remaining teeth, was decidedly different from the anterior one; they are all very wide as compared with their length and set transversely in the jaw as in *Diadectes* and *Telerpeton*. The third, fourth and fifth teeth are complete and perfect. The sixth tooth is represented by the base only and posterior to this there is evidence of a socket for a seventh tooth. All the teeth posterior to the first increase regularly in size from the second to the seventh.

The three perfect teeth are so similar that a description of one will suffice for all. The short base rises sharply from the surface of the bone without change of size to the base of the enameled crown. The crown swells out sharply but slightly and then the anterior and posterior faces incline inwards and upwards to meet in the sharp cutting edge at the top. Each face is marked by two shallow grooves which divide the crown into three parts and the cutting edge into three cusps. The cutting edge is inclined slightly inward and downward; the outer cusp is somewhat higher than the median and inner ones. The bases of the second and sixth teeth and small perforations of the bases of the third and fifth show that there was a large pulp cavity.

The method of attachment of the teeth to the jaw is most puzzling. Viewed from above the teeth are apparently acrodont in their attachment; the fibres of bone passing from the base of the tooth to the adjacent bone are clearly seen under a low magnification, but the bases of the second and sixth teeth, showing the continuation of the root into the bone, and the evident socket for the root of a seventh tooth indicate that the method of attachment was thecodont. The peculiarity of the method of attachment led to an examination of the literature of the members of the *Procolophonidae* and it was discovered that most of the writers upon these forms had been puzzled by the same thing and that the shape of the crowns of the maxillary teeth was for long misunderstood because in most of the specimens the teeth are presented only in lateral profile.

In 1867 Huxley wrote of *Telerpeton elginense:* "I have carefully examined into the mode of implantation of these (maxillary) teeth, and I have been unable to satisfy myself that they are lodged in true alveoli. They appear to be anchylosed to the edges of the jaw bone, as in many modern lizards with an 'acrodont' dentition."

In 1904 Boulenger wrote on the same form correcting some of Huxley's observations. He said: ". . . the exaggerated length of the anterior mandibular tooth is owing to the roots being made to project beyond the bone, the thecodont nature of the Reptile not having been recognized. As to the side teeth, Huxley only described them from the lateral aspect, without mentioning that on the right side of the specimen the fourth maxillary tooth may distinctly be seen, in a transverse section, to have been transversely expanded and molar like." In the same article he said: "Each maxillary bore six large bilobate teeth, the roots hourglass-shaped in horizontal section, the transverse diameter at least double the longitudinal and with large pulp cavities. The teeth are implanted in sockets, not acrodont as believed by Huxley. Three conical teeth were present in each premaxillary, the first the largest. . . .

"The dentition therefore bears a close resemblance to that of *Procolophon*, differing, however, in the molar teeth being wider still, in this respect agreeing with *Diadectes* and *Empedias* as figured by Cope."

In 1917 Huene gave a description of the skull of *Telerpeton elginense* in which he mentions a large conical premaxillary tooth separated by a diastema from a second smaller premaxillary tooth. The maxillary teeth are described as having two conical cusps connected by a somewhat lower cutting edge. The outer cusps of the mandibular teeth are somewhat higher than the inner ones.

In 1876 Owen described the maxillary teeth of *Procolophon* as conical in form with a pulp cavity: "The base of the tooth seems to be confluent with substance of the jaw; there is no appearance of an alveolar cavity."

In 1878 Seeley, describing three species of *Procolophon*, speaks of the maxillary teeth as subcylindrical with a large pulp cavity. He says of the species *cuniceps* that the teeth are "anchylosed to the jaw with no trace of fangs or sockets," and of the species *laticeps*: "The bases of the fangs invested by bone, so as to have the aspect of being in sockets." In 1889, in another paper, he describes the teeth of *Procolophon* as cylindrical, terminating in sharp conical points, and as having "slightly expanded bases, which are in close union with the jaws."

In 1903 Broom gave a catalogue of the specimens of *Procolophon* in the Albany Museum in Grahamstown, South Africa; the following notes are extracted from his descriptions.

Specimen 16.— "The teeth of the maxillary bone are seen to be seven in number; of which the last two are smaller than the others. The teeth which may be regarded as molars have broad flattened crowns."

Specimen 24.— "The cast of the broad tops of the molar teeth are well shown."

Specimen 27.— "In the lower jaw there are seen to have been 9 teeth, the anterior ones pointed, the posterior with broad crowns."

Specimen 30.— "There have been three premaxillary pointed teeth, and six maxillary teeth, of which the first only is somewhat pointed."

Specimen 32.— "There have been only eight teeth in each mandible, of which the first three are pointed."

Specimen 37.— "The molars have broad though fairly sharp crowns."

Specimen 42.— "Left mandible showing the thorough anchylosis of the teeth to the dentary."

In 1905 Broom described the two species of *Thelegnathus* with the teeth "closely resembling those of *Procolophon*," and with "little doubt that the remains belong to a member of the *Procolophonia*." "In the maxilla are six and possibly seven molar teeth, and as in *Procolophon* they are anchylosed to the jaw. In structure and shape the teeth also resemble closely those of *Procolophon*. They differ, however, in becoming steadily larger on passing back." Of a second specimen of the same species, he says: "The crown [in the maxillary teeth] is distinctly constricted antero-posteriorly in the middle, and though worn is fairly sharp."

In 1912 Huene described *Koiloskiosaurus* and said of the maxillary teeth that they are apparently conical, but an exploration of the cast of a tooth with a fine point shows that there was a transverse cutting edge and that the cross-section of the base was not round but oval, with the greatest diameter transverse to the length of the jaw.

In the same paper Huene described the teeth of *Sclerosaurus* as sharply conical in the anterior portions and the maxillary and opposing mandibular teeth as bluntly conical, flattened above and with the base oval and obliquely transverse to the jaw.

It is apparent from these citations that the members of the family *Procolophonidae* had a tendency to the development of teeth set transversely in the jaw, with a transverse cutting edge and the edge divided into cusps. These characters show in varying degree in the various forms and reach their greatest development in *Telerpeton* and the form here described. The peculiar attachment of the teeth is apparently also a common character.

The discussion of fragmentary material is often of doubtful value, but when a fragment indicates the presence of a new form or group previously unknown in a certain geological horizon or locality, the geological and paleogeographical implication transcends in importance the morphological description. The members of the family *Procolophonidae* are all Triassic. *Telerpeton* occurs in the upper Keuper, according to Jukes-Brown, in the middle Triassic, according to Broili (Zittel); *Thelegnathus* occurs in the middle Triassic of South Africa; *Procolophon* in the lower Triassic of the same region; *Sclerosaurus* and *Koiloskiosaurus* occur in the upper Bunter of Germany.

The present form is from the upper Triassic, though Huene has suggested that the beds may be middle Triassic. The advanced stage of specialization of the teeth is confirmatory evidence of the upper Triassic age of the beds in which it was found.

The second specimen, number 7506, Museum of Paleontology, University of Michigan, was figured in outline and briefly described in 1922 by the author. It was then described as a form of unknown relationships. Since that time the author has searched for further light upon the specimen but has been unsuccessful; it is here further described and figured, Plate I, Figures 6, 7, 8, in the hope that some worker may recognize it or may make some helpful suggestion. This specimen was found in the matrix enclosing the skull described by the author as *Leptosuchus crosbiensis* in 1922; it is from the same geological horizon as the other specimens recovered from Crosby County, Texas, but from another layer than the specimen described above and from a locality a few miles away.

The form of the tooth is clearly shown in the figures. The bone is evidently that of a fish and from the shape of the bone and of a depression on the inner side, becoming narrower anteriorly, the broken base is toward the anterior end. The teeth were typically acrodont in attachment. The broken base of the anterior tooth is 5.5 mm. from the anterior end of the fragment; it is separated from the complete tooth by less than 1 mm. and the second tooth is 3.5 mm. from the posterior end. As there is no indication of teeth anterior or posterior to the two and as they are set very close together, it is probable that they were the only teeth in the jaw or were isolated in position. The two teeth have their greatest diameter parallel to the axis of the jaw and the anterior one is much smaller than the second, as indicated by its broken base.

The complete tooth is 10 mm. long and 4.5 mm. in its greatest breadth. The crown is slightly swollen at the base but contracts toward the top from all sides. One side, believed to be the outer, is concave. The top of the tooth is a narrow ridge, probably originally a sharp cutting edge but worn by use into a narrow facet sloping slightly toward the concave side. The facet is 4.8 mm. long. A portion of the crown is broken away from the convex side showing characteristic dentine structure. As the author has sought unsuccessfully for any form to which this may be compared or referred, the name *Xenognathus obscurus* is tentatively proposed.

The exposures of Upper Triassic in northern Arizona have been consistently treated by all workers in the region as revealing the typical section but there has been no consensus of opinion as to the limits of the Triassic, above and below, of the exact horizon of the component beds, or of their areal extent. Gregory [2] and Darton [3] have given résumés of the observations and conclusions of various workers, and recently Lee [4] and Branson [5] have discussed the Red Beds of the whole Rocky Mountain region. Increasing knowledge of the fauna of the Triassic beds in western Texas permits this contribution to an appreciation of the extent of the Shinarump conglomerate member.

A very brief statement of the character of the beds in the type region is:

Chinle. Essentially red sandstone and shale, frequently very massive. *Limestone conglomerate* a characteristic feature.

Shinarump. Cross-bedded, lenticular conglomerate or coarse sandstone, the pebbles chiefly *quartz and quartzite.*

Moenkopie. Chocolate red and banded arenaceous shales and clays, and thin sandstones. Variable in stratification and color.

Gregory correlates the Shinarump with the Dolores of southwestern Colorado, which is characterized by the abundance of "limestone conglomerate" recognizing both types, calcareous and siliceous, of conglomerate as a feature of the Shinarump. The Shinarump, in both its phases, is accompanied by a distinct vertebrate fauna and by this means can be traced from its type occurrence to its final, easternmost, appearance on the east side of the Staked Plains of Texas. Chart I, compiled from the literature and the observations of the author, shows the possible correctness of this suggestion. The chart differs from that of

[2] Gregory, H. E., *Geology of the Navajo Country,* Professional Paper No. 93, U. S. G. S., 1917.

[3] Darton, N. H., *Resumé of Arizona Geology,* Arizona Bureau of Mines, Bulletin No. 119, 1926.

[4] Lee, W. T., *Correlation of Geologic Formations between East Central Colorado, Central Wyoming, and Southern Montana,* Professional Paper No. 149, U. S. G. S., 1927.

[5] Branson, E. B., "Triassic-Jurassic 'Red Beds' of the Rocky Mountain Region," *Journal of Geology,* Vol. XXXV, No. 7, p. 607, 1927.

CHART I

L. Trias. and U. Trias. refer solely to the relation of the beds to the conglomerate. All are Upper Triassic in age. (F) Indicates the position of the vertebrate fauna.

Age	Tanner's Crossing Tucker's Springs Adamana	Ft. Wingate	Abiquiu Chama Rv.	S. W. Colorado	Carthage	Carlsbad Guadaloupe	Santa Rosa Cuervo Tucumcari San Jon Endee	Dickens Spur	Big Springs
Tertiary							Tertiary	Tertiary	Tertiary
Comanchean							Comanchean	Comanchean	Comanchean
Jurassic	Wingate ss.	Wingate ss.	Jura (?)	La Plata	Wingate ss. (?)	Absent	Wingate ss. (?)	Absent	Absent
Triassic	Chinle (F)	Chinle (F)	U. Trias.	U. Dolores	Absent ?	Absent	U. Trias. Dockum 3	U. Trias. ? Dockum 3	Absent
	Shinarump	Shinarump	Poleo ss.	L. Dolores Ls. Cong.	Cong. (F)	Absent	Cong. Dickens Dockum 2	Dickens Dockum 2	Dickens Dockum 2
	Moenkopi	Moenkopi	L. Trias. (F)	L. Dolores	L. Trias. (F)	Absent	L. Trias. (F) Dockum 1	L. Trias. (F) Dockum 1	L. Trias. (F) Dockum 1
Permian	Kaibab	Shale and clay	Shale and sandstone	Cutler	Chupadero (Gypsum at top)	Castile Gypsum Rustler dol.		Gypsum Double Mt.	Gypsum Double Mt.
					Chupadero	Delaware ls.		Clear Fork	Clear Fork
			Abo ss.						

CHART II

Stegocephalia are all either Metoposaurus or Buettneria. Desmatosuchus possibly equals Episcoposaurus. It is probable that many of the genera of Parasuchians should be united in the genus Phytosaurus.

Tanner's Crossing Tucker's Springs Adamana	Fort Wingate	Abiquiu Chama River	S. W. Colorado	Carthage	Santa Rosa Cuervo Tucumcari San Jon Endee	Dickens Spur (Walker's Tank and Cedar Mountain)	Big Springs (Otischalk)
Stegocephalia Metoposaurus	Stegoceph. fragments	Stegoceph. fragments	Stegoceph. fragments	Stegoceph. fragments	Stegoceph. fragments	Metoposaurus Buettneria	Stegoceph. fragments
Parasuchia Macheroprosopus Angistorhinus	Accompsosaurus Paleorhinus	Phytosaurus Episcoposaurus	Phytosaur fragments Heterodontosuchus	Phytosaur fragments	Phytosaur fragments	Phytosaurus Episcoposaurus Leptorhinus Desmatosuchus Promystriosuchus	Phytosaurus (?)
Dinosauria		Coelophysis			Dinosaur fragments	Coelophysis Dinosaur indet.	
Forms indet. Placerias		Typothorax				Trilophosaurus	Forms nov.

Branson mainly in that the Dockum beds are considered as equivalent to the Shinarump rather than to the Chinle.

The evidence for the position assigned to the Dockum, other than that of the vertebrate fauna, is the apparent actual continuity and the similarity of association with other beds. The evidence from the contained fossil wood has not yet been evaluated and the known invertebrates are too few to be of great value.

The vertebrate fauna associated with the Shinarump (Chart II) is a highly specialized and characteristic one which bears internal evidence of response to a fixed and peculiar environment. The author has more than once insisted that correlation of beds by the contained fossils is more a correlation of life conditions than a measure of exact time equivalence. In this consideration is to be found an answer to one objection that might be made to the proposed correlation. In the extreme western part the vertebrate fauna is found in the Shinarump and above it, in the Chinle; in the eastern exposures, in western Texas, it is found in the Shinarump and below it, in the Moenkopi or its cquivalents. Future discoveries may necessitate alteration of this statement, but as yet there is no evidence to the contrary. The fauna is an easily recognized one of giant Stegocephalians and Phytosaurs of bizarre form, of Cotylosaurs (as reported in this paper), of Dinosaurs, and of Ganoid and Dipnoan fishes.

Chart II shows the location of the better known members of the fauna. It can be traced from Tanner's Crossing north of Winslow and Holbrook, in Arizona, to Fort Wingate, in western New Mexico, and as far east as Carthage, in central New Mexico. It occurs in the exposures on the Chama River, in northwestern New Mexico near Abiquiu, and as abundant fragments in the Dolores Formation of southwestern Colorado. It can be traced across southern Colorado from the San Miguel River to the Purgatory River, Animas County, in extreme southeastern Colorado. It occurs near Folsom, New Mexico, and from there can be traced southward on the western side of the Staked Plains through Endee, San Jon, Tucumcari, Cuervo, Montoya, and Santa Rosa. The beds at Folsom and on the

Purgatory River are the same as those which appear in the valley of the Red River north of Amarillo, Drake's Dockum beds, and these can be traced south with the same fauna, on the eastern side of the Staked Plains from Clarendon to Big Springs, in Texas. The absence of Upper Triassic beds in central New Mexico is due to the elevation of the Chupadero (Permian) in that region; probably thin outliers containing the fauna will be found between Santa Rosa and Carthage.

Branson has very justly remarked upon the uncertainty of identification, generically and specifically, of the few known vertebrates and upon the danger of using the material in exact correlation of the beds, but there can be little doubt that the specimens are all parts of one fauna, perhaps with local geographical differences, and reflect a similar stage of evolution under similar conditions of environment. The fauna as a whole is so closely associated with the Shinarump that it may safely be used as an indication of equivalence of conditions and perhaps of time.

In all the exposures on both sides of the Staked Plains the fauna lies in and below a conglomerate or heavy, coarse sandstone. In the localities directly west of the Plains the conglomerate lies directly beneath the Tertiary cap or within a few feet of it, but at Tucumcari, Montoya, Cuervo, etc., there is a heavy bed of red sandstone, the equivalent of the Chinle, above the conglomerate, which has been repeatedly searched unsuccessfully for the vertebrate fauna. These upper beds extend a long distance southward toward Roswell and westward toward Las Vegas.

On the east side of the Plains the conglomerate lies directly beneath the Tertiary cap. It is typically exposed at Dickens, in Dickens County, and is referred to in the author's notes as the Dickens member of the Dockum; it is probably the same as Gould's Trujillo member and Drake's Santa Rosa member. The author has described this conglomerate from several localities and has published sections. The upper layers of conglomerate, or coarse sandstone, are heavy, brown or red, cross-bedded and siliceous; immediately below this is a bluish or whitish bed of similar material but with many included limestone pebbles and a high calcareous content, a "limestone conglomerate." These beds thin rapidly toward the east and probably did not extend a great distance beyond their present exposure. The clay and sandstone below, the equivalent of the Moenkopi extend farther east and are probably separated from the Permian below by the Quartermaster or Blaine gypsum which Gould has traced from Oklahoma into Texas.

The author does not believe that the fact that the fauna occurs above the conglomerate in the west and below it in the east militates against the proposed correlation for the following reasons:

1. The conglomerate on both sides of the Staked Plains is dominantly siliceous above and dominantly calcareous below; both phases of the Shinarump are present.

2. Water-worn fragments of the fauna are found in the conglomerate wherever the fauna occurs in immediately adjacent beds, either above or below.

3. The occurrence of the remains of the fauna is peculiar. Literally hundreds of square miles of exposure of beds that might be expected to contain the fauna are utterly barren. The bones occur in what are obviously old river channels or in areas of wash of rapid water. It is extremely rare that any association of skeletal parts is found.

It seems probable that great areas of barren, fine-grained Triassic sediment were deposited in immense shallow lakes, of such extent that only locally could anything but the finest sediment reach the major part of their bottom. Bordering the lakes were lands that afforded a most favorable habitat for the aquatic and semi-aquatic vertebrates. The land was covered by a mantle of deeply decayed residual soil, in places red from the effect of alternately high and low water table. Some slight physiographic or meteorologic change, insufficient to affect the life, might permit the unreduced pebbles and fragments to be borne far out into the lakes and deposited as the conglomerate. An equally slight change in the opposite direction would restore the previous conditions. Whatever the process really was, such a succession of events would account for the extent and continuity of the beds over large areas and for the persistence of the fauna. The dominantly siliceous or calcareous content of the beds is easily accounted for by the supposition of different areas of degradation affording the material. The stratigraphic location of the fossils already discovered may very easily be but the accidents of the collectors' fortune or of original deposition.

DESCRIPTION OF PLATE I

FIGURES 1 to 5, *Trilophosaurus buettneri*. All figures × 1.7

FIG. 1. Inner surface

FIG. 2. Outer surface

FIG. 3. Upper surface

FIG. 4. Posterior surface

FIG. 5. Obliquely from above and in front

FIGURES 6 to 8, *Xenognathus obscurus*. All figures × 1.8

FIG. 6. Outer surface

FIG. 7. Inner surface

FIG. 8. Upper surface

PLATE I

1

3

2

4

5

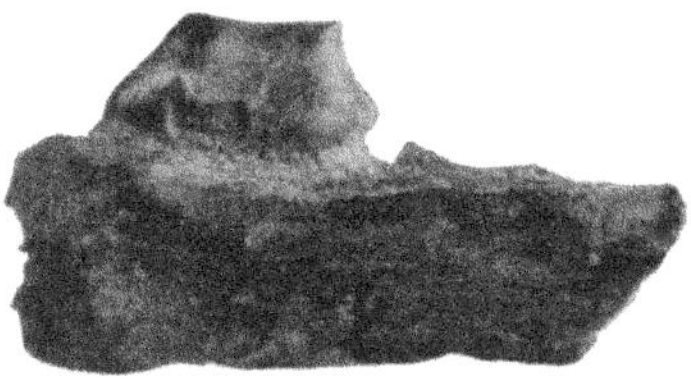

6

8

7

Vol. III, No. 2, pp. 15–18 (1 pl.) November 10, 1928

FOSSIL FISHES FROM THE TRIASSIC OF TEXAS

By ALDRED S. WARTHIN, Jr.

INTRODUCTION

THE expedition from the Museum of Paleontology of the University of Michigan in the summer of 1925 recovered considerable material from the Triassic beds of western Texas. A few fragments of fossil fish were included in the lot and these were turned over to the writer for examination by Dr. E. C. Case. Much of the material is too fragmentary to admit of identification, but two specimens of interest are here noted and figured. These fragments gain in significance when it is borne in mind that the Triassic of this part of the continent has thus far yielded almost no evidence of fish life. The great fauna of fishes which flourished in eastern North America during Newark time seems to have been unable to penetrate to the west.

DESCRIPTION OF MATERIAL

CERATODUS CROSBIENSIS, n. sp.

(Plate I, Fig. 1)

Description. — The specimen is a left palatal tooth plate. Its outline is nearly that of a right-angled triangle, the long face being the internal one. There are four main denticles with a smaller fifth on the internal border of the plate. The first denticle is low and has a knob on the outer border, probably an abnormal state due to injury, as the others are long and straight. They are narrow and sharp and are separated from one another by deep angular furrows. The second, third, and fourth denticles extend all the way to the angulus internus, and are discernible as separate ridges on its extreme point. The first denticle does not reach to the angulus internus.

When viewed from the edge the denticles are seen to be made of narrow folds of dentine covered with a thin coat of enamel. This enamel surface is preserved only in the younger portions of the furrows between the denticles, where it presents a smooth shiny surface with faint lines of growth. Its thickness here is 0.10 mm. On the remainder of the tooth plate the enamel is missing and the dentine is exposed. On the surface of the dentine are numerous small openings, round or oval, and averaging 0.11 mm. in diameter. They are spaced according to no apparent plan but each opening is about 0.35 mm. distant from its neighbors. Weathering has affected the type specimen so much that many of these pores are further exposed and lie revealed as part of a complex canal system which pervades all of the dentine. *Ceratodus* teeth from widely varying localities have been sectioned by numerous scientists and all show this canal system. The pulp cavity beneath extends up into the denticles only in those forms having a thin layer of dentine.

The length of the tooth plate along the internal border is 21 mm., its width is 12 mm. and its greatest thickness (near the angulus internus) is 4.5 mm.

Occurrence. — The type was collected from the Dockum beds of Upper Triassic age at Walker's Tank, Crosby County, Texas, by Doctor R. C. Hussey. It occurred in association with plant and reptilian remains which indicate a continental environment.

Type. — The holotype of *Ceratodus crosbiensis* bears the number 9634 and is preserved in the Museum of Paleontology of the University of Michigan.

Remarks. — No trace of the genus *Ceratodus* has as yet been recorded from the Triassic of eastern North America, and only one other species, *C. dorothea* Case, has been collected in Texas. The type of *C. crosbiensis* differs from that of *C. dorothea* in being thinner, in having the second, third and fourth denticles separate for their entire length and radiating straight from the angulus internus, and also in not having the second denticle arise from the first. In *C. crosbiensis* the denticles have a greater horizontal length and slope down at the end only gently, instead of dropping off almost vertically as in *C. dorothea.*

All these differences indicate the more primitive character of *C. crosbiensis.*

MACROPOMA SP.

(Plate I, Figs. 2–3)

Description. — In the same region which yielded the preceding form was collected a fragment which appears to be the first evidence of the presence of fishes of the family *Coelacanthidae* in the Triassic of Texas. This fragment consists of a quadrate bone with a portion of the pterygoid bone attached to it. They are from the left side of the skull.

The quadrate is thin, with a rib on its posterior edge, and closely appressed to the pterygoid. On its lower end it carries two well-developed condyles. The inner one is nearly hemispherical, about 8 mm. in diameter, and was probably somewhat anterior to the outer condyle which is more conical and is only 6 mm. in diameter. Both structures have slightly flaring margins. There is a long narrow groove on the back of the quadrate above the condyles.

The pterygoid is flat and joined to the quadrate by a suture. On its posterior edge it has a rib which increases in size from below upward, in proportion as the rib on the quadrate decreases. On its outer surface about 20 mm. above its base it also develops another rib which curves anteriorly. The interior surface of the pterygoid is gently curved and is covered with numerous denticles, mammillary in shape, with obscure radial striae. The denticles are set in shallow depressions on the surface of the pterygoid and they range up to 0.2 mm. in diameter. Portions of the interior surface of the pterygoid are also ornamented with fine striae in irregularly arranged groups.

Total length of the fragment, 40 mm.; length of the quadrate alone, 35 mm.

Occurrence. — The specimen was collected by Dr. E. C. Case from the Dockum beds, Upper Triassic in age, of Crosby County, Texas. It is now preserved in the Museum of Paleontology of the University of Michigan where it bears the catalogue number 9630.

Remarks. — This fragment has a detailed resemblance to the specimen of *Macropoma mantelli* figured by Watson [1] in an article on the morphology of the coelacanth skull. The only observable differences are a slight discrepancy in the position of the anterior rib of the pterygoid, and some slight variation in the shape of the condyles. The genus *Macropoma* is typically Cretaceous, and has not previously been recorded from the Triassic so that it is with some hesitation that this specimen is so identified. A survey of the Triassic coelacanth genera already known, including the forms described from Spitzbergen by Stensio,[2] failed to discover any form with a comparable structure.

None of the characters observable seem to be other than generic in nature, so no specific name is here applied to the specimen.

[1] Watson, D. M. S., "On the Coelacanth Fish," *Ann. Mag. Nat. Hist.*, Ser. 9, Vol. 8, 1921, pp. 320–337.

[2] Stensio, E. A., *Triassic Fishes from Spitzbergen*, Part 1, 1921. Vienna, Adolf Holzhausen.

EXPLANATION OF PLATE I

Ceratodus crosbiensis, n. sp.

FIG. 1. Left palatal tooth plate, showing canals in dentine, and straight denticles radiating from the angulus internus. × 2
(No. 9634, Museum of Paleontology, University of Michigan)

Macropoma sp.

FIG. 2. View of left quadrate and pterygoid, seen from outer side. The black line marks the suture between the bones. × 1.5

FIG. 3. View of the same specimen from the internal side. × 1.5
(No. 9630, Museum of Paleontology, University of Michigan)

PLATE I

1

2

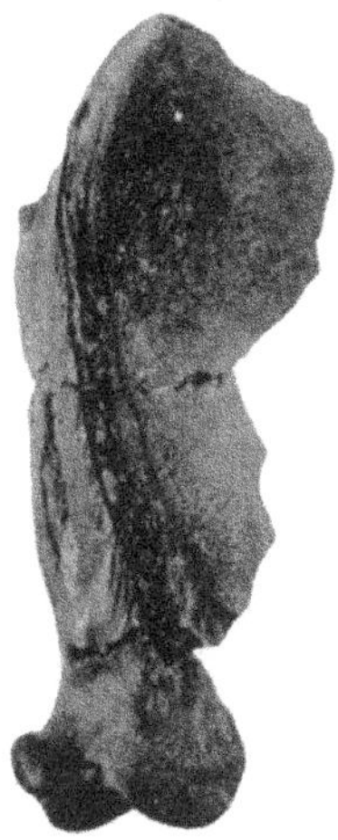

3

Vol. III, No. 3, pp. 19–76 (17 pls., 1 text fig., 1 map) November 10, 1928

CONTRIBUTIONS TO THE GEOLOGY OF FOXE LAND, BAFFIN ISLAND

By LAURENCE M. GOULD, AUG. F. FOERSTE AND RUSSELL C. HUSSEY

Part I: The Rocks and Their Distribution, By Laurence M. Gould

Part II: The Cephalopods of Putman Highland, By Aug. F. Foerste

Part III: Corals, Brachiopods, Gastropods and Ostracods from Putnam Highland, By Russell C. Hussey

Part I

THE ROCKS AND THEIR DISTRIBUTION

By LAURENCE M. GOULD

THE data and the fossils which constitute the basis for the following studies were collected by the author as geographer of the Putnam Baffin Island Expedition of 1927. This expedition was organized and directed by George Palmer Putnam[1] of New York and was sponsored primarily by the American Geographical Society, since its main purpose was geographic exploration.

The name "Foxe Land" has long been used in a very general way to indicate the southwestern portion of Baffin Island. In the following studies the peninsula which projects toward the southwest and adjacent parts of the main island proper will be

[1] Putnam, George Palmer, "The Putnam Baffin Island Expedition," *The Geographical Review*, Vol. XVIII, No. 1, pp. 1–40, January, 1928.

included in this term. These relationships are clearly indicated on the geological map and on the small insert with it (Map 1).

The physiographic characteristics of this region have been elsewhere sufficiently described.[2]

Pre-Cambrian crystalline rocks, mainly gneisses and schists with varying amounts of intrusive granites, comprise the rocks of the peninsula proper. Intense denudation beneath a continental ice-sheet has ironed these old rocks into a landscape which shows but little change from place to place and is ever one of dead monotony. Its bleakness is further accentuated by the almost complete absence of vegetation of any appreciable size. Along the north coast of Foxe Land for more than a hundred miles eastward from Cape Dorchester the general level of the land is but a few feet above the sea and even near Bowman Bay, where it becomes most nearly rugged, the general level is scarcely a hundred feet above the same level. This landscape with its flat or at least flatly-convex profiles of glacial origin furnishes a most startling contrast to the physiographic expression of the sediments of the region which are included in the Putnam Highlands. Here to the south of Bowman Bay rises a great northward-facing scarp of typical mesa character which stretches inland to the east and northeast farther than one can see. The general altitude of this scarp is from 600 to 700 feet above sea-level but the relatively low level and flatness of the adjacent pre-Cambrian rocks make it stand out sharply against the sky (Plate I, Fig. 1), even from very great distances. A nearer approach demonstrates that the Highland consists essentially of a great thickness of shales with a cap rock of limestone. The mesa character of the whole is as perfect as one might expect to find in the Colorado Desert, yet here it exists in a climate which is evidently decidedly humid. Two factors appear in large measure to have caused and preserved these sharp profiles: first, the total absence of vegetation, except for negligible amounts of lichens and mosses, has allowed

[2] Gould, L. M., "Report on the Physical Geography of the Putnam Baffin Island Expedition," *The Geographical Review*, Vol. XVIII, No. 1, pp. 27–40, January, 1928.

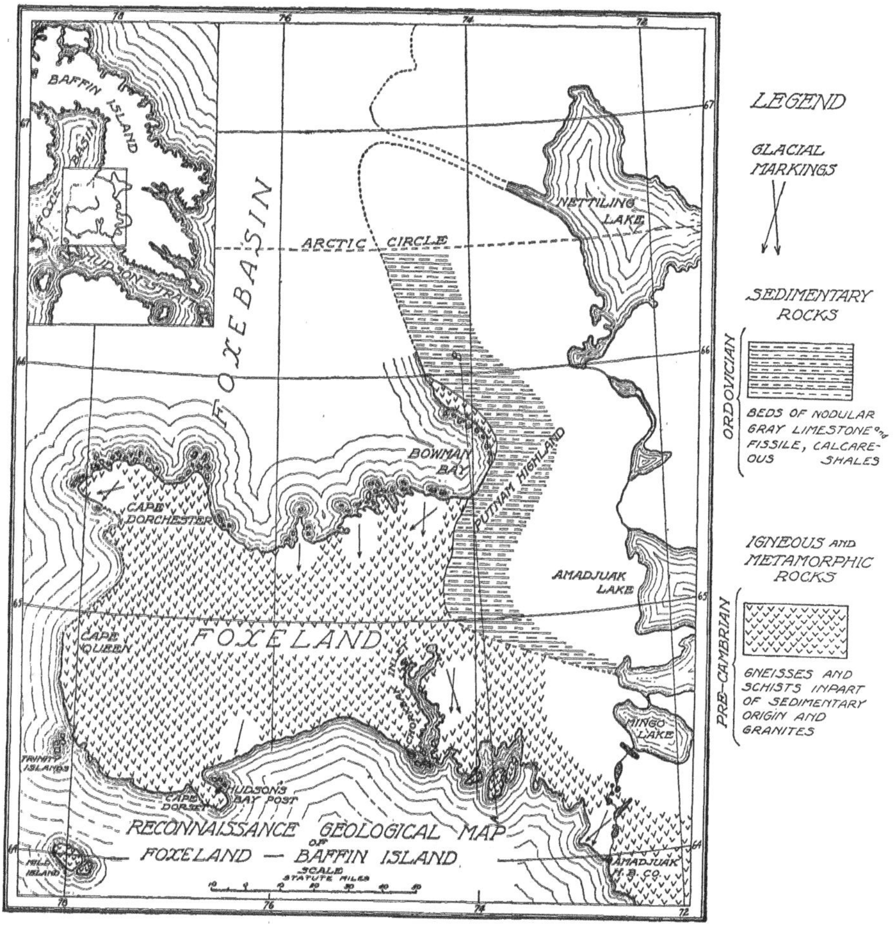

LEGEND
GLACIAL MARKINGS
SEDIMENTARY ROCKS
ORDOVICIAN
BEDS OF NODULAR GRAY LIMESTONE and FISSILE, CALCAREOUS SHALES
IGNEOUS AND METAMORPHIC ROCKS
PRE-CAMBRIAN
GNEISSES AND SCHISTS IN PART OF SEDIMENTARY ORIGIN AND GRANITES
BAFFIN ISLAND
HUDSON STRAIT
FOXE BASIN
ARCTIC CIRCLE
NETTILING LAKE
BOWMAN BAY
PUTNAM HIGHLAND
AMADJUAK LAKE
MINGO LAKE
CAPE DORCHESTER
CAPE QUEEN
FOXE LAND
TRINITY ISLANDS
CAPE DORSET
HUDSON'S BAY POST
MILL ISLAND
AMADJUAK H.B. CO.
RECONNAISSANCE GEOLOGICAL MAP OF FOXELAND — BAFFIN ISLAND
SCALE
STATUTE MILES

Map 1

frost action unlimited opportunity to work; and secondly, the cap rock itself is of such a character as to be peculiarly subject to such action. It is a porous, nodular limestone which is conspicuously jointed; a very friable sort of rock which shatters and crumbles when struck with a hammer. In Figure 2 of Plate I this character of the rock is indicated. This non-resistive sort of rock maintains itself as the rim or cap rock of the mesa only because the underlying shales appear to offer no more resistance to weathering and erosion than would totally unconsolidated materials. The recession of this mesa escarpment has been so rapid in post-glacial time that all evidences of the Highlands having ever been glaciated have entirely disappeared.

Rather than a single distinct stratum, the mesa cap rock is found to consist of a series of beds which in most places total less than fifty feet. Considerable variation in both the thickness and the character of this series is noticeable in a horizontal direction. Figure 1, Plate II, shows it in its most nearly massive development, while Figure 2 of the same plate illustrates the same horizon not more than three miles distant. This latter exposure shows a typical manner in which the limestone grades into the fissile calcareous shales associated with it. Both are of the same blue-gray color and it is sometimes scarcely possible to delimit the boundaries between the two kinds of rocks.

These limestones and shales exhibit their greatest development along the north face of Putnam Highland. As determined by aneroid the maximum thickness here is 700 feet, of which about 600 feet consist of shales. The strata are all flat-lying and throughout this great thickness there is no evidence of any angular unconformities. Limestone beds like the cap rock are found interbedded with the shales far below the mesa rim. In Plate I, Figure 1, it will be noted that such strata form an imperfect shelf part way down the slope.

The limestone members of this series are highly fossiliferous and only the fact that but two or three hours were available for collecting and that the materials collected had to be carried out to our boat on our backs, a distance of about forty miles, prevented us from securing a much more comprehensive collection. With the exception of an ostracod elsewhere described by Dr. Hussey all the fossils figured by him and by Dr. Foerste were collected from the limestone strata at or near the rim of the mesa or from the bottoms of washes where they had accumulated after weathering out of the rocks above.

North of Putnam Highland and up the coast north of Bowman Bay as far as we voyaged and as far as we could see, the Putnam Highland Formation is represented by only a few feet of shale. Figure 1, the structure section along line A–B of the geological map, illustrates the relationships between the Highland and the terrain to the north. A more completely monotonous landscape than that exhibited where the shales are the country rock could hardly be imagined. For instance at latitude 66.2 degrees north, where the ostracod described by Dr. Hussey was collected, one can look inland in any direction across a landscape without any perceptible change of level. So flat is this terrain that where it passes beneath the waters of Foxe Basin there are ex-

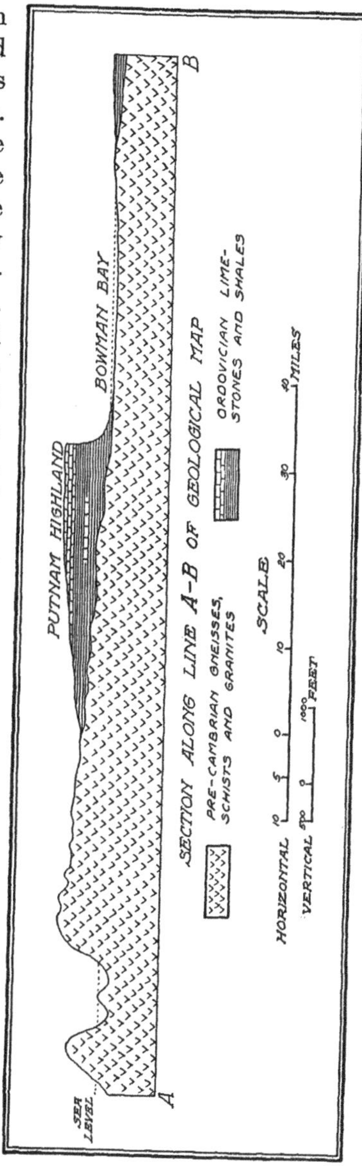

FIGURE 1

posed at low tide great mud flats in places at least ten miles wide.

On the geological map of North America compiled by Bailey Willis and George W. Stose and published by the United States Geological Survey in 1911, Silurian rock is mapped as occurring as a broad band from the head of Frobisher Bay northwestward to Foxe Basin, including a narrow strip along the north coast of Foxe Land. The same rock is also mapped as recurring across Foxe Basin on Melville Peninsula and on islands in Fury and Hecla Strait. Such a relationship seems to have been first suggested by Bell, who, writing in 1898, says: "The scanty information we possess goes to show that Foxe Basin is partly bordered on both sides by comparatively low land occupied by undisturbed fossiliferous Silurian limestone." [3]

It is at once evident that the rocks along the western coast of Baffin Island, believed by Bell to be Silurian, are in large part, and perhaps entirely, Ordovician. This is assuredly true of the regions studied by the Putnam Expedition and in the light of this information the existence of Silurian strata beyond our farthest north becomes more problematical than ever. Even greater uncertainty still exists as to the exact age of the rocks bordering the western side of Foxe Basin. Freuchen and Mathiassen,[4] who, as members of the Fifth Thule Expedition under the leadership of Knud Rasmussen, mapped portions of the eastern side of Melville Peninsula and the western part of northern Baffin Island, refer to the surface as being covered by rocks of Cambro-Silurian age. This designation appears to have been made before any of the fossils collected were studied.

[3] Bell, Robert, "Report of an Exploration on the Northern Side of Hudson Strait," *Canada Geol. Survey, Annual Report*, Vol. XI (N.S.), 1898, p. 17 M.

[4] Freuchen and Mathiassen, "Contributions to the Physical Geography of the Region North of Hudson Bay — Preliminary Report of the Fifth Thule Expedition," *Geographical Review*, Vol. XV, No. 4, p. 552, October, 1925.

PLATE I

Fig. 1

Fig. 2

PLATE II

Fig. 1

Fig. 2

Part II

THE CEPHALOPODS OF PUTNAM HIGHLAND

By AUG. F. FOERSTE

THE cephalopods here discussed were collected by Professor Laurence M. Gould, geographer and geologist of the Putnam Baffin Island Expedition of 1927, and were kindly submitted to the writer for study.

They belong to a widely ranging Upper Ordovician fauna, first described from southern Manitoba, chiefly by Whiteaves. Here they occur on the west shore and on the islands of Lake Winnipeg, and in the adjacent part of the Red River Valley. A closely similar fauna occurs on the Nelson and Shammattawa rivers, southwest of Hudson Bay, and the cephalopods of this fauna have been studied by Professor W. A. Parks, and more recently by Foerste and Savage. A much more abundant and varied development of the southern Manitoba cephalopod fauna is that recently described by Gustaf Troedsson from Cape Calhoun, on the northern shore of Kane Basin, in the northwestern part of Greenland. It was believed long ago by John Richardson that the Lake Winnipeg fauna is represented also on Igloolik Island, on the northwestern margin of Foxe Basin, at the eastern end of Fury and Hecla Strait, and recent studies appear to confirm this opinion.

The wide geographical range of this cephalopod fauna is indicated by the following statements. From the Putnam Highland southeast of Foxe Basin, on the west shore of Baffin Land southwestward to the exposures southwest of Hudson Bay is a distance of 830 miles, while the total distance to the southern end of Lake Winnipeg, in the same direction, is 1200 miles. The distance of the Baffin Land locality from Cape Calhoun, in northern Greenland, almost directly northward, is 950 miles. Igloolik Island is only 320 miles northwest of the Baffin Land locality.

In his Report on the geology of the west shore and islands of Lake Winnipeg, published in 1900, Dowling divides the Ordovician strata of this area into the following formations, listed in descending order:

Stony Mountain formation
Upper Mottled limestone
Cat Head limestone
Lower Mottled limestone
Winnipeg sandstone.

Of these, the *Stony Mountain* formation is definitely correlated with the Richmond group, while the *Upper Mottled* limestone, *Cat Head* limestone, and *Lower Mottled* limestone were regarded by Dowling as of Galena or Trenton age. Regarding the *Winnipeg* sandstone, Dowling stated that "The fossils give no definite information as to the age of the beds, but suggest a passage from Black River to Trenton."

For the Upper Mottled limestone the term *Selkirk* limestone here is proposed, in view of the numerous fossils collected from this horizon at East Selkirk and Lower Fort Garry in the Red River valley. For the Lower Mottled limestone, the name *Dog Head* limestone is selected, from Dog Head, on the west shore of Lake Winnipeg, this being another important fossil locality. For the combined Selkirk, Cat Head and Dog Head limestone section the term *Red River* formation is proposed, which will result in the following list of strata, in descending order:

Stony Mountain formation		100+ feet
Red River formation	Selkirk limestone	130 feet
	Cat Head limestone	70 feet
	Dog Head limestone	70 feet
Winnipeg sandstone		100+ feet

In the area southwest of Hudson Bay, the Ordovician is divided by Savage and Van Tuyl, in descending order, into the

Shammattawa limestone, and
Nelson limestone.

Of these, the Shammattawa limestone is correlated with the Stony Mountain limestone, since it contains, in common with the latter, such characteristic species as *Streptelasma trilobatum*

(Whiteaves), the coarsely striated form of *Dinorthis* which resembles *Dinorthis subquadrata*, and the large and coarse form of *Rhynchotrema* which resembles *Rhynchotrema capax*. The Nelson limestone is correlated with the Red River formation, owing to the presence of *Cyrtogomphoceras nutatum*, which is closely related to *Cyrtogomphoceras magnum* (Whiteaves) and *Cyrtogomphoceras whiteavesi* (Miller) from that formation.

In the Kane Basin area in northern Greenland, Troedsson has divided the Ordovician into the

Cape Calhoun series, and
Gonioceras Bay formation.

Of these groups of strata, the Cape Calhoun series is regarded by Troedsson as including several distinct horizons: the Richmond, Trenton, and Black River, the fossils of which, unfortunately, were not collected separately, owing to the vertical exposures of these strata at Cape Calhoun, and owing to the absence of readily recognizable lithologic distinctions among the formations. However, those cephalopods which are listed by Troedsson from the Richmond part of his Cape Calhoun series suggest affinities with the Red River formation of the southern Manitoba area.

The cephalopods here studied, from the Putnam Highland, southeast of Foxe Basin, on the western side of Baffin Island, also suggest affinity with those of the Red River formation of southern Manitoba. This is true especially of *Cyrtogomphoceras baffinense*, which is a close relative of *Cyrtogomphoceras whiteavesi* (Miller), and which is listed by Dowling from all three members of the Red River formation, but which is most abundant from its lower, or Dog Head, member. The specimens here described as *Westonoceras putnami* and *Westonoceras gould* evidently show affinities with *Westonoceras manitobense* (Whiteaves), also from the Dog Head member, but how close the relationship may be, is not certain. It is noted, for instance, that both the Baffin Land species differ from the Manitoba type of the genus in having the sutures of the septa virtually straight across the lower part of the phragmacone, while they curve distinctly downward laterally in *Westonoceras manitobense*. In this respect they differ also from *Thuleoceras ornatum* Troedsson, in which the lateral lobes are distinct, but much less pronounced than in the Manitoba species. The Putnam Highland specimens here referred to *Gonioceras* may belong to *Lambeoceras*, but, if so, this cannot be determined from their present state of preservation.

The Putnam Highland specimen here described as *Leurorthoceras* (?) *baffinense* is closely related to *Leurorthoceras* (?) *ruedemanni* Troedsson, described from the Gonioceras Bay section, along the northern margin of Kane Basin in northern Greenland. The species of *Cyrtogomphoceras* from that basin, however, are listed by Troedsson from the Richmond part of the Cape Calhoun series, while *Westonoceras* and *Ephippiorthoceras* are listed from the Trenton, with the possibility that the *Ephippiorthoceras* may be from a horizon as low as Black River. Unfortunately, Troedsson does not state on what basis he separated the specimens from the Cape Calhoun series into those of Richmond, Trenton, or Black River origin, considering the fact that they were not so discriminated in the field by Lauge Koch, the collector.

As stated in the earlier part of this paper, the strata included in the Red River formation of southern Manitoba were regarded by Whiteaves and Dowling as chiefly of Trenton age, with the basal parts possibly of Black River age. The first suggestion of the possibility of these strata being of Richmond age is given in Bassler's *Bibliographic Index of American Ordovician and Silurian Fossils*,[1] where, on page 1458, the Red River cephalopods are cited under the heading "Black River or Richmond (Manitoba)," with the supplementary statement: "This list includes the species from the Cat Head and associated formations in Manitoba. Although recorded as of Mohawkian age by Whiteaves, it is probable that many of them were derived from Richmond strata." Unfortunately there is no indication in this list as to which species gave rise to this correlation with the Richmond.

Troedsson, in his very important contribution "On the Middle and Upper Ordovician Faunas of Northern Greenland;

[1] U. S. National Museum, Bull. 92, Vol. 2, 1915.

I, Cephalopods"[2] (1926), refers more than half of the cephalopods studied by him to the Richmond division of the Cape Calhoun series. Among these are such characteristic Red River genera as *Narthecoceras* and *Cyrtogomphoceras.* Among other genera listed as Richmond are *Apsidoceras, Charactoceras, Lambeoceras, Huronia,* and *Diestoceras.* Among these, *Apsidoceras* is represented also in the Richmond of Anticosti, in the Shammattawa limestone of the Hudson Bay area, and in the Stony Mountain representative of the Richmond in southern Manitoba. *Charactoceras* occurs in the Richmond of Anticosti, and also in that of Ohio and Indiana; also in the Red River formation of southern Manitoba. *Lambeoceras* occurs in the Selkirk member of the Red River formation of Manitoba, and also in the Richmond of Indiana and Ohio. *Huronia* occurs not only in the Silurian, but also in the Shammattawa limestone of the Hudson Bay area, and in the Red River formation of southern Manitoba; it occurs also in the Richmond of Mocassin Mountain, in the Big Horn range of Wyoming. *Diestoceras* occurs in the Richmond of Anticosti, Indiana, and Ohio, and also in the Nelson limestone of the Hudson Bay locality.

Among the genera not listed by Troedsson from the Richmond part of the Cape Calhoun series, but of interest in this connection, are *Ephippiorthoceras* and *Westonoceras,* of which the former occurs also in the Shammattawa limestone of the Hudson Bay area, while the latter occurs also in the Nelson limestone and in the Dog Head member of the Red River formation.

Among the species found in the Dog Head member of the Red River formation in southern Manitoba is *Billingsites costulatum* (Whiteaves), representing a genus the typical members of which are known elsewhere only from the Richmond; for instance, in the Shammattawa limestone of the Hudson Bay area, in the English Head, Vaurial, and Ellis Bay formations of Anticosti, in the Stonington member of the Richmond in the northern peninsula of Michigan, and in the basal part of the Whitewater member of the Richmond in Ohio.

[2] Jubilaeumsekspeditionen Nord om Grønland 1920–1923. Nr. 1., Medd. om Grønland, 71. 1926.

The preceding notes indicate that the age of the Red River formation requires further study, and that there are arguments in favor of its Richmond, rather than Trenton, age.

When the Trenton age of the Red River formation first was accepted, this determination was based largely upon the presence of a large species of *Receptaculites*, and on large specimens of gasteropoda usually identified as *Maclurea magna*. Later, however, this gasteropod was described by Whiteaves as *Maclurea manitobensis*, and still later this species was made the genotype of the new genus *Maclurina* of Ulrich and Scofield.

The *Receptaculites* of the Red River formation was identified first with *Receptaculites neptuni* Hall (not Defrance), described as though from the Trenton at Carlisle, Pennsylvania. Later it was identified with *Receptaculites occidentalis* Salter, when the former name was abandoned on account of synonomy. The latter is a Black River species described originally from the Pauquette Rapids on the Ottawa River, in eastern Ontario, Canada. Later it was identified with *Receptaculites oweni* Hall, from the Galena of Illinois and Wisconsin. Dr. Ulrich, however, informs me that this Manitoba Red River species is distinct from the Trenton and Black River forms. Possibly it is related to *Receptaculites arcticus*, described by Etheridge from the west side of Kane Basin. Evidently both *Maclurina* and *Receptaculites* may occur in Richmond as well as in Trenton and Black River strata.

In fact, the possibility of the Red River formation belonging to the Richmond, instead of to the Trenton, is suggested by the presence also of such corals as *Calapoecia cribriformis*, *Columnaria* (*Palaeophyllum*) *stokesi*, and *Halysites gracilis*, all three of which are present in the Big Hill member of the Richmond in the peninsula east of Escanaba, on the northern peninsula of Michigan.

From this point of view it now becomes desirable to restudy the Ordovician exposures on the west side of the Kane Basin, along the eastern shore of Ellesmereland. Here, within a length of about fifty-five miles, occur four exposures, all apparently exposing Richmond strata. The most southern of these is Cape

Harrison, northwest of Norman Lockyer Island, on the northern coast of Princess Marie Bay, in latitude 79° 19′ north. The next, Cape Louis Napoleon, at the northern edge of the mouth of Dobbin Bay, is in latitude 79° 38′ north. Cape Hilgard is at 79° 41′; and Cape Frazer is at 79° 45′, about half way between Dobbin Bay and Scoresby Bay. From Cape Harrison, Whitfield described *Receptaculites pearyi, Halysites agglomeratiformis, Calapoecia borealis,* and *Heliolites perelegans.* From Cape Louis Napoleon, Etheridge described *Receptaculites arcticus* and cited *Maclurea magna.* From Cape Hilgard he described *Halysites catenulatus feildeni,* and cited *Favistella reticulata, Favistella franklini,* and *Maclurea magna.* From Cape Frazer he described *Receptaculites arcticus* and *Halysites catenulatus harti.* Fortunately all three of the forms of *Halysites* here mentioned are figured in the publications including their original descriptions, and all evidently belong to the Ordovician group of species in which the corallites are more quadratic than elliptical in cross-section, and do not alternate with narrow tubules, as in some Silurian species. The species of *Favistella* cited apparently are species of *Columnaria.* The *Calapoecia borealis* is an excellent representative of the genus *Calapoecia.* The figure of *Heliolites perelegans,* published by Whitfield, lacks every essential detail, and the presence of this genus at Cape Harrison needs verification.

It is evident that progress in knowledge of Arctic geology depends on the accumulation of data from all possible sources. To this progress every new locality makes some contribution. The fossils collected by Professor L. M. Gould very materially extend our knowledge of the geographic distribution of the Red River fauna, originally described from southern Manitoba, then identified in the Hudson Bay area and in northern Greenland, and now known also from the western side of Baffin Land. It now becomes necessary to determine how this equivalent of the Red River fauna in western Baffin Island is related to the fauna discovered at the head of Frobisher Bay, on the eastern side of Baffin Land, and only 200 miles southeast of the western Baffin Land exposure. This fauna at the head of Frobisher Bay has been identified as Trenton, but, according to Bassler, probably the material collected includes some Richmond species.

It is evident that the progress of our knowledge of Arctic Ordovician faunas is greatly retarded by the uncertainty which lingers as to the Black River, Trenton, or Richmond age of the various Red River formation faunas here discussed. Our present knowledge of the cephalopods and corals appears to favor the Richmond age of the entire Red River formation, as exposed in southern Manitoba, and of the combined Shammattawa-Nelson section, as exposed in the area southwest of Hudson Bay. At least the greater part of the Cape Calhoun series of northern Greenland appears to be of Richmond age. In the Greenland area, stratigraphical details are lacking.

As the fauna most typical of the doubtful part of the upper Ordovician of the American Arctic, that of the Red River formation of southern Manitoba here is selected. Of the areas of exposure known at present, it is the most easy of access. Moreover, it contains an abundant and diversified fauna. The list of fossils from this formation published by Dowling [3] includes 142 items, consisting of 5 species of algae, 3 *Receptaculitidae,* 2 sponges, 3 graptolites, 9 corals, 2 unclassified, 1 unsorted series of crinoid fragments, 1 cystoid, 1 starfish, 2 annelids, 9 bryozoans, 28 brachiopods, 12 pelecypods, 22 gasteropods, 1 Conularia, 26 cephalopods, 3 ostracods, and 13 trilobites.

Since Dowling's report has long been out of print, and is missing in many important libraries, that part pertaining to the cephalopods here is appended. In this list, various species have been identified with characteristic Trenton and Black River forms, in my opinion erroneously. For others, new generic affiliations have been proposed in recent years, chiefly by the present writer. These here are adopted, but the generic terms employed by Dowling in his list are added in brackets, for ready reference to the original literature. The most important part of this list is the indication of the horizon at which the individual species are found.

Regarding the Richmond age of the Stony Mountain formation, immediately above the Red River formation, there is no question. From this Stony Mountain formation Dowling lists 42

[3] *Geol. Surv. Canada, Ann. Rept.*, XI, pt. F, pp. 48–53 (1900).

STRATIGRAPHICAL DISTRIBUTION OF ORDOVICIAN CEPHALOPODS IN SOUTHERN MANITOBA.

List of Species	Red River Formation			Stony Mountain Formation
	Dog Head limestone	Cat Head limestone	Selkirk limestone	
Cyclendoceras (Endoceras) subannulatum (Whitfield)	*	*	*	....
Narthecoceras (Endoceras) crassisiphonatum (Whiteaves)	....	....	*	....
Narthecoceras (Endoceras) simpsoni (Billings)	*	*	....	....
Armenoceras (Actinoceras) richardsoni (Stokes)	....	....	*	....
Actinoceras bigsbyi (?) Bronn	*	*	*	....
Ormoceras (Actinoceras) allumettense (Billings)	*	....	*	....
Deiroceras (Actinoceras) python (Billings)	....	....	*	....
Paractinoceras (Actinoceras) canadense (Whiteaves)	*	....	....	....
Orthoceras winnipegense Whiteaves	*	....	....	....
Kionoceras (Orthoceras) magnisulcatum (Billings)	....	....	*	....
Cycloceras (Orthoceras) selkirkense (Whiteaves)	....	....	*	*
Spyroceras (Orthoceras) anellus (Conrad)	....	....	*	....
Lambeoceras (Tripteroceras) lambi (Whiteaves)	....	....	*	....
Garryoceras (Tripteroceras) semiplanatum (Whiteaves)	....	....	*	....
Billingsites (Ascoceras) costulatum (Whiteaves)	*	....	....	....
Billingsites (Ascoceras) sp.	....	....	....	*
Diestoceras (Poterioceras) nobile (Whiteaves)	*	....	*	....
Diestoceras (Poterioceras) apertum (Whiteaves)	*	*	....	....
Dowlingoceras (Poterioceras) gracile (Whiteaves)	*	....	....	....
Cyrtogomphoceras (Oncoceras) magnum (Whiteaves)	....	....	*	....
Cyrtogomphoceras (Oncoceras) intermedium (Whiteaves)	*	....	....	....
Cyrtogomphoceras (Oncoceras) whiteavesi (Miller)	*	*	*	....
Westonoceras (Cyrtoceras) manitobense Whiteaves)	*	....	....	....
Winnipegoceras (Cyrtoceras) laticurvatum (Whiteaves)	*	....	....	....
Charactoceras (Eurystomites) plicatus (Whiteaves)	*	....	....	....
Discoceras canadense Whiteaves	*	....	....	....
Trochoceras (?) mccharlesi Whiteaves	*	....	*	....
Apsidoceras (Litoceras) insigne Whiteaves	....	....	....	*

species and varieties, of which 18 species of bryozoans, 2 species of *Beatricea*, various corals, and one species each of *Kionoceras*, *Billingsites*, and *Apsidoceras* suggest affinities with the Richmond of Anticosti. It is the Richmond age of the underlying Red River formation which here is asserted.

LIST OF SPECIES DESCRIBED

1. Endoceras sp. (A)
2. Endoceras sp. (B)
3. Cyclendoceras sp. (A)
4. Cyclendoceras sp. (B)
5. Orthoceras sp.
6. Ephippiorthoceras sp. (A)
7. Ephippiorthoceras sp. (B)
8. Garryoceras semiplanatum (Whiteaves)
9. Dowlingoceras gracile (Whiteaves)
10. Leurorthoceras (?) baffinense Foerste
11. Gonioceras (?) sp. (A)
12. Gonioceras (?) sp. (B)
13. Westonoceras manitobense (Whiteaves)
14. Westonoceras putnami Foerste
15. Westonoceras gouldi Foerste
16. Westonoceras cf. gouldi Foerste
17. Winnipegoceras laticurvatum (Whiteaves)
18. Winnipegoceras dowlingi Foerste
19. Cyrtogomphoceras magnum (Whiteaves)
20. Cyrtogomphoceras whiteavesi (Whiteaves)
21. Cyrtogomphoceras baffinense Foerste
22. Cyrtogomphoceras cf turgidum Troedsson
23. Cyrtogomphoceras intermedium (Whiteaves)

NEW GENERA

Garryoceras; genotype: **Orthoceras semiplanatnm** Whiteaves.
Dowlingoceras; genotype: **Poterioceras gracile** Whiteaves.
Winnipegoceras; genotype: **Cyrtoceras laticurvatum** Whiteaves.

ENDOCERAS Hall

Endoceras Hall, Pal. New York, 1, p. 58, 207, footnote (1847). See also Foerste, Journ. Sci. Lab. Denison Univ., 20, p. 208 (1924).

First described species: *Endoceras subcentrale* Hall, Pal. New York, 1, p. 59, pl. 17, fig. 4 (1847).

Accepted genotype: *Endoceras proteiforme* Hall, Pal. New York, 1, p. 208, pls. 48, 49; see also Foerste, loc. cit., pls. 21, 22, 23, 25 (1924).

Conch containing long, funnel-like endocones within the siphuncle. According to Hyatt, the septal funnels or necks extend posteriorly only for the length of one camera, or from one septum to the next septum to that from which they originated. In the type of *Endoceras subcentrale*, however, they extend for the length of two camerae.

1. ENDOCERAS SP. (A)

(Plate II, Fig. 3; Plate X, Fig. 1)

Specimen 110 mm. long, with a lateral diameter of 58 mm., not enlarging to an appreciable extent within the limit of the length of the specimen. Almost five camerae occupy a length equal to the diameter of the conch. The sutures of the septa are directly transverse. The concavity of the septa is at least 10 mm. Apparently the siphuncle was 22 mm. in diameter and 5 or 6 mm. from contact with the ventral wall of the conch, but the specimen is slightly crushed dorso-ventrally, so that the original distance of the siphuncle from the ventral wall cannot be determined with confidence.

Locality and Horizon. — Putnam Highland, Baffin Island; in strata equivalent to the Red River formation of Manitoba.

Specimen No. 10280 in the Museum of Paleontology, University of Michigan.

Remarks. — Compared with typical *Endoceras proteiforme* Hall, from the Trenton at Middleville, New York, the number of camerae in a length equal to the diameter of the conch is less, the diameter of the siphuncle is considerably smaller, and its distance from the ventral wall of the conch is much greater. The number of camerae, moreover, is less than in *Endoceras nelsonense* Foerste and Savage, from the Nelson limestone west of Hudson Bay, in *Endoceras chidleyense* Foerste, from erratic blocks west of Cape Chidley in northern Labrador, probably of Black River age, and in *Endoceras baffinense* Foerste, from Frobisher Bay, on the east side of Baffin Island, probably of Trenton age, though possibly Richmond. Compared with *Endoceras nelsonense*, its siphuncle is of about the same relative size, but it is nearer the ventral wall of the conch. Compared with *Endoceras chidleyense,* the siphuncle is at about the same distance from the ventral wall of the conch, and its size is only slightly greater. Compared with *Endoceras baffinense,* the siphuncle is of about the same size, and it is located at about the same distance from the ventral wall of the conch. Evidently this Putnam Highland specimen is related most closely to *Endoceras baffinense,* but it is not known to be identical.

2. ENDOCERAS SP. (B)

(Plate II, Fig. 2; Plate X, Fig. 2)

Specimen 110 mm. long, 75 mm. in diameter, not appreciably enlarging within the length of the specimen. Nearly six camerae in a length equal to the diameter of the conch. Sutures of septa directly transverse. Concavity of septa about 20 mm. Siphuncle apparently indicated by a downward flexure of the septum at the base of the specimen, suggesting a siphuncle about 30 mm. in diameter and only moderately eccentric in location, but both size and location are too indefinite to serve for exact specific discrimination of the specimen.

Locality and Horizon. — Putnam Highland, Baffin Island; in strata corresponding to the Red River formation of southern Manitoba.

Specimen No. 10279 in the Museum of Paleontology, University of Michigan.

Remarks. — This specimen might possibly belong to the same species as the preceding, here called *Endoceras sp. (A),* but the size and location of the siphuncle cannot be determined with sufficient accuracy to verify this.

CYCLENDOCERAS Grabau and Shimer

Cyclendoceras Grabau and Shimer, North American Index Fossils, 2, p. 43 (1910).

Genotype: *Endoceras annulatum* Hall, Pal. New York, 1, p. 207, pl. 44, figs. 1a, b (1847); see also Journ. Sci. Lab. Denison Univ., 19, p. 299, pl. 31, fig. 3 (1921); also 23, pl. 3, fig. 1 (1928).

Cyclendoceras included the annulated species otherwise having the structure of *Endoceras*. In the genotype, these annulations are relatively distant from each other. In the species figured by Whiteaves (*Royal Soc. Canada, Trans.*, 9, iv, pl. 5, figs. 2, 2a, 1892) from the upper or Selkirk member of the Red River formation at East Selkirk, in southern Manitoba, under the name of *Endoceras subannulatum* (not Whitfield), the annulations are much more numerous. Species of this type occur in the Maquoketa of Iowa, and in the Richmond of Idaho, Montana, and Colorado.

3. CYCLENDOCERAS SP. (A)

(Plate I, Fig. 1; Plate X, Fig. 3)

Specimen 110 mm. long, with a lateral diameter of 45 mm., not enlarging appreciably within the length of the specimen. Six camerae occupy a length equal to the diameter of the conch. The sutures of the septa apparently are directly transverse. The concavity of the septa equals 10 mm. The septal necks are exposed, but not sufficiently well to determine whether their length equals that of one or two camerae. The siphuncle is 18 mm. in diameter, and it is estimated that it was about 2 mm. distant from the ventral wall of the conch. The cast of the interior of the conch is annulated transversely, there being 10 annulations in a length equal to that of 11 camerae. These annulations vary from one fourth to one half of a millimeter in elevation. Ventro-laterally they slope downward toward the median part of the ventral side of the conch at an angle of about 33 degrees with a horizontal plane.

Locality and Horizon. — Putnam Highland, Baffin Island; in strata corresponding to the Red River formation of southern Manitoba.

Specimen No. 10266 in the Museum of Paleontology, University of Michigan.

Remarks. — This Baffin Land specimen is similar to *Cyclendoceras annulatum* (Hall), from the Trenton at Watertown, New York, in the relative number of its camerae and annulations, compared with the diameter of the conch, and also in the relative size of its siphuncle and in the distance of the latter from the ventral wall of the conch. The annulations, however, curve much more strongly downward along the ventral side of the conch than in the New York species.

4. CYCLENDOCERAS SP. (B)

(Plate I, Fig. 2; Plate X, Fig. 4.)

Specimen 83 mm. long, 53 mm. in lateral diameter, not appreciably enlarging within the length of the specimen. Sutures of the septa apparently directly transverse. Four septa are exposed at intervals of 21 to 23 mm.; if intermediate septa were present originally, no definite trace remains in the specimen at hand. The concavity of the septa is at least 16 mm. The siphuncle is 17.5 mm. in diameter at the top of the specimen, and is located about 5 mm. from the ventral wall of the conch; the length of its septal necks cannot be determined. Nearly six annulations cross the conch in a length equal to its lateral diameter. These annulations curve downward ventro-laterally toward the median part of the ventral side of the conch, at an angle of about 33 degrees with a horizontal plane.

Locality and Horizon. — Putnam Highland, Baffin Island; in strata corresponding to the Red River formation in southern Manitoba.

Specimen No. 10274 in the Museum of Paleontology, University of Michigan.

Remarks. — This Baffin Land specimen is similar to typical *Cyclendoceras annulatum* (Hall) in the number of its annulations, and in the size and location of its siphuncle, compared with the diameter of the conch. It agrees with the New York species, moreover, in the small downward curvature of its annulations ventrally. The distance between the septa cannot, however, be determined with confidence, and there is a possibility of this distance being considerably greater than in the New York species.

ORTHOCERAS Breynius

Orthoceratites Breynius, Dissertatio Physica de Polythalmiis, pp. 12, 19, 31; pl. 3, figs. 1–7 (1732); not a generic name in accordance with the Linnaean system.

First published species: *Orthoceras regulare* Schlotheim, not illustrated, but with citation of figures differing greatly in form, and apparently in structure. See also Foerste, Journ. Sci. Lab. Denison Univ., 20, p. 218 (1924).

Typical *Orthoceras*, as originally figured by Breynius, had a shell with a smooth surface, the septa were distant, and the segments of the siphuncle were cylindrical, instead of expanded within the camerae. This genus is credited to Breynius only by courtesy, the latter not having used a binomial system of nomenclature. Schlotheim, who first adopted the binomial system in connection with this genus, included under *Orthoceratites regulare* very divergent types of orthocones, illustrating none of these himself. Hyatt was the first to confine this term to conchs with a smooth surface ("Genera of Fossil Cephalopods," *Boston Soc. Nat. Hist., Proc.* 22, p. 275, 1884), and Foerste restricted this name to those with distant septa and cylindrical segments of the siphuncle (*Journ. Sci. Lab. Denison Univ.*, 20, p. 218, 1924).

5. ORTHOCERAS SP.

(Plate II, Fig. 1; Plate X, Fig. 5)

Specimen 150 mm. long, enlarging from a lateral diameter of 54 mm. to one of 63 mm. in a length of 80 mm. The conch is depressed owing to pressure previous to fossilization. At its base, the shortest diameter is 46 mm. Two camerae occupy a total length equal to the maximum lateral diameter of the conch. The sutures of the septa are directly transverse. The concavity of the septa equals 15 to 18 mm. The siphuncle is not preserved. The surface of the conch apparently was smooth.

Locality and Horizon. — Putnam Highland, Baffin Island; in strata corresponding to the Red River formation.

Specimen No. 10261 in the Museum of Paleontology, University of Michigan.

Remarks. — In the absence of the siphuncle it is impossible to determine the generic relationship of this specimen beyond doubt, but the relatively long distance between the septa suggests that it is a typical species of *Orthoceras,* with cylindrical segments of the siphuncle.

EPHIPPIORTHOCERAS Foerste

Ephippiorthoceras Foerste, Geol. Surv. Canada, Memoir 145, p. 71, pl. 11, fig. 11, also text fig. 7.

Genotype: *Orthoceras formosum* Billings, Geol. Surv. Canada, Rept. of Progress for 1853–56, p. 317 (1857); also Foerste, as cited above.

Conchs orthoconic, with sutures of septa curving distinctly downward laterally, and with the segments of the siphuncle enlarging to vertically elliptical or spherical outlines, but not vertically depressed and disk-like, or nummuloidal.

6. EPHIPPIORTHOCERAS SP. (A)

(Plate IV, Fig. 3; Plate X, Fig. 7)

Specimen 75 mm. long, consisting of the ventro-lateral half of the conch. To judge from the outline of the cross-section at its base, its dorso-ventral diameter there is estimated at 35 mm., and its lateral one at 30 mm. The dorso-ventral diameter at the top of the specimen is estimated at 40 mm. Seven camerae occur in a length equal to the estimated dorso-ventral diameter of the conch at the top of the series counted. The sutures of the septa curve downward laterally about 7 mm. The saddles are conspicuous. The concavity of the septa is slight in a lateral direction, but conspicuous dorso-ventrally. The passage of the siphuncle through the septum at the base of the specimen appears to be obscurely indicated. This passage apparently is 2 mm. in diameter, and its center is about 10 mm. from the ventral wall of the conch.

Locality and Horizon. — Putnam Highland, Baffin Island; in strata corresponding to the Red River formation of southern Manitoba.

Specimen No. 10272 in the Museum of Paleontology, University of Michigan.

Remarks. — This Putnam Highland specimen closely resembles the *Ephippiorthoceras dowlingi* Foerste and Savage, from the Shammattawa River, on the western side of Hudson Bay, as far as can be determined from the fragment at hand. Compared with *Ephippiorthoceras baffinense* Foerste, the sutures of the septa curve more strongly downward along the lateral side of the conch.

7. EPHIPPIORTHOCERAS SP. (B)

(Plate IV, Fig. 4; Plate X, Fig. 8)

Specimen 65 mm. long, enlarging from a dorso-ventral diameter of 28.5 mm. at its base to 33 mm. at its top. The conch apparently is compressed laterally, and its lateral diameter is estimated at 26 mm. at its base, but the poor preservation of the specimen does not admit of an accurate determination of this diameter. Six camerae occur in a length equal to the dorso-ventral diameter of the conch. The sutures of the septa curve downward laterally about 3 mm. The concavity of the septa in a lateral direction is much less than dorso-ventrally. The center of the siphuncle is located 12 mm. from the ventral side of the conch at the top of the specimen, where the dorso-ventral diameter of the conch is 33 mm. The diameter of the siphuncle at its passage through the septum is 1.8 mm.

Locality and Horizon. — Putnam Highland, Baffin Island; in strata corresponding to the Red River formation in southern Manitoba.

Specimen No. 10275 in the Museum of Paleontology, University of Michigan.

Remarks. — This specimen resembles *Ephippiorthoceras dowlingi* Foerste and Savage in the small rate of enlargement of the conch, in the number of its camerae, and in the amount of downward curvature of the sutures of the septa laterally. The conch appears to have been more compressed laterally.

GARRYOCERAS, gen. nov.

Genotype: *Orthoceras semiplanatum* Whiteaves, Royal Soc. Canada, Trans., 9, iv, p. 81, pl. 8, figs. 3, 3a (1892).

Conchs with siphuncle on the more convex side of the conch, as in *Mixosiphonoceras* Hyatt (see Foerste, *Journ. Sci. Lab. Denison Univ.*, 21, p. 306, pl. 32, figs. 3 A–D, 1926), but with the dorsal side more strongly flattened, and with the sutures curved more strongly downward on this dorsal side. Moreover, the siphuncle is relatively much smaller, is cylindrical, and shows no trace of an actinosiphonate structure.

8. GARRYOCERAS SEMIPLANATUM (Whiteaves)

(Plate XI, Fig. 7 A, B)

Orthoceras semiplanatum Whiteaves, Royal Soc. Canada, Trans., 9, iv, p. 81, pl. 8, figs. 3, 3a (1892).

Original description: Shell compressed subcylindrical, increasing very slowly in thickness (at the rate, so far as can be ascertained, of three millimeters and a half in a length of five centimeters), nearly plano-convex in transverse section, one side being broadly and very gently convex and the other nearly flat, the shorter of the two diameters of the tube being about one third less than the longer, and the lateral margins narrowly rounded. Surface markings unknown; sutures of the septa shallowly concave on the flattened side, slightly convex on the other, and closely approximated, the six anterior chambers together measuring half an inch, on the median line of the flattened side; siphuncle small, cylindrical, placed close to the margin of the convex side.

Locality and Horizon. — Lower Fort Garry, in southern Manitoba; in the Selkirk member of the Red River formation.

Specimen No. 1834 in the Victoria Memorial Museum, at Ottawa, Ontario, Canada.

DOWLINGOCERAS, gen. nov.

Genotype: *Poterioceras gracile* Whiteaves, Royal Soc. Canada, Trans. 9, iv, p. 87, pl. 11, figs. 4, 4a, 4b (1892).

Conch somewhat similar to *Diestoceras* in form (see Foerste, *Journ. Sci. Lab. Denison Univ.*, 20, p. 262, pls. 25–28, 1924), but more slender, much more strongly compressed laterally, with a relatively much smaller siphuncle, enlarging but slightly within the camerae.

9. DOWLINGOCERAS GRACILE (Whiteaves)

(Plate I, Fig. 5; Plate III, Figs. 2 A, B)

Poterioceras gracile Whiteaves, Royal Soc. Canada, Trans., 9, iv, p. 87, pl. 11, figs. 4, 4a, 4b, 1892).

Original description: Shell fusiform, strongly compressed, straight and rather slender, flattened conical and obtusely pointed posteriorly, thickest at mid-length, where it is very gently convex, thence narrowing gradually and very slightly toward the aperture, which apparently is simple and broadly truncated; immediately behind the aperture there is a faint annular constriction; siphonal and antisiphonal regions narrowly rounded; sides compressed, somewhat expanded, especially at the mid-length and anteriorly; chamber of habitation occupying about one third of the entire length, which is more than twice but less than three times the maximum breadth; outline of transverse section, in the thickest part, elliptical, with the longer axis of the ellipse not quite twice the length of the shorter; surface showing indications of fine transverse costae, though the exterior of the test is not very well preserved. Septa rather closely approximated and averaging about 3 millimeters apart; siphuncle nearly marginal, moniliform, and slightly inflated between the septa, the maximum breadth of each siphuncular segment being one third less than its height or depth.

Locality and Horizon. — Black Island, near Swampy Island, Lake Winnipeg; in the Dog Head member of the Red River formation.

Specimen No. 1841 in the Victoria Memorial Museum, Ottawa, Ontario, Canada.

LEURORTHOCERAS Foerste

Leurorthoceras Foerste, Journ. Sci. Lab. Denison Univ., 19, p. 278 (1921).

Genotype: *Leurorthoceras hanseni* Foerste, loc. cit., p. 278, pls. 30, 31, 32, 34 (1921).

Conch flattened along its ventral side, the sutures of the septa with broad ventral lobes. The location of the siphuncle is distinctly ventral, but not in contact with the ventral wall of the conch. The septal necks are distinct and relatively long. The connecting rings enlarge within the camerae, and vary from vertically elliptical to approximately globular in outline, but they are not enlarged sufficiently in a lateral direction to secure a flattened disk-like appearance, as in typical *Actinoceras*.

10. LEURORTHOCERAS (?) BAFFINENSE, sp. nov.

Plate III, Fig. 1; Plate X, Figs. 6 A, B)

Specimen 170 mm. long, and 93 mm. wide at its broadest part. It consists of three fragments. It is estimated that the conch originally was about 140 mm. in diameter, but this cannot be determined with any accuracy. It is estimated also that nearly eleven camerae occur in a length equal to the diameter of the conch. The sutures of the septa curve downward at least 12 mm. on the ventral side of the conch, producing broad ventral lobes. The septa slope downward on the ventral side of the siphuncle at an angle of 40 degrees with the ventral vertical outline of the conch. From this it is assumed that the concavity of the septa was considerable. The siphuncle is about 5 mm. distant from the ventral wall of the conch. Its diameter is 16 mm., at least dorso-ventrally. The septal necks are 4 or 5 mm. long where the septa are 11 or 12 mm. distant from each other vertically. These septal necks curve concavely inward a distance of 4 or 5 mm. on the ventral side of the siphuncle and about 2 mm. on its dorsal side. In consequence, the lower half of each segment of the siphuncle appears to be prominently annulated at a strongly oblique angle, the slope being toward the interior of the conch. No other details are preserved.

Locality and Horizon. — Putnam Highland, Baffin Island; in strata corresponding to the Red River formation of southern Manitoba.

Specimen No. 10283 in the Museum of Paleontology, University of Michigan.

Remarks. — The specimen here described as *Leurorthoceras baffinense* is closely related to the second specimen figured by Troedsson under the name *Leurorthoceras* (?) *ruedemanni* ("On the Middle and Upper Ordovician Faunas of Northern Greenland; I, Cephalopods," pl. 31, figs. 3, 4; 1926). This specimen was numbered 137; and it came from the Gonioceras Bay formation, in the lower part of the Black River series of strata as exposed at Cape Calhoun, along the northern part of Kane Basin, on the coast of Greenland.

Compared with the latter, the Putnam Highland specimen here described has longer segments of the siphuncle, compared with the maximum dorso-ventral diameter of the latter, and the septal necks are relatively longer. Unfortunately, not enough of the transverse outline of the conch of the Putnam Highland specimen remains to insure that its outline is similar to that of typical *Leurorthoceras*, in which the ventral outline is distinctly flattened compared with the remainder of the outline of the conch, the latter forming almost a semicircle. Under these circumstances, its reference to the genus *Leurorthoceras* can be only tentative.

GONIOCERAS Hall

Gonioceras Hall, Pal. New York, 1, p. 54 (1847).

Genotype: *Gonioceras anceps* Hall, loc. cit., p. 54, pl. 14, figs. 1a–d (1847).

Conch strongly depressed dorso-ventrally, with its ventral face flat or only slightly convex, and with its dorsal face gently convex, these two faces being separated by the narrowly acute lateral margins of the conch. On approaching the lateral margin of the conch, the dorsal face of the conch tends to become slightly concave in some specimens. The sutures of the septa curve strongly downward along the median part of both the dorsal and ventral faces of the conch. Along the lateral fourths or fifths of both of these faces the sutures curve strongly upward, forming very conspicuous saddles, and then they curve strongly downward as far as the acute margins of the conch, where they form very deep and narrowly angulated lateral lobes, which embrace both the dorsal and the ventral faces of these margins. The siphuncle is located slightly nearer the flat ventral face of the conch. Its segments are strongly nummuloidal, of small height, but broad.

Lambeoceras Foerste (*Journ. Cincinnati Soc. Nat. Hist.*, 22, p. 44, pl. 1, figs. 3 A–D; pl. 3, fig. 2, 1917) differs in having the dorsal and ventral faces of the conch equally convex, and its lateral margins less acute. The sutures of the septa curve downward along the entire width of both faces, but near the lateral margins they curve more directly outward. The siphuncle is located much nearer the ventral wall of the conch. The segments of the siphuncle are nummuloidal, and similar in form to those of *Gonioceras*.

The two specimens here described from western Baffin Land are referred to *Gonioceras* because their ventral sides present no evidence of distinct convexity, as in typical *Lambeoceras*; it is possible, however, that more complete and better preserved specimens might lead to an opposite opinion. Unfortunately, these specimens do not preserve their lateral parts, so that the course of their septa near the lateral margins cannot be determined.

11. GONIOCERAS (?) SP. (A)

(Plate I, Fig. 4)

Specimen 37 mm. in length along the median line of the specimen. Ventral side apparently nearly flat, and hence the specimen is assumed to be a *Gonioceras*, rather than a *Lambeoceras*. In this length of 37 mm. there are five camerae. The sutures of the septa curve strongly downward both ventrally and dorsally, but only the ventral side of the specimen is preserved fairly well. Since the lateral parts of the conch are not preserved, it cannot be determined whether the sutures here recurved strongly, as in *Gonioceras*, or only moderately, as in *Lambeoceras*. In a dorso-ventral direction the septa are nearly horizontal along the median part of the conch, but slope distinctly downward from the ventral toward the dorsal side along the more lateral parts of the specimen. The siphuncle is faintly indicated along the septum at the base of the specimen, and here is located about as close to the ventral wall as in *Lambeoceras*.

Locality and Horizon. — Putnam Highland, Baffin Island; in strata corresponding to the Red River formation in southern Manitoba.

Specimen No. 10289 in the Museum of Paleontology, University of Michigan.

Remarks. — Not enough of the specimen remains to determine whether is it a *Gonioceras* or a *Lambeoceras*. It is characterized by the slightly angular concavity of the sutures of its septa along the median part of the ventral side of the conch.

12. GONIOCERAS (?) SP. (B)

(Plate V, Fig. 4)

Specimen 57 mm. in length along the median part of its ventral side, which is the only part preserved. This ventral part is nearly flat, suggesting the reference of the specimen to *Gonioceras*, rather than to *Lambeoceras*. Unfortunately, the lateral parts of this ventral side are not preserved, so that it is not possible to determine whether the lateral extensions of the sutures of the septa are strongly reversed in curvature, as in typical *Gonioceras*, or are only moderately reversed and for only a short distance, as in *Lambeoceras*. Ten camerae occupy a length of 52 mm. along the median line of the ventral side. The sutures of the septa curve downward distinctly, evidently producing lobes both on the ventral and dorsal sides of the flattened conch. To judge from that part of the conch preserved, the original width of these lobes was at least 65 mm., but may have been considerably greater. The siphuncle is obscurely indicated along the ventral side of the specimen, apparently almost in contact with the latter, but nothing definite is known about its segments.

Locality and Horizon. — Putnam Highland, Baffin Island; in strata corresponding to the Red River formation of southern Manitoba.

Specimen No. 10264 in the Museum of Paleontology, University of Michigan.

Remarks. — Not enough of the specimen remains to determine whether it is a *Gonioceras* or a *Lambeoceras*. Compared with the preceding specimen (*A*), the camerae appear to be relatively more numerous.

WESTONOCERAS Foerste

Westonoceras Foerste, Journ. Sci. Lab. Denison Univ., 20, p. 253 (1924).

Genotype: *Cyrtoceras manitobense* Whiteaves.

Conch faintly concave along the dorsal outline, but more or less distinctly convex along the upper part of the phragmacone and the lower part of the living chamber, becoming faintly concave again toward the aperture. Ventral outline convex along its entire length, its maximum convexity being located five or six camerae beneath the base of the living chamber. Cross-section laterally compressed, especially at the point of the maximum diameter of the conch. From this point the cross-section of the conch contracts along the upper part of the phragmacone and along the entire living chamber. The sutures of the septa curve distinctly downward laterally in the genotype. The siphuncle is located near the ventral side of the conch but is not in contact with the latter. The inner margins of the septa, at the passage of the siphuncle through these septa, do not curve downward into septal necks, but end abruptly. The intervening connecting rings are in contact with the septa, both above and beneath. These connecting rings are conspicuously larger in diameter than the passages of the siphuncle through the septa, but their lateral vertical outlines are only moderately convex.

The two species from western Baffin Land, here described

under *Westonoceras,* differ from the genotype in having the sutures of the septa nearly straight, except along the upper part of the ventral outline of their phragmacones, where these sutures curve strongly upward.

13. WESTONOCERAS MANITOBENSE (Whiteaves)

(Plate V, Fig. 1; Plate XI, Fig. 5)

Cyrtoceras manitobense Whiteaves, Trans. Royal Soc. Canada, 7, sec. 4, p. 80, pl. 13, figs. 3, 4, 5; pl. 15, fig. 4 (1890).

Type specimen. — Specimen 127 mm. long, obliquely depressed dorso-laterally, so that the dorsal and left lateral sides of the conch are exposed equally well. In general, the dorsal outline of the conch is slightly convex, the upper part of the phragmacone, for a length of 65 mm., being almost straight. The ventral outline is much more strongly convex; the maximum convexity is somewhere between 30 and 40 mm. beneath the top of the phragmacone, where there is a tendency toward angularity. Owing to the oblique depression of the conch, it is not possible to determine exactly the length of the dorso-ventral diameters of the conch, but it is estimated that between eight and nine camerae occupied a length equal to the dorso-ventral diameter along the upper part of the phragmacone, a similar number occurring at its lower end. The sutures of the septa curve distinctly downward laterally, even at the lower end of the specimen, where its dorso-ventral diameter is only 13 mm. Along the lower part of the specimen the ventral saddles are on the same level as the dorsal ones. Along the upper part of the phragmacone, however, the ventral saddles attain a distinctly higher elevation. The dorsal saddles are broadly rounded; the ventral ones, on the contrary, are distinctly angular, at least at the base of the specimen. The center of the siphuncle is 3.5 mm. from the ventral wall of the conch, where its dorso-ventral diameter is 13 mm. The diameter of the siphuncle here is 3 mm., and its distance from the ventral wall is 2 mm. The segments of the siphuncle are oblong in outline, their vertical outlines being straight, except in the immediate vicinity of the septa, where they curve abruptly inward. The concavity of the septa is relatively slight, and their inner margins, at the passage of the siphuncle, end abruptly, without any downward deflection forming a neck. This passage of the siphuncle through the septum is about 1 mm. in diameter where the diameter of the siphuncle is 3 mm.

The specimen is a cast of the interior of the conch, and the surface of this cast is ornamented by low, broad, vertical ribs, separated by narrow grooves, numbering about forty within the circumference of the conch at mid-height of the phragmacone. These vertical ribs are very distinct along the lower part of the phragmacone, but disappear farther up.

The maximum dorso-ventral diameter of the conch is about six or seven camerae beneath the base of the living chamber. From this point the conch narrows both dorso-ventrally and laterally as far as the aperture. This aperture slopes moderately downward from the ventral toward the dorsal side of the conch, without any indication of a hyponomic sinus, as far as known. The cast of the interior of the living chamber is transversely striated, the coarser markings numbering about five in a length of 5 mm., and the finer ones numbering about five in a length of 2 mm. In addition there appear to be distinct vertical striae, numbering about eight in a width of five mm. Some of these vertical marks appear like file-marks, so that the presence of these marks upon the specimen in its original state is in doubt.

Locality and Horizon. — Big Island, in Lake Winnipeg, in southern Manitoba; from the Lower Mottled or Dog Head limestone, beneath the Cat Head limestone of Dowling's Report;[4] in the Red River formation.

Specimen No. 1879 in the Victoria Memorial Museum, Ottawa, Canada.

[4] Dowling, D. B. "Report on the Geology of the West Shore and Islands of Lake Winnipeg," *Geological Survey of Canada, Annual Report,* Vol. XI, part F, p. 53 (1900).

14. WESTONOCERAS PUTNAMI, sp. nov.

(Plate V, Fig. 3; Plate XI, Figs. 3 A, B)

Specimen 105 mm. long, relatively straight along its dorsal outline, as far as preserved, moderately humped ventrally, the maximum elevation of the hump being at the fifth or sixth camera beneath the living chamber. Above and below this hump, the ventral outline of the conch appears to be only slightly convex, or nearly straight. The conch enlarges dorso-ventrally from a diameter of 11 mm. at its base to 32 mm. at the hump, and then narrows to about 27 mm. at the base of the living chamber, and to 22 mm. at the top of this chamber, as far as preserved. It is assumed that in its original state the specimen was moderately compressed laterally, but at present it is compressed obliquely by pressure, thus giving the conch a wider aspect dorsally than it probably had originally. At the level of the hump on its ventral side, the lateral diameter of the conch originally may have been approximately 28 mm. Along the lower half of the phragmacone the sutures of the septa are straight along their entire course around the conch; but, beginning at about the ninth suture beneath the living chamber, the sutures rise increasingly on approaching the median part of the ventral side of the conch, forming saddles about 5 mm. in height along the upper four or five sutures. No downward curvature of the sutures is present along the median part of their downward course. The sutures are at right angles to the vertical dorsal outline of the conch along the upper half of the phragmacone, but slope increasingly downward in a ventrad direction on approaching the lower end of the specimen.

At the lower end of the specimen, where its diameter is estimated at 13 mm., the diameter of the siphuncle is 3.5 mm. The concavity of the septa, at least in a lateral direction, is relatively small. The inner margin of these septa is not curved downward into a septal neck at the passage of the siphuncle through the septa. This passage is 1.5 mm. in diameter where the diameter of the segment of the siphuncle is 3.5 mm. A vertical section through the lower part of the siphuncle in a ventro-lateral direction suggests that the ventral outline of its segments was almost straight, while its dorsal vertical outline appears to have been straight along the lower part of each segment, but moderately curved inward along the upper two fifths of each segment. No actinoceroid deposits occur within this siphuncle. The distance of the siphuncle from the ventral wall of the conch cannot be determined with exactness, but it is estimated at about 1 mm.

The upper part of the conch narrows from the hump on its ventral side as far as the aperture. This aperture is not preserved well enough to establish its characteristics, but it appears to have been similar to that of *Westonoceras manitobense.*

Locality and Horizon. — Putnam Highland, Baffin Island; in strata corresponding to the Red River formation in southern Manitoba.

Specimen No. 10273 in the Museum of Paleontology, University of Michigan.

Remarks. — This species is distinct from typical *Westonoceras manitobense* in the absence of distinct lateral lobes or downward curvatures along the sutures of its septa. The structure of its siphuncle appears to be similar.

15. WESTONOCERAS GOULDI, sp. nov.

(Plate V, Fig. 2; Plate XI, Figs. 1 A, B, C)

Specimen 120 mm. long, nearly straight along its dorsal outline, but conspicuously humped along its ventral side, the maximum convexity of this hump being five camerae beneath the base of the living chamber. Above and below this hump the ventral vertical outline of the conch is nearly straight. At the base of the specimen its dorso-ventral diameter is 22 mm., and its lateral one is 21.5 mm. At the level of the ventral hump mentioned above the corresponding diameters are 56 mm. and 43 mm., but the original lateral diameter here may have been a little greater. At the base of the living chamber these diameters are 49 and 38 mm. At the top of the specimen, about 34 mm. above the base of the living chamber, they are 38 mm. and 32 mm. Along the lower part of the specimen, below the level of the ventral hump, eleven camerae occur in a length equal to the dorso-ventral diameter of the conch at the top of the series counted. Along the lower half of the phragmacone the sutures of the septa are straight, but along the upper half of the specimen they rise distinctly along the ventral side of the specimen, forming conspicuous ventral saddles which, in some specimens, rise fully 5 mm. above the remaining portion of the same sutures laterally and dorsally. Along the median part of the dorsal side of the phragmacone, the upper five or six sutures curve narrowly downward, the upper two or three lobes having a depth of nearly 3 mm. The fifth suture below the living chamber is at right angles to the dorsal outline of the conch. Below this level the sutures slope increasingly downward in a ventral direction. Above this level, they rise slightly until close to the ventral saddles, where they curve strongly upward. At the base of the specimen, the concavity of the septa is slight.

At the base of the specimen, where its dorso-ventral diameter is 22 mm., the center of the siphuncle is 6 mm. from the ventral wall of the conch. The passage of the siphuncle through the basal septum here is 2.5 mm. in diameter. The lunular deposit of calcareous material embracing the inner margin of this lowest septum, at the passage of the siphuncle, is 5 mm. in diameter. The total diameter of the segments probably was at least 6.5 mm., but the connecting rings forming these segments no longer are preserved.

The specimen is a cast of the interior of the conch, and along the phragmacone the surface of this cast is marked by low, broad, vertical ribs, separated by narrow grooves. Ten of these ribs occupy a width of 27 mm. at the tenth camera above the base of the specimen. Evidently, the upper part of the living chamber is not preserved.

Locality and Horizon. — Putnam Highland, Baffin Island; in strata corresponding to the Red River formation of southern Manitoba.

Specimen No. 10263 in the Museum of Paleontology, University of Michigan.

Remarks. — This specimen evidently is related generically to *Westonoceras putnami,* from the same locality and horizon. It has the same straight course of the sutures of the septa dorsally and laterally, changing to an upward curvature ventrally, especially along the upper part of the phragmacone. In this respect both species differ conspicuously from typical *Westonoceras,* in which the sutures curve distinctly downward laterally.

16. WESTONOCERAS cf. GOULDI Foerste

(Plate I, Figs. 3 A, B; Plate XI, Figs. 2 A, B)

The specimen consists of seven camerae from the lower part of a phragmacone, their total length being 25 mm. The dorso-ventral diameter of the specimen enlarges from 22 mm. at its base to 35 mm. at its top. The corresponding lateral diameters are 21 mm. and 33 mm. The number of camerae in a length equal to the dorso-ventral diameter at the top of the series counted is nine. The sutures of the septa curve slightly downward laterally, the low ventral saddles being slightly angulated, while the median part of the much broader dorsal saddles is slightly curved downward into a faint lobe. The distance of the center of the siphuncle from the ventral wall of the conch varies from nearly 7 mm. at the base of the specimen to 10 mm. at its top. The corresponding diameters of the siphuncle vary from 6 mm. to 8 mm., and the corresponding diameters of the passages of the siphuncle through the septa vary from 2.8 mm. to slightly over 3 mm. The vertical outlines of segments of the siphuncle are slightly convex, the convexity increasing in the immediate vicinity of the septa. The inner margin of the septa at the passage of the siphuncle is not curved downward into a septal neck, but ends abruptly without any change in the slight general concavity of the septa. This inner margin of the septa is embraced by whitish calcareous deposits which, in vertical sections, appear lunulate, but depressed in a vertical direction. At the base of the specimen these lunulate deposits close the passage of the siphuncle centrally, but at its top they leave a passage 3 mm. in diameter.

The specimen is a cast of the interior of the conch. The surface of this cast is marked by fifty-five to sixty vertical, low, broad ribs, separated by narrow grooves, which are better defined laterally than along the median part of its ventral and dorsal sides.

Locality and Horizon. — Putnam Highland, Baffin Island; in strata corresponding to the Red River formation of southern Manitoba.

Specimen No. 10270 in the Museum of Paleontology, University of Michigan.

Remarks. — This specimen is assumed to be a fragment of a specimen of *Westonoceras gouldi.* In favor of such an interpretation are the rate of enlargement of the specimen, the location of its siphuncle, and the character of the vertical ribbing on its surface. The size of the passage of the siphuncle through the septa corresponds, and apparently the size of the segments of the siphuncle was similar. It differs from the type of *Westonoceras gouldi*, however, in the distinct, though slight, downward curvature of the sutures of its septa laterally. In this respect it resembles *Westonoceras manitobense,* in which the sutures of the septa curve much more strongly downward.

WINNIPEGOCERAS, gen. nov.

Genotype: *Cyrtoceras laticurvatum* Whiteaves, Geol. Surv. Canada, Pal. Foss., 3, pt. 3, p. 224, text fig. 14 (1897).

Conch strongly curved in a lengthwise direction, with the ventral side convex and the dorsal side strongly concave, differing in this respect from *Westonoceras,* in which the lengthwise curvature is relatively small, the dorsal outline being slightly gibbous along the upper part of the phragmacone and the lower part of the living chamber. In *Winnipegoceras* the conch is humped along the upper part of its ventral outline at a point five or six camerae below the base of the living chamber, and this chamber is long and attenuated toward the aperture, as in the genus named. Although the vertical dorso-ventral section through the genotype passes directly through the center of the siphuncle for most of its length, the details of structure of this siphuncle still remain vague.

17. WINNIPEGOCERAS LATICURVATUM (Whiteaves)

(Plate IX, Figs. 1, 2)

Cyrtoceras laticurvatum Whiteaves, Geol. Surv. Canada, Pal. Foss., 3, pt. 3, p. 224, text fig. 14 (1897).

Type. — Specimen (Plate IX, Fig. 1) 330 mm. long, measured along its convex ventral outline. The convexity of this outline increases moderately from the smaller end of the specimen toward the upper part of the phragmacone, reaching a maximum at 50 mm. beneath the base of the living chamber. This chamber is 70 mm. long and is relatively straight. The dorso-ventral diameter of the conch increases from 20 mm. at a point 15 mm. above the base of the specimen to 55 mm. at a point 50 mm. beneath the base of the living chamber. At the base of this chamber the diameter is estimated at 45 mm., and near the aperture it diminishes to 36 mm. Where the dorso-ventral diameter of the conch is 55 mm., its lateral diameter is estimated at 38 mm. The sutures of the septa curve strongly downward laterally. The concavity of the septa is correspondingly great in a dorso-ventral direction, but is relatively smaller at right angles to this direction. The siphuncle is near the convex outline of the conch, but not in contact with the ventral wall. Where the dorso-ventral diameter of the conch is 50 mm., the corresponding diameter of the siphuncle is 8 mm., and its distance from the ventral wall of the conch is slightly over 2 mm. Owing to the strong concavity of the septa, the upper and lower faces of the segments of the siphuncle are correspondingly oblique. Along the lower half of the phragmacone, the vertical outlines of the segments are distinctly convex, and their general form is nummuloidal. Along the upper half of the phragmacone, the vertical outlines are less convex, the segments are relatively taller, and their general form is more oblong. In both parts of the phragmacone, the ventral outlines of the segments project in a ventral direction beyond the corresponding outlines of the segments beneath, in a scalariform manner. The structure of the siphuncle is not clear. In some parts of the siphuncle the segments appear to be in contact with the septa both above and below, at least along the ventral side of the base and the dorsal side of the top of the segments. In those cases where a constriction of the siphuncle appears to intervene between the septa and the adjacent part of the segments of the siphuncle, this constriction is located on the ventral side of the top and on the dorsal side of the base of the segments. In no part of the specimen, however, has any septum been observed to curve downward into a septal neck.

Locality and Horizon. — Little Black Island, Swampy Harbor, Lake Winnipeg, in southern Manitoba; in the Lower Mottled limestone of Dowling's *Report on the Geology of the West Shore and Islands of Lake Winnipeg* (1900), here called the basal or Dog Head member of the Red River formation.

Specimen No. 7148 in the Victoria Memorial Museum, Ottawa, Canada.

In another specimen, from the same horizon and locality, the surface of the cast of the interior of the conch is marked longitudinally by low ribs, separated by narrow grooves, eight ribs occupying a width of 28 mm. where the dorso-ventral diameter of the conch is 40 mm. Specimen No. 7148a in the Victoria Memorial Museum.

A third specimen (Plate IX, Fig. 2) was found at the Point off Moose Creek, on Lake Winnipeg. It is of the same size as the type specimen and preserves almost as great a length of the phragmacone. Its most striking feature, however, is the living chamber, which is 100 mm. in length, and almost straight, except at the top where it curves slightly in a dorsal direction. At its base the dorso-ventral diameter of the chamber is 36 mm., narrowing to 33 mm. at the aperture. At the fourteenth septum beneath the living chamber the suture of this septum curves 3 mm. downward laterally; at the third septum this downward curvature equals 8 mm. At the largest dorso-ventral diameter, five camerae occur within a length equal to this diameter, when counted along the ventral outline. Along the lower part of the phragmacone seven camerae occur within a corresponding length. Specimen No. 7153, Victoria Memorial Museum, Ottawa, Canada. From the Lower Mottled or Dog Head limestone.

18. WINNIPEGOCERAS DOWLINGI, sp. nov.

(Plate VIII, Fig. 1)

Conch attaining a dorso-ventral diameter of 65 mm. at the hump along the convex ventral outline of the conch. Apparently not fully mature, since the uppermost camera is not conspicuously shorter than those directly beneath. It is assumed that at later stages of growth the living chamber was elongated and relatively narrow, as in the genotype *Winnipegoceras laticurvatum* (Whiteaves), and that at gerontic stages of growth the base of the living chamber was 25 to 30 mm. above the base as exhibited in the specimen here described. Of course, there is a possibility of a difference of size, within the same species, corresponding to sex, but of that nothing is known at present.

Locality and Horizon. — Little Tamarack Island, in Lake Winnipeg; in the Lower Mottled or Dog Head member of the Red River formation.

Specimen No. 7152 in the Victoria Memorial Museum, Ottawa, Ontario, Canada.

CYRTOGOMPHOCERAS Foerste

Cyrtogomphoceras Foerste, Journ. Sci. Lab. Denison Univ., 20, p. 267 (1924).

Genotype: *Oncoceras magnum* Whiteaves, Royal Soc. Canada, Trans. 7, iv, p. 79, pl. 15, fig. 1 (1890).

Conch distinctly curved lengthwise, endogastric, and compressed laterally. The dorsal outline is conspicuously convex; the ventral outline is concave along the lower part of the phragmacone and along the upper part of the living chamber, but is more or less distinctly convex along the upper part of the phragmacone and the lower part of the living chamber. The sutures of the septa are nearly straight or curve slightly downward laterally. Along the lower part of the specimen they are nearly directly transverse to the curving vertical axis of the conch, but toward the top of the phragmacone these sutures rise at an increasing angle from the ventral toward the dorsal side of the conch. The aperture of the living chamber has a corresponding obliquity, and this chamber narrows conspicuously toward the aperture. The siphuncle is relatively large; it is located near the ventral wall of the conch, and its segments are strongly nummuloidal.

19. CYRTOGOMPHOCERAS MAGNUM (Whiteaves)

(Plate VI, Fig. 1)

Oncoceras magnum Whiteaves, Trans. Royal Soc. Canada, 7, sec. 4, p. 79, pl. 15, fig. 1 (1890).

Specimen 187 mm. long. Its ventral side is moderately concave along the lower part of the phragmacone, but is distinctly gibbous for a length of 75 mm. along the upper part of the phragmacone and the lower part of the living chamber. At a point 43 mm. above the ventral margin of the base of the living chamber this ventral outline becomes concave again, as far as the aperture. The amount of the gibbosity equals 4 mm. The dorsal outline of the conch is much more strongly convex. Directly opposite the ventral margin of the base of the living chamber the radius of convex curvature of this vertical dorsal outline is 80 mm., changing to 180 mm. along the lower part of the phragmacone, and to 140 mm. along the top of the phragmacone and the entire length of the living chamber. The dorso-ventral diameter enlarges from 52 mm. at the base of the specimen to 106 mm. at the fourth camera beneath the dorsal margin of the base of the living chamber, and then decreases to 65 mm. at the aperture. Nine camerae occupy a length equal to the maximum dorso-ventral diameter of the conch, when measured along its dorsal outline, immediately beneath this diameter. The sutures of the septa are relatively straight. The ninth septum beneath the living chamber has a suture almost at right angles with the general vertical outline of the ventral side of the conch.

Below this level the sutures slope increasingly downward in a dorsal direction, the lowest septum at an angle of about 65 degrees with this outline. Above this level, the sutures rise at an increasing angle, until the suture at the base of the living chamber forms an angle of about 60 degrees in the opposite direction. Along the median part of the ventral side of the conch, the sutures tend to be angulate. In addition there is a slight tendency toward an upward deflection of the suture at the base of the living chamber at points about 22 mm. on each side of the median line of the conch. This results in an apparent downward curvature of the median part of the suture for a total width of 44 mm., but this downward deflection is slight. The margin of the aperture slopes downward in a ventral direction at about the same angle as the suture at the base of the specimen. The concave vertical outline along the upper part of the ventral side of the living chamber diminishes laterally and disappears along the dorsal side of the chamber. The attachment ring at the base of the living chamber is represented by a deep groove on the cast of the interior of the conch, varying from 5 to 7 mm. in height. The surface of this cast is marked also by low vertical ribs, especially along the upper part of the lateral sides of the phragmacone. Much fainter indications occur along the lower parts of the living chamber. The siphuncle is about 13 mm. in diameter at the base of the specimen, and apparently is within 1 mm. from the ventral wall of the conch. It is exposed relatively close to the ventral wall also half-way between the base and the top of the phragmacone.

Locality and Horizon. — East Selkirk, in southern Manitoba; in the Upper Mottled or Selkirk limestone, above the Cat Head limestone of Dowling's *Report on the Geology of the West Shore and Islands of Lake Winnipeg* (1900), where it forms the upper part of the Red River formation of southern Manitoba.

Specimen No. 1875 in the Victoria Memorial Museum, Ottawa, Ontario, Canada.

20. CYRTOGOMPHOCERAS WHITEAVESI (Miller)

(Plate VII, Figs. 1, 2, 3)

Oncoceras gibbosum Whiteaves (not Hall, 1847), Trans. Royal Soc. Canada, 7, sec. 4, p. 80, pl. 15, figs. 2, 3 (1890).

Oncoceras whiteavesi Miller, North Amer., Geol. Pal., 1st App., p. 697 (1892).

Type. — Specimen (Plate VII, Fig. 1) 102 mm. long, enlarging in a dorso-ventral direction from an estimated diameter of 32 mm. at the base of the specimen to 62 mm. at the base of the living chamber, and then diminishing to an estimated diameter of 31 mm. at the aperture of this chamber. The ventral vertical outline of the conch is distinctly gibbous for a length of 46 mm., the gibbosity equalling 5 mm. and extending about equal distances above and below the ventral margin of the base of the living chamber. Toward the aperture and along the lower part of the phragmacone this ventral outline is moderately, but distinctly, concave. The siphuncle, therefore, has an endogastric location, near, but not in contact with the ventral wall of the conch. The dorsal outline of the conch is strongly convex, having a radius of curvature of 60 mm. along the upper half of the phragmacone and along the living chamber. Along the lower part of the phragmacone this radius may have equalled 80 mm. About eight camerae occur in a length equal to the maximum diameter of the conch when counted immediately below this diameter along the dorsal outline of the conch. At the base of the specimen the sutures of the septa are almost directly transverse to the vertical ventral outline of the conch, but toward the top of the phragmacone these sutures rise at an increasing rate in a ventral direction, the uppermost septum forming an angle of about 60 degrees with the general ventral outline of the conch. The concavity of the septa is slight at the base of the specimen, but increases to 12 mm. at its top. The sutures of the septa are relatively straight. The three upper segments of the siphuncle are 15 mm. in diameter, the next lower segment has a diameter of 16 mm., and the two directly beneath measure 14 mm. The distance of these segments from the ventral wall of the conch, in the same sequence, decreases from 4 to 1 mm. These segments appear strongly nummuloidal, being separated by constrictions at which the passage of the siphuncle through the septa is greatly narrowed. At the top of the siphuncle this constriction equals 8 mm. in diameter where the diameter of the segments of the siphuncle equals 15 mm. At the top of the siphuncle the septum appears to curve downward into a septal neck 1 mm. in length, the upper surface of the immediately underlying segment being that distance below the septum discussed. Elsewhere the septa appear to have sagged downward, until in contact with the top of the underlying segments.

The living chamber contracts strongly toward the aperture, laterally as well as dorso-ventrally. There is a faint concave depression in its vertical outlines about 10 mm. beneath the aperture. This aperture slopes downward in a ventral direction, apparently at about the same angle as the suture at the base of the living chamber. The annular ring at the base of this chamber is about 4 mm. in height, and is crossed by short, low, flat, vertical ribs, as in numerous other gomphoceroid cephalopods.

The shell of the conch is relatively thick, fully equalling 2 mm. on some parts of the conch. The shell is striated transversely, five raised lines occurring in a length of 5 mm. at a point 10 mm. above the base of the specimen, along its ventral side.

Locality and Horizon. — Pike Head, Lake Winnipeg, in southern Manitoba; in the Lower Mottled or Dog Head limestone, beneath the Cat Head limestone of Dowling's *Report on the Geology of the West Shore and Islands of Lake Winnipeg* (1900); in the Red River formation in southern Manitoba.

Specimen No. 1877, Victoria Memorial Museum, Ottawa, Canada.

Swampy Island specimen. — The specimen represented by figure 3 on plate 15 accompanying the original description of this species by Whiteaves is numbered 1882 in Victoria Memorial museum. Along the basal part of the specimen five transverse, low and broad, raised lines occur in a length of 7 mm.; farther up they becomes less numerous. Only the basal part of the specimen was figured.

Lower Moose Island specimen. — The transverse raised lines are shown much better by a specimen from Lower Moose Island, 90 mm. long, including the living chamber. Here the crests of six lines occur in a length of 5 mm. At a point 50 mm. above the base of the specimen six lines occupy a length of 9 mm. The margin of the aperture of the living chamber slopes strongly downward in a ventral direction, but there is no evidence of a hyponomic sinus, as supposed when studying the specimens of *Cyrtogomphoceras* found by Professors Savage and Van Tuyl in the area west of Hudson Bay. Specimen No. 7145 in the Victoria Memorial Museum, Ottawa, Canada. Plate VII, Fig. 3.

Black Island specimen. — A specimen preserving the natural dorsal and ventral outlines of the conch better than the type was found on Black Island, at Swampy Harbor, in Lake Winnipeg. It is represented by Figure 2 on Plate VII, and is numbered 7146 in the Victoria Memorial Museum, Ottawa, Canada.

21. CYRTOGOMPHOCERAS BAFFINENSE, sp. nov.

(Plate IV, Fig. 1; Plate XI, Fig. 4)

Specimen 115 mm. long. Its ventral side is relatively straight. From a point 27 mm. beneath the base of the living chamber to a point 28 mm. above this base the ventral vertical outline is slightly gibbous; this gibbosity does not exceed 3 mm. Above and below this gibbosity, the ventral outline is slightly concave. It is assumed that the lower part of the phragmacone, missing in the specimen at hand, was slightly concave along its length, so that the location of the siphuncle is endogastric. The dorsal side of the conch is conspicuously convex, the radius of curvature of its dorsal outline being 70 mm. On a level with the ventral margin of the base of the living chamber the dorso-ventral diameter of the conch is 65 mm., and the lateral one is about 50 mm. At the base of the specimen the corresponding diameters are estimated at 39 mm. and 30 mm.; and at the aperture of the living chamber they are approximately 41 mm. and 30 mm. Eight camerae occupy a length along the dorsal side of the conch equal to the dorso-ventral diameter of the latter, when they are counted immediately below the level of the maximum dorso-ventral diameter of the conch. At the base of the specimen the sutures of the septa are almost directly transverse to the curving vertical axis of the conch. On approaching the base of the living chamber, however, the sutures rise at increasing rates from the ventral toward the dorsal side of the conch, the suture at the base of the living chamber rising at an angle of about 30 degrees with the horizontal plane. At the base of the specimen the sutures of the septa are nearly straight; but, at the top of the phragmacone the sutures curve downward about 5 mm., forming broad and shallow lateral lobes. The margin of the aperture appears to be approximately parallel to the base of the living chamber. The living chamber is constricted strongly from the base of the chamber to within 5 or 6 mm. from the margin of the aperture, above which the constriction is distinctly less, especially ventrally. This specimen is a cast of the interior of the conch, and the exterior of this cast is ribbed vertically, ten ribs occupying a width of 22 mm. ventro-laterally. The siphuncle is not exposed, but is assumed to be located near the relatively straight side of the conch, as in other species of *Cyrtogomphoceras*.

Locality and Horizon. — Putnam Highland, Baffin Island; in strata corresponding to the Red River formation of southern Manitoba.

Specimen No. 10262 in the Museum of Paleontology, University of Michigan.

Remarks. — This specimen differs from typical *Cyrtogomphoceras whiteavesi* (Miller) chiefly in the somewhat smaller curvature of its vertical dorsal outline, and in its moderately greater size. It evidently is closely related to the latter. From *Cyrtogomphoceras nutatum* Foerste and Savage it differs in the greater crowding of the camerae along the upper part of the ventral side of the phragmacone. Moreover, along the median part of this ventral side the sutures of the septa are relatively straight for a width of nearly 15 mm., with a tendency toward angulation on each side of this straightened part, a feature similar to that observed in some specimens of *Cyrtogomphoceras whiteavesi*, but not in *Cyrtogomphoceras nutatum*.

Second specimen. — A second specimen (Plate IV, Fig. 2; Plate XI, Fig. 6) of *Cyrtogomphoceras*, numbered 10278, consists of a fragment of the upper part of a phragmacone. Its siphuncle is 15 mm. in diameter, where that of the conch is 65 mm. The distance of this siphuncle from the ventral wall of the conch is 4 mm. The sutures of the septa do not rise as strongly dorsally as along the upper part of the phragmacone in the type of *Cyrtogomphoceras baffinense.*

22. CYRTOGOMPHOCERAS cf. TURGIDUM Troedsson

(Plate VII, Fig. 4)

Cyrtogomphoceras turgidum Troedsson, On the Middle and Upper Ordovician Faunas of Northern Greenland; I, Cephalopods, p. 97, pl. 2, fig. 2; pl. 57, figs. 1, 2; pl. 58, figs. 2–4.

Conch similar to *Cyrtogomphoceras whiteavesi* (Miller), but enlarging more rapidly at early stages of growth, and curving more strongly at the hump along the dorsal outline, and more distinctly concave along the ventral outline of the lower part of the phragmacone.

Locality and Horizon. — North end of Big Island, in Lake Winnipeg; in the lower part of the Lower Mottled or Dog Head member of the Red River formation.

Specimen No. 7144 in the Victoria Memorial Museum, Ottawa, Ontario, Canada.

23. CYRTOGOMPHOCERAS INTERMEDIUM (Whiteaves)

(Plate VIII, Fig. 2)

Oncoceras (*magnum* ? *var.*) *intermedium* Whiteaves, Palaeozoic Fossils, Geol. Surv. Canada, 3, pt. 3, pp. 221, 222, text fig. 13 (1897).

Specimen 200 mm. long, curved lengthwise. The ventral vertical outline is slightly concave, especially along the upper part of the specimen, where the radius of curvature of this outline is 275 mm., instead of 300 mm., as along the lower half of the specimen. The dorsal outline is conspicuously convex, the maximum curvature, with a radius of 120 mm., being at the seventh camera beneath the dorsal side of the base of the living chamber. Above and below this point of maximum convexity the radius of curvature is nearer 200 mm. At the extreme top of the dorsal side of the specimen this outline becomes more convex again, but for how great a distance is unknown, since the aperture of the living chamber is not preserved. The dorso-ventral diameter of the conch increases from 37 mm. at the base of the specimen to 69 mm. at the maximum convexity of the dorsal outline; then it decreases to 56 mm. at the highest point at which this diameter can be measured, 40 mm. above the ventral part of the base of the living chamber. The lateral diameter, at the point of maximum diameter of the conch, is estimated at 60 mm., the specimen, in its present condition being sectioned vertically in a dorso-ventral direction. Along the larger part of the conch six camerae occur in a length equal to the dorso-ventral diameter of the conch, when the camerae are counted along the dorsal outline. Toward the lower part of the specimen this number increases to eight. The sutures of the septa are nearly straight. Compared with the slightly curved ventral outline of the conch, the sutures of the septa are nearly directly transverse at the base of the specimen, but rise at an increasing rate on approaching the upper part of the phragmacone. The septa are only moderately concave at the base of the specimen, but at the point of its maximum dorso-ventral diameter their concavity equals 16 mm. The largest segments of the siphuncle, between 28 and 29 mm. in diameter, are those between the fifth and ninth from the top of the phragmacone. Within these limits the dorso-ventral diameter of the conch varies from 60 mm. at the base to 67 mm. at the top of the series of segments under investigation. The maximum dorso-ventral diameter of the conch is at the third segment of the siphuncle; here the dorso-ventral diameter of the conch is 69 mm., while that of the siphuncle is only 26 mm. The second segment from the top has a diameter of 24.5 mm., and the topmost segment is still smaller. The thirteenth segment from the top of the phragmacone has a diameter of 17.5 mm., the dorso-ventral diameter of the conch here being approximately 50 mm. The segments of the siphuncle are strongly nummuloidal. Where the diameter of the segments is between 28 and 29 mm., that of the septal necks is between 17 and 18 mm., so that the constrictions of the siphuncle at the septal necks equal about 5 mm. Most of the septal necks equal 1 mm. in height, or only slightly exceed this height. The septa are adnate to the lower face of the connecting rings, becoming free along the basal part of their convex exterior vertical outlines. This structure is seen most distinctly along the ventral side of the siphuncle, where the width of the septum intervening between the siphuncle and the nearest part of the wall of the conch is short; it is shown best by the unfigured half of the type here described. The same type of structure is shown also by the structure at the base of the seventh camera from the base of the specimen, and, to a less degree, also by that at the base of the eighth camera. This structure is obscured along the greater part of the length of the dorsal side of the siphuncle by the sagging of the soft deposit of matrix, filling the interior of the camerae, in a downward direction for a distance of about 1 mm., thus producing an appearance as though the septa had been in contact with the upper face of the underlying connecting rings of the siphuncle, instead of with the lower face of the overlying rings. The septa apparently consisted either of an extremely thin calcareous deposit, or it was more or less membranous and not sufficiently impregnated with calcareous material to crack into distinct fragments, rather than to sag.

Locality and Horizon. — Black Island, Lake Winnipeg, in southern Manitoba; from the Lower Mottled or Dog Head limestone, beneath the Cat Head limestone, of Dowling's *Report on the Geology of the West Shore and Islands of Lake Winnipeg* (1900); in the Red River formation of southern Manitoba.

Specimen No. 7143, Victoria Memorial Museum, Ottawa, Canada.

Remarks. — This specimen agrees with typical *Cyrtogomphoceras* in the nummulitic structure of its siphuncle and in the endogastric location of this siphuncle. It agrees also in the conch attaining its greatest diameter near the level of the top of the siphuncle, and tapering thence upward along the living chamber. The sutures of the septa rise in a similar manner in a dorsal direction, the rate of rise increasing toward the top of the phragmacone. It differs, however, in its ventral outline, the latter being slightly concave along its entire length, instead of gibbous along the upper part of the phragmacone and the basal part of the living chamber. Moreover, there is a possibility of the aperture of this living chamber differing from that of typical *Cyrtogomphoceras*. In the specimen at hand, this living chamber appears to be more slender than in the genus named, and nothing is known of its aperture.

BIBLIOGRAPHY

DOWLING, D. B. Notes on the Stratigraphy of the Cambro-Silurian Rocks of Eastern Manitoba. Ottawa Naturalist, 9, pp. 65–74 (1895).

DOWLING, D. B. Report on the Geology of the West Shore and Islands of Lake Winnipeg. Geol. Surv. Canada, Annual Report, 11, F, 100 pp. and map (1900).

ETHERIDGE, R. Paleontology of the Coasts of the Arctic Lands Visited by the Late British Expedition. Geol. Soc. London, Quarterly Journal, 34, pp. 568–636, ill. (1878).

FOERSTE, AUG. F. Notes on Richmond and Related Fossils. Journ. Cincinnati Soc. Nat. Hist., 22, pp. 42–55 (1917).

FOERSTE, AUG. F. Actinosiphonate, Trochoceroid, and Other Cephalopods. Journ. Sci. Lab. Denison Univ., 21, pp. 285–383 (1926).

FOERSTE, AUG. F. American Arctic and Related Cephalopods. Journ. Sci. Lab. Denison Univ., 23, pp. 1–106 (1928).

FOERSTE, AUG. F., AND SAVAGE, T. E. Ordovician and Silurian Cephalopods of the Hudson Bay Area. Journ. Sci. Lab. Denison Univ., 22, pp. 1–107 (1927).

WHITEAVES, J. F. On Some Silurian and Devonian Fossils from Manitoba and the Valleys of Nelson and Churchill Rivers. Geol. Surv. Canada, Report of Progress for 1878–1879, C, pp. 45–51 (1880).

WHITEAVES, J. F. Description of Eight New Species of Fossils from the Cambro-Silurian Rocks of Manitoba. Royal Soc. of Canada, Transactions, 7, iv; pp. 75–83, for 1889 (1890).

WHITEAVES, J. F. The Orthoceratidae of the Trenton Limestone of the Winnipeg Basin. Royal Soc. of Canada, Trans., 9, iv, pp. 77–90, for 1891 (1892).

WHITEAVES, J. F. The Fossils of the Galena-Trenton and Black River Formations of Lake Winnipeg and Its Vicinity. Geol. Surv. Canada, Palaeozoic Fossils, 3, pt. 3, pp. 129–242 (1897).

WHITFIELD, R. P. Observations on and Descriptions of Arctic Fossils. Amer. Mus. Nat. Hist., Bull. 13, pp. 23–24, ill. (1900).

PLATE I

FIG. 1. *Cyclendoceras sp.* (*A*). Ventral side of conch with annulations curving strongly downward ventrally; median parts weathered so as to expose the siphuncle, the entire width of the latter being indicated at the base of the specimen. Putnam Highland, in western Baffin Island. No. 10266, Mus. Paleontol., Univ. Michigan. See also Pl. X, Fig. 3.

FIG. 2. *Cyclendoceras sp.* (*B*). Ventral side of conch, with annulations curving slightly downward ventrally. Putnam Highland, in western Baffin Island. No. 10274, Mus. Paleontol., Univ. Michigan. See also Pl. X, Fig. 4.

FIG. 3. *Westonoceras cf. gouldi* Foerste. *A*, ventral side of conch, showing slight upward angulation of sutures of septa along median line; *B*, lateral view, with ventral outline on left. Putnam Highland, in western Baffin Island. No. 10270, Mus. Paleontol., Univ. Michigan. See also Pl. XI, Figs. 2 *A*, *B*.

FIG. 4. *Gonioceras sp.* (*A*). Ventral side, surface weathered. Putnam Highland, in western Baffin Island. No. 10289, Mus. Paleontol., Univ. Michigan.

FIG. 5. *Dowlingoceras gracile* (Whiteaves). Ventral side, part of specimen filed away so as to expose the siphuncle. Swampy Island, Lake Winnipeg. Figure copied from Royal Soc. Canada, Trans. 9, iv, pl. 11, fig. 4 a (1892). No. 1841, Victoria Memorial Museum. See also Pl. 3, Figs. 2 *A*, *B*.

PLATE I

1

2

3A

3B

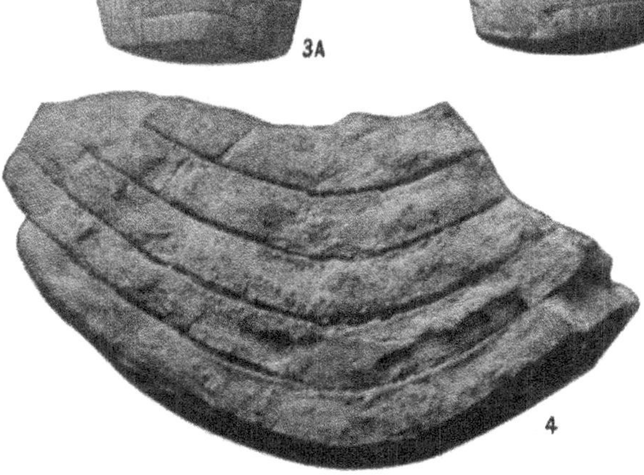

4

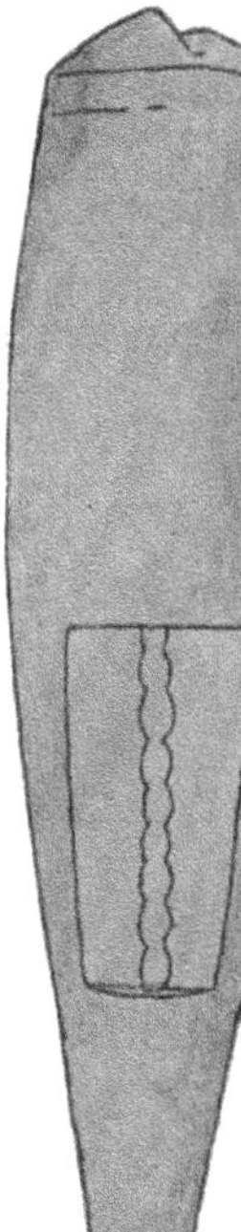

5

PLATE II

All figures on this plate are reduced to nine tenths of their natural size.

Fig. 1. *Orthoceras sp.* Lateral view, showing distant sutures of septa. Putnam Highland, in western Baffin Island. No. 10261, Mus. Paleontol., Univ. Michigan. See also Pl. X, Fig. 5.

Fig. 2. *Endoceras sp.* (*B*). Lateral view, showing sutures of septa. Putnam Highland, in western Baffin Island. No. 10279, Mus. Paleontol., Univ. Michigan. See also Pl. X, Fig. 2.

Fig. 3. *Endoceras sp.* (*A*). Ventral view showing sutures of septa. Putnam Highland, in western Baffin Island. No. 10280, Mus. Paleontol., Univ. Michigan. See also Pl. X, Fig. 1.

PLATE II

1

2

3

PLATE III

FIG. 1. *Leurorthoceras baffinense* Foerste. Ventral side, showing downward curvature of sutures of septa. Putnam Highland, in western Baffin Island. No. 10283, Mus. Paleontol., Univ. Michigan. See also Pl. X, Figs. 6 *A*, *B*.

FIG. 2. *Dowlingoceras gracile* (Whiteaves). *A*, lateral view, with ventral outline on right, showing location of section exposing the siphuncle. *B*, cross-section showing location of siphuncle. Swampy Island, Lake Winnipeg. Figures copied from Royal Soc. Canada, Trans. 9, iv, pl. 11, figs. 4, 4b. See also Pl. I, Fig. 5.

PLATE III

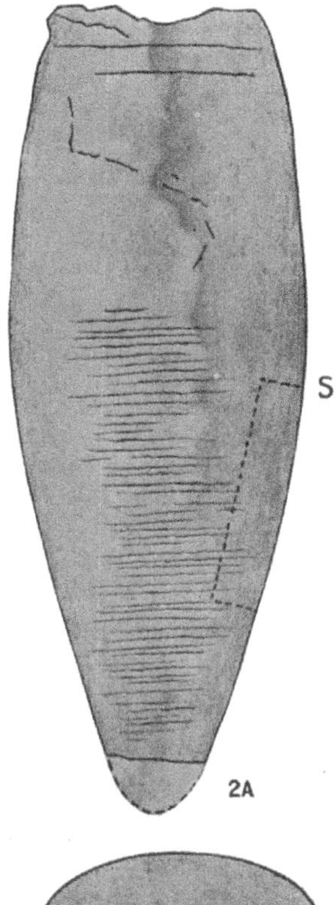

2A

2B

1

PLATE IV

FIG. 1. *Cyrtogomphoceras baffinense* Foerste. Lateral view, with ventral outline on right. Putnam Highland, in western Baffin Island. No. 10262, Mus. Paleontol., Univ. Michigan. See also Pl. XI, Fig. 4.

FIG. 2. *Cyrtogomphoceras cf. baffinense* Foerste. Lateral view, with ventral outline on right. Putnam Highland, in western Baffin Island. No. 10278, Mus. Paleontol., Univ. Michigan. See also Pl. XI, Fig. 6.

FIG. 3. *Ephippiorthoceras sp. (A)*. Ventro-lateral view, showing ventral saddles and lateral lobes of sutures of septa. Putnam Highland, in western Baffin Island. No. 10272, Mus. Paleontol., Univ. Michigan. See also Pl. X, Fig. 7.

FIG. 4. *Ephippiorthoceras sp. (B)*. Lateral view, with ventral outline on left. Putnam Highland, in western Baffin Island. No. 10275, Mus. Paleontol., Univ. Michigan. See also Pl. X, Fig. 8.

PLATE IV

1

2

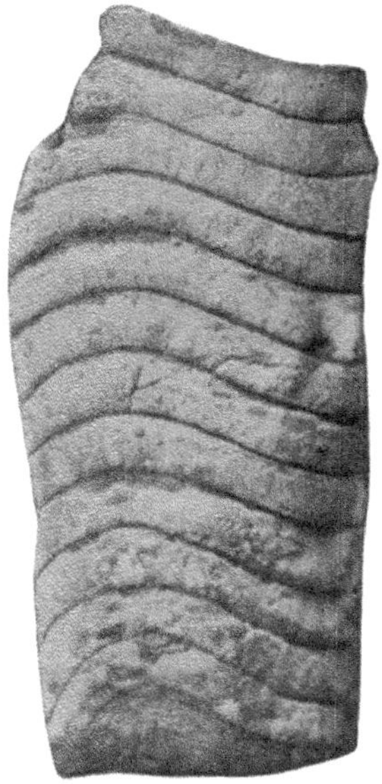

3

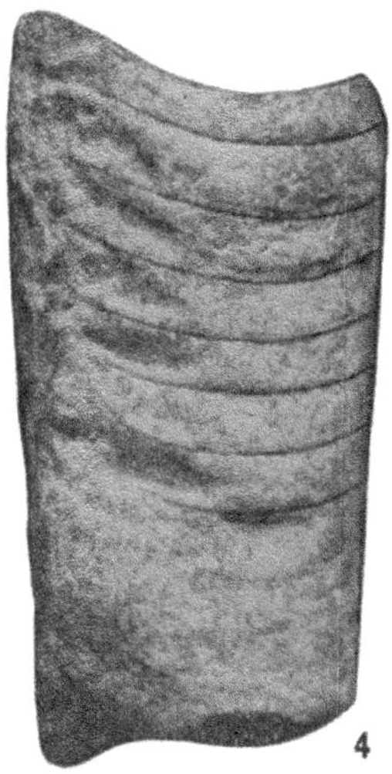

4

PLATE V

FIG. 1. *Westonoceras manitobense* (Whiteaves). Dorso-lateral view, the conch being distorted by oblique lateral pressure during fossilization. Showing the dorsal side of the conch along the right half of the figure, and most of its lateral side along its left half. The median line of the dorsal side passes vertically about 10 mm. left of the most gibbous part of the right outline of the figure. The lateral lobes of the sutures of the septa are most conspicuous along the upper part of the phragmacone, but are distinct also near its base. Big Island, Lake Winnipeg, in southern Manitoba. No. 1879, Victoria Memorial Museum. See also Pl. XI, Fig. 5. Same specimen as Royal Soc. Canada, Trans., 7, iv, pl. 13, fig. 3 (1890).

FIG. 2. *Westonoceras gouldi* Foerste. Lateral view, with dorsal outline on left. Sutures of septa nearly straight, except along upper half of ventral side of phragmacone. Putnam Highland, in western Baffin Island. No. 10263, Mus. Paleontol., Univ. Michigan. See also Pl. XI, Figs. 1 *A*, *B*, *C*.

FIG. 3. *Westonoceras putnami* Foerste. Lateral view, with dorsal outline on right. Putnam Highland, in western Baffin Island. No. 10273, Mus. Paleontol., Univ. Michigan. See also Pl. XI, Figs. 3 *A*, *B*.

FIG. 4. *Gonioceras sp.* (*B*). Ventral side, strongly weathered. Putnam Highland, in western Baffin Island. No. 10264, Mus. Paleontol., Univ. Michigan.

PLATE V

PLATE VI

FIG. 1. *Cyrtogomphoceras magnum* (Whiteaves). Lateral view, with ventral outline on left. East Selkirk, in southern Manitoba. No. 1875, Victoria Memorial Museum. Same specimen as Royal Soc. Canada, Trans., 7, iv, pl. 15, fig. 1 (1890).

PLATE VI

1

PLATE VII

FIG. 1. *Cyrtogomphoceras whiteavesi* (Miller). Vertical dorso-ventral section through type specimen, exposing the siphuncle, but passing through its center only along the upper three segments. Pike Head, in Lake Winnipeg. No. 1877, Victoria Memorial Museum. Same specimen as Royal Soc. Canada, Trans., 7, iv, pl. 15, fig. 2.

FIG. 2. *Cyrtogomphoceras whiteavesi* (Miller). Conch showing the typical ventral and dorsal outlines of this species. Black Island, Swampy Harbor, in Lake Winnipeg. No. 7146, Victoria Memorial Museum.

FIG. 3. *Cyrtogomphoceras whiteavesi* (Miller). Dorsal view of specimen which has been strongly depressed dorso-ventrally by pressure previous to fossilization; showing the transverse striae or bands. Lower Moose Island, in Lake Winnipeg. No. 7145, Victoria Memorial Museum.

FIG. 4. *Cyrtogomphoceras cf. turgidum* Troedsson. Lateral view, showing dorsal and ventral outlines. North end of Big Island, in Lake Winnipeg. No. 7144, Victoria Memorial Museum.

PLATE VII

PLATE VIII

Both figures on this plate are reduced to four fifths of their natural size.

FIG. 1. *Winnipegoceras dowlingi* Foerste. Lateral view, with ventral outline on left. Little Tamarack Island, in Lake Winnipeg. No. 7152, Victoria Memorial Museum.

FIG. 2. *Cyrtogomphoceras intermedium* (Whiteaves). Dorso-ventral vertical section through the siphuncle, with ventral outline of conch on right. Black Island, in Lake Winnipeg. No. 7143, Victoria Memorial Museum. Same specimen as Geol. Surv. Canada, Pal. Foss., 3, pt. 3, p. 221, text fig. 13 (1897).

PLATE VIII

1

2

PLATE IX

Both figures on this plate are reduced to four fifths of their natural size.

FIG. 1. *Winnipegoceras laticurvatum* (Whiteaves). Dorso-ventral vertical section through siphuncle, with ventral outline on right. Little Black Island, in Lake Winnipeg. No. 7148, Victoria Memorial Museum. Same specimen as Geol. Surv. Canada, Pal. Foss., 3, pt. 3, p. 224, text fig. 14 (1897).

FIG. 2. *Winnipegoceras laticurvatum* (Whiteaves). Lateral view, weathered so as to show the sutures of the septa; omitting the apical part of the specimen, 50 mm. in length. Including this apical part, the lengthwise curvature of this specimen is closely similar to that of the type, represented by Fig. 1 on this plate. Point off Moose Creek, on west shore of Lake Winnipeg. No. 7153, Victoria Memorial Museum.

PLATE IX

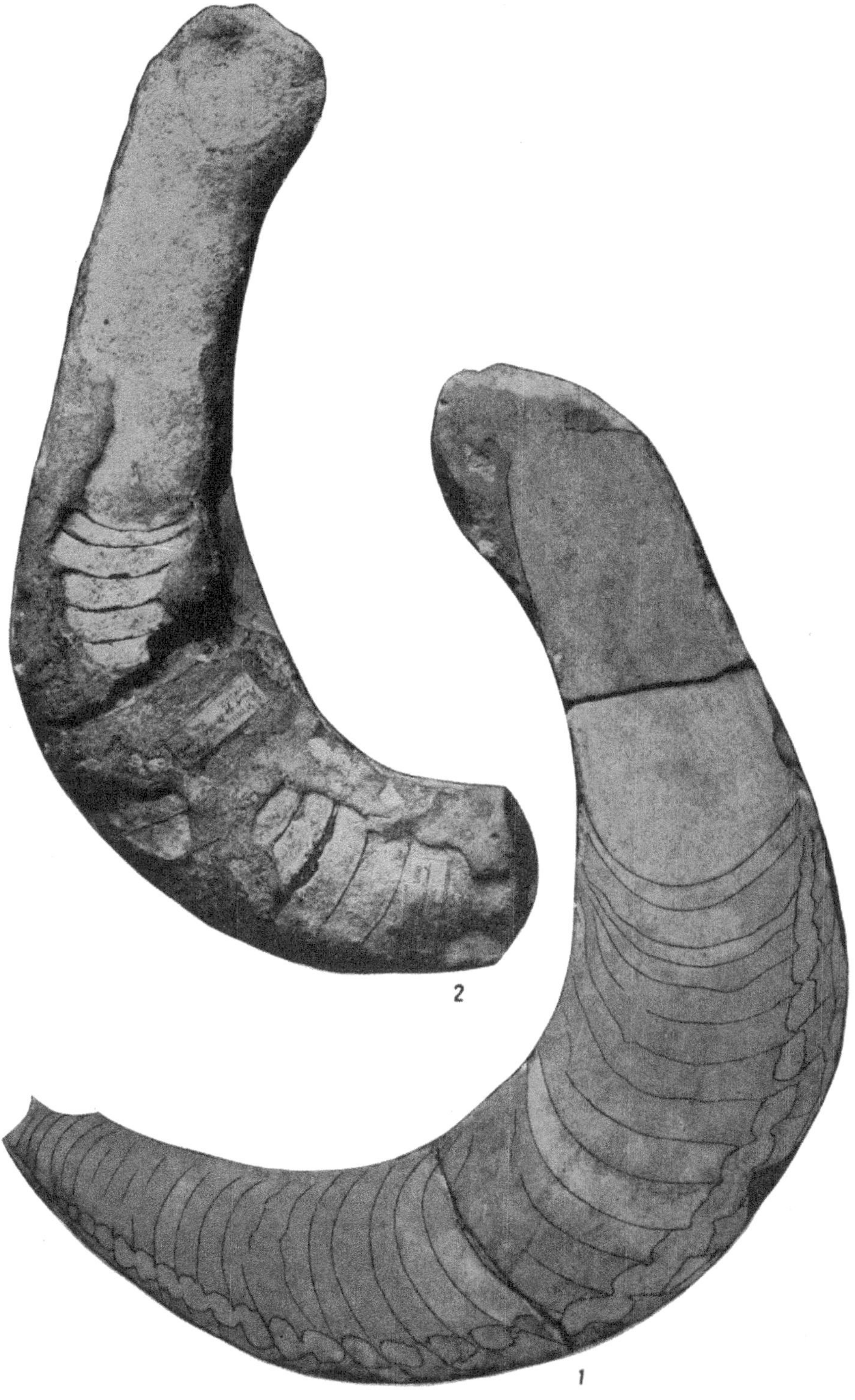

2

1

PLATE X

FIG. 1. *Endoceras sp.* (*A*). Cross-section, showing location of siphuncle. See also Pl. II, Fig. 3.

FIG. 2. *Endoceras sp.* (*B*). Cross-section, with an attempt to show the size and general location of the siphuncle. See also Pl. II, Fig. 2.

FIG. 3. *Cyclendoceras sp.* (*A*). Cross-section showing location of siphuncle. See also Pl. I, Fig. 1.

FIG. 4. *Cyclendoceras sp.* (*B*). Cross-section showing location of siphuncle. See also Pl. I, Fig. 2.

FIG. 5. *Orthoceras sp.* Cross-section, location of siphuncle unknown. See also Pl. II, Fig. 1.

FIG. 6. *Leurorthoceras baffinense* Foerste. *A*, dorso-ventral vertical section through the siphuncle, with the ventral outline of the conch on the left. *B*, cross-section of conch, showing location of siphuncle. See also Pl. III, Fig. 1.

FIG. 7. *Ephippiorthoceras sp.* (*A*). Cross-section, showing location of siphuncle. See also Pl. IV, Fig. 3.

FIG. 8. *Ephippiorthoceras sp.* (*B*). Cross-section showing location of siphuncle. See also Pl. IV, Fig. 4.

PLATE X

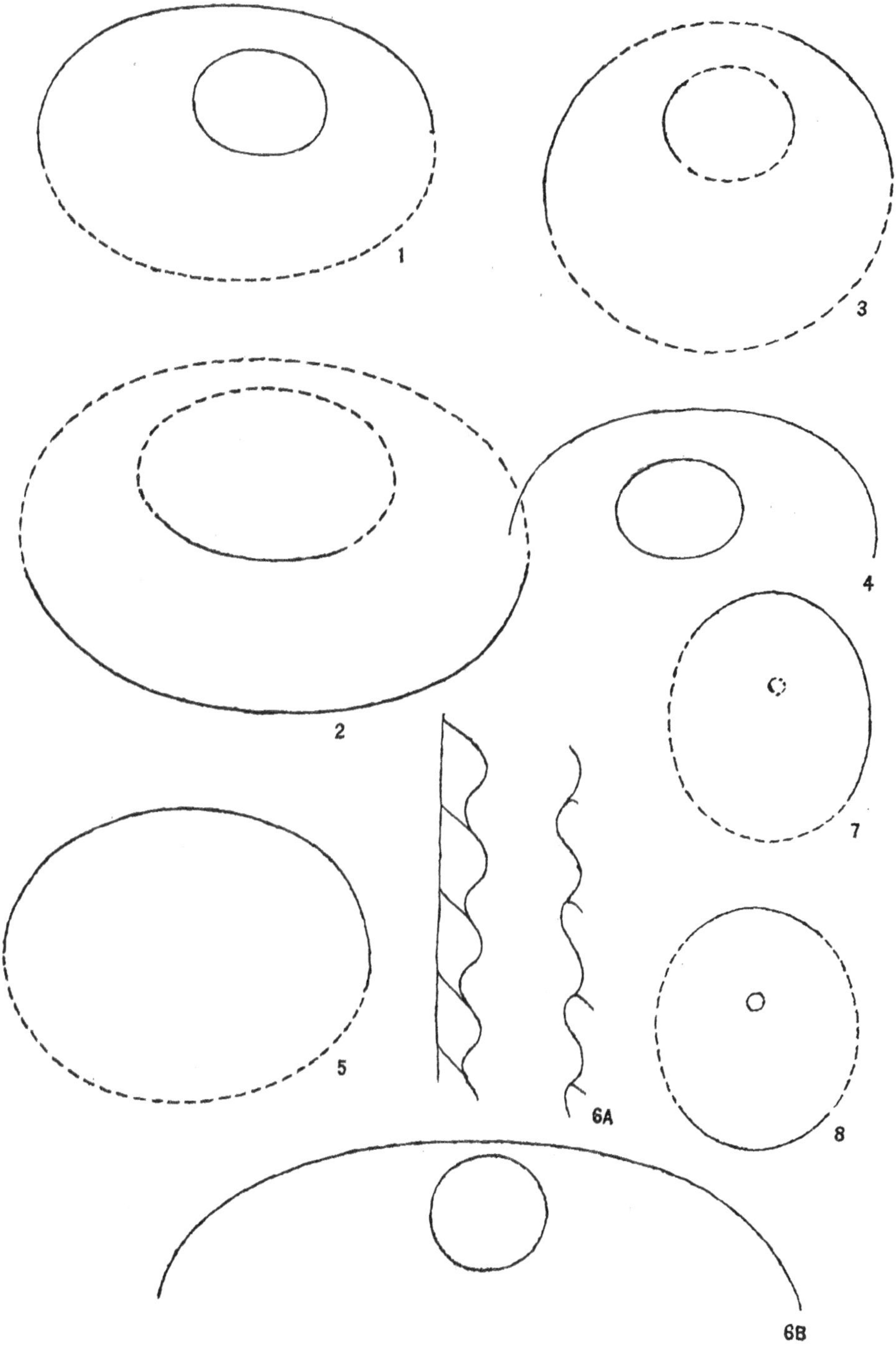

PLATE XI

FIG. 1. *Westonoceras gouldi* Foerste. *A*, cross-section at maximum diameter of conch; *B*, dorso-ventral vertical section at base of specimen, showing calcareous deposit embracing inner margin of septum at base of specimen, at passage of siphuncle; *C*, cross-section at base of specimen, showing location of siphuncle. See also Pl. V, Fig. 2.

FIG. 2. *Westonoceras cf. gouldi* Foerste. *A*, cross-section of specimen at its maximum diameter, showing location of siphuncle; *B*, dorso-ventral vertical section of basal part of specimen, magnified two diameters, showing siphuncle, and form of calcareous deposits embracing the inner margin of the septa at the passage of the siphuncle. See also Pl. I, Figs. 3 *A*, *B*.

FIG. 3. *Westonoceras putnami* Foerste. *A*, cross-section at maximum diameter of specimen; *B*, vertical section through specimen in a lateral direction, passing through center of siphuncle, but not through the center of the conch; figure magnified two diameters. See also Pl. V, Fig. 3.

FIG. 4. *Cyrtogomphoceras baffinense* Foerste. Cross-section of conch at its maximum diameter. See also Pl. IV, Fig. 1.

FIG. 5. *Westonoceras manitobense* (Whiteaves). Vertical section through the center of the siphuncle in a lateral direction, but not through the center of the conch. Magnified two diameters. See also Pl. V, Fig. 1.

FIG. 6. *Cyrtogomphoceras cf. baffinense* Foerste. Cross-section at upper end of specimen, showing location of passage of siphuncle. See also Pl. IV, Fig. 2.

FIG. 7. *Garryoceras semiplanatum* (Whiteaves). *A*, dorsal view of conch; *B*, cross-section, showing location of siphuncle. Lower Fort Garry, in southern Manitoba. No. 1834, Victoria Memorial Museum. Copied from Royal Soc. Canada, Trans., 9, iv, pl. 8, figs. 3, 3a.

PLATE XI

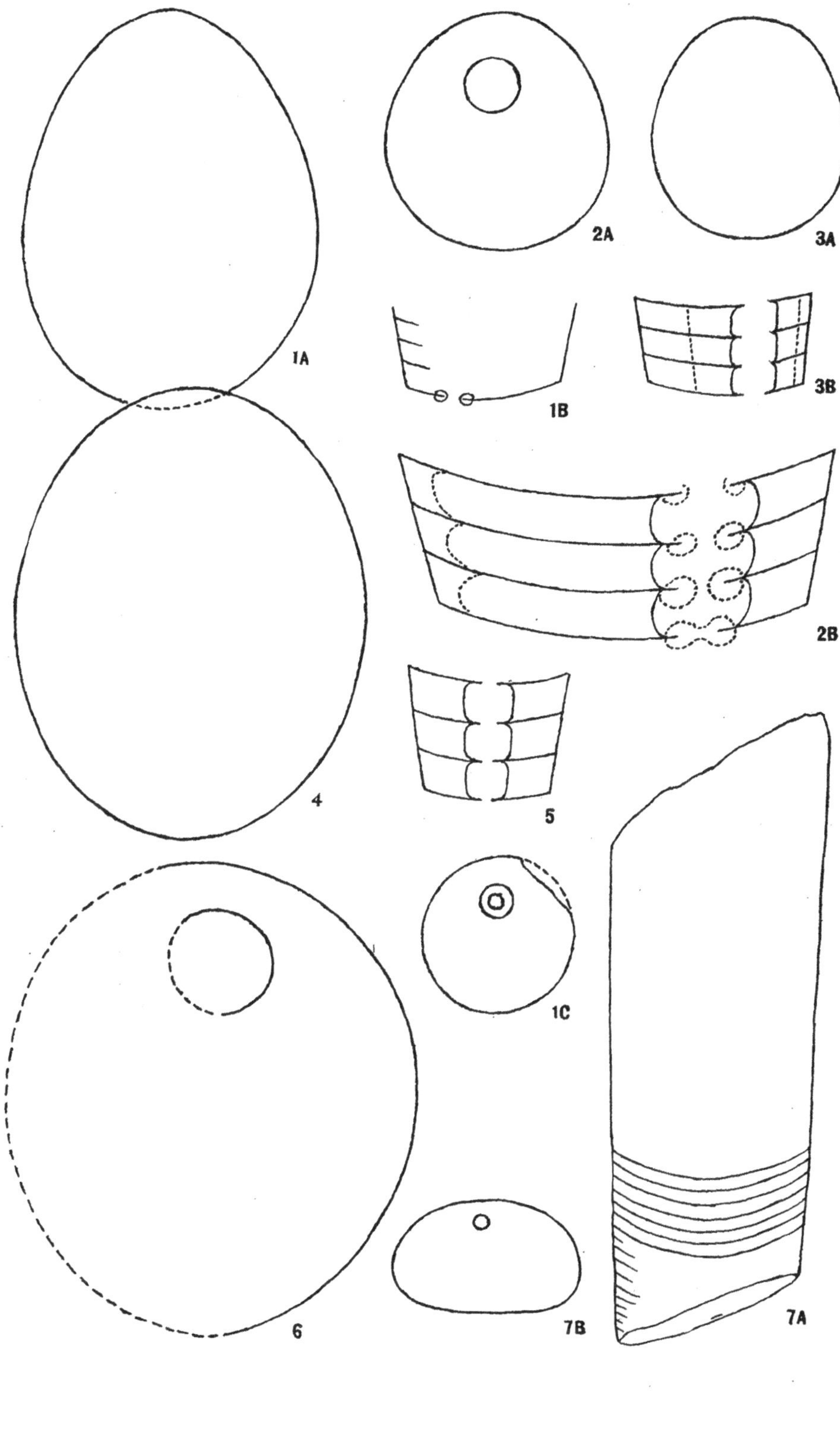

Part III

CORALS, BRACHIOPODS, GASTROPODS AND OSTRACODS FROM PUTNAM HIGHLAND

By RUSSELL C. HUSSEY

INTRODUCTION

THE specimens here described and figured were collected by Dr. L. M. Gould during the summer of 1927 while a member of the Putnam Baffin Island Expedition. This party explored and mapped parts of Baffin Land which were totally unknown, and during the course of these explorations Dr. Gould found an entirely new locality where fossils of Ordovician age are very abundant. Time was not available for making an extensive collection, but the material is exceptionally valuable for purposes of correlation, the Arctic Richmond horizon being clearly determined. This new locality is located on Putnam Highland.

Class ANTHOZOA

Genus HALYSITES Fischer

HALYSITES GRACILIS (Hall)

(Plate I, Fig. 4)

Several unusually well preserved specimens of this coral were found. There are two varieties of the species at the Putnam Highland locality, one in which the longer diameter of the corallites is about one-half millimeter, and another in which the diameter ranges from one to one and one-half millimeters. The shape of the corallites is typically quadrangular.

Horizon and Locality. — Richmond. Putnam Highland, Baffin Land. No. 10208, Museum of Paleontology, University of Michigan.

Class BRACHIOPODA

Genus RHYNCHOTREMA Hall

RHYNCHOTREMA SP.

(Plate II, Figs. 1–3)

The specimen has been considerably weathered, but the general form can be determined. Outline broadly triangular. Brachial valve inflated, particularly towards the umbo. One slope of this valve shows five strong, rounded, plications; a sixth short one may have been present. Pedicle valve less inflated than the brachial. Sinus well-developed, deepening and widening towards the front, with three plications. One slope of the pedicle valve shows five strong, rounded, plications and traces of a sixth, which is the maximum number that could have been present. All traces of lamellose lines, which were probably present, have been removed by weathering. The general shape of this specimen and its association with *Halysites gracilis* suggest that it may be *Rhynchotrema perlamellosum.*

Horizon and Locality. — Richmond. Putnam Highland, Baffin Land. No. 10103, Museum of Paleontology, University of Michigan.

Class GASTROPODA

Genus LOPHOSPIRA Whitfield

LOPHOSPIRA SP.

(Plate III, Figs. 3, 4)

Three volutions are preserved, the two upper ones being somewhat weathered. Lower volution tricarinate. Slope from the lower carina to the peripheral angle concave. Middle carina very prominent, angular. Upper carina removed about one half of the biconcave upper slope of the volution from the suture; the lower half of this slope is more strongly concave than the upper. No surface markings preserved.

Horizon and Locality. — Richmond. Putnam Highland, Baffin Land. No. 10211, Museum of Paleontology, University of Michigan.

GENUS MACLURINA ULRICH AND SCOFIELD

MACLURINA CUNEATA (Whitfield)

(Plate II, Fig. 6; Plate III, Fig. 2; Plate IV, Fig. 2)

This species is distinguished from *Maclurina manitobensis*, found in the same beds, by the smaller umbilicus. The specimens are preserved as casts of the interior. A small fragment of the test, preserved on the inner side of an inner whorl in one individual, shows no traces of surface markings, being smooth.

Horizon and Locality. — Richmond. Putnam Highland, Baffin Land. No. 10215, Museum of Paleontology, University of Michigan.

MACLURINA MANITOBENSIS (Whiteaves)

(Plate III, Figs. 1, 5; Plate IV, Fig. 4)

This species is very common at the Putnam Highland locality where it reaches a diameter of eight inches. The specimens found are all casts of the interior. The general shape is very much like that of *Maclurina cuneata* from which it differs in having a larger umbilicus.

Horizon and Locality. — Richmond. Putnam Highland, Baffin Land. No. 10207, Museum of Paleontology, University of Michigan.

MACLURINA MANITOBENSIS (?) (Whiteaves)

(Plate II, Figs. 4, 8, 9; Plate IV, Fig. 3)

Only one specimen of this form is available for description. The shell is moderately large, and the oval outline is the result of distortion. About four and one-half volutions are preserved; these expand very rapidly after the second volution. The test is thick on the inside of the whorls. Whorls very flat on the base. Upper surface of whorls very steep, broadly convex. Umbilicus moderately large, deep, descending abruptly. The last whorl overlaps more than one half of the one below. Two patches of the test showing surface markings have been preserved; these markings consist of strong, rounded, revolving ridges about one millimeter wide, separated by rounded grooves.

Horizon and Locality. — Richmond. Putnam Highland, Baffin Land. No. 10214, Museum of Paleontology, University of Michigan.

Genus MACLURITES Lesueur

MACLURITES CRASSUS (Ulrich and Scofield)

(Plate I, Figs. 1, 2; Plate IV, Fig. 1)

Lower surface flat, the outer angle slightly obtuse. Inner whorls rounded. Umbilicus large, exposing a little more than one half of each of the inner whorls. Margin of umbilicus abruptly rounded to angular. Slopes of umbilicus abrupt, convex.

Horizon and Locality. — Richmond. Putnam Highland, Baffin Land. No. 10212, Museum of Paleontology, University of Michigan.

Genus TROCHONEMA Salter

TROCHONEMA SP. aff. UMBILICATUM (Hall)

(Plate II, Figs. 5, 7)

The general form of this species is like that of *Trochonema umbilicatum* or *Trochonema beachi,* although it is possibly a little lower-spired than either of these two. Peripheral band nearly vertical, flat. Above the peripheral band is a gently concave slope leading up to the shoulder-like space bordering the suture; the outer margin of this space is elevated a little above the suture.

Horizon and Locality. — Richmond. Putnam Highland, Baffin Land. No. 10210, Museum of Paleontology, University of Michigan.

Class EUCRUSTACEA

Genus LEPERDITIA Rouault

LEPERDITIA SP.

(Plate I, Fig. 3)

A single right valve was found. Outline somewhat oval. Valve strongly convex, greatest convexity about the middle.

Posterior end broadly rounded, shell narrowing a little anteriorly. Hinge-line straight. Eye spot relatively large, showing as a dark, pigmented area in the antero-dorsal fourth. Differs from *Leperditia fabulites* in being shorter, and in having the greatest height of the shell somewhat nearer the middle line. Resembles *Leperditia alta* in general outline.

Horizon and Locality. — Horizon uncertain. Found several miles north of Putnam Highland, Baffin Land. No. 10201, Museum of Paleontology, University of Michigan.

Faunal list of invertebrates, other than cephalopods, from Putnam Highland, Baffin Land.

Receptaculites sp.
Streptelasma corniculum Hall
Halysites gracilis (Hall)
Rhynchotrema sp.
Lophospira sp.
Maclurina cuneata (Whitfield)
Maclurina manitobensis (Whiteaves)
Maclurina sp.
Maclurites crassus (Ulrich and Scofield)
Trochonema sp.
Fusispira aff. nobilis Ulrich and Scofield
Leperditia sp.
Illaenus sp.

EXPLANATION OF PLATE I

FIGURES

1–2. *Maclurites* crassus (Ulrich and Scofield)
   1. View of the flat side
   2. View showing the umbilicus
3. *Leperditia sp.* Right valve. ×2
4. *Halysites gracilis* Fischer. ×2

PLATE I

1

2

3

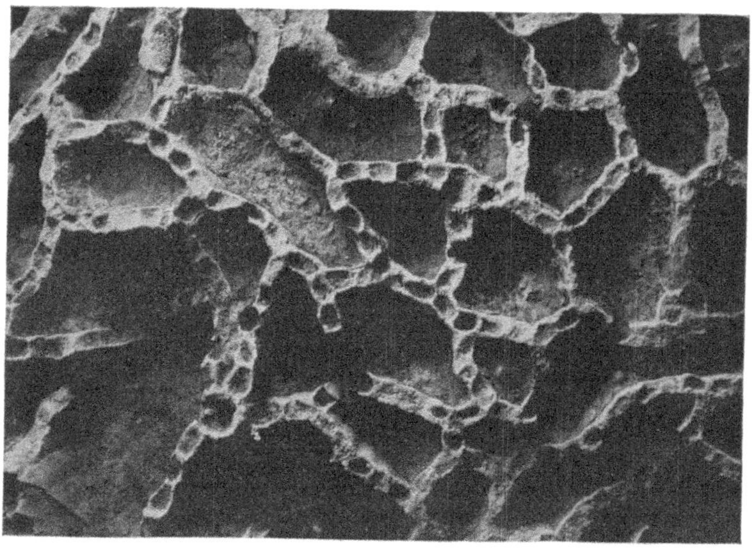

4

EXPLANATION OF PLATE II

FIGURES

1–2. *Rhynchotrema sp.* Probably *R. perlamellosum*
- 1. Brachial valve
- 2. Side view
- 3. Pedicle valve

4, 8–9. *Maclurina manitobensis* (?) (Whiteaves)
- 4. A portion of the surface showing the surface markings
- 8. Side view
- 9. View showing the umbilicus

5, 7. *Trochonema sp. aff. umbilicatum* (Hall)
- 5. Apical view
- 7. Side view

6. *Maclurina cuneata* (Whitfield)

PLATE II

EXPLANATION OF PLATE III

FIGURES

1, 5. *Maclurina manitobensis* (Whiteaves)

2. *Maclurina cuneata* (Whitfield)

3–4. *Lophospira sp.*

3. Apical view

4. Side view

PLATE III

EXPLANATION OF PLATE IV

FIGURES

1. *Maclurites crassus* (Ulrich and Scofield)
2. *Maclurina cuneata* (Whitfield). View showing the umbilicus
3. *Maclurina manitobensis* (?) (Whiteaves). View of the flat side
4. *Maclurina manitobensis* (Whiteaves). View showing the umbilicus

PLATE IV

1

2

3

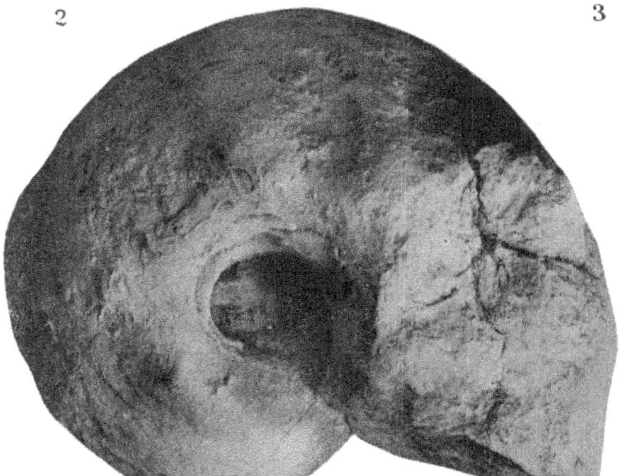

4

Vol. III, No. 4, pp. 77–79 (1 pl.) November 10, 1928

CYSTOIDS FROM THE TRENTON ROCKS OF MICHIGAN

By RUSSELL C. HUSSEY

THE cystoids described and illustrated in this paper were collected during the summer of 1927 from an outcrop of Trenton limestone exposed along the Escanaba River, three miles north of the city of Escanaba, Michigan. The rock at this locality is very fossiliferous but well-preserved specimens of cystoids are exceedingly rare.

Class CYSTOIDEA VON BUCH

Genus PLEUROCYSTITES Billings

PLEUROCYSTITES SQUAMOSUS Billings

(Plate I, Figs. 1, 3)

Description. — The specimen shows only the antanal side, the anal side being embedded in the hard limestone matrix. The calyx has been somewhat distorted by crushing but nearly all the plates are well shown and the sutures can easily be traced. The column is round, tapering distally, with the projecting edges of the segments vertically striated; the segments are separated from each other by deep grooves which become shallow distally where the segments are also thicker. The surface of the plates is granular. On a few plates there are traces of concentric lines running parallel with the margin. Basal pectinorhomb narrow, about 1.5 mm. wide and 5 mm. long, with the longer axis running in the same direction as the length of the calyx. The longer axes of the two upper pectinorhombs are transverse to the length of the calyx. The chief distinction between *Pleurocystites filitextus* and *P. squamosus* is in the plates on the anal side. In the former the anal area is covered with forty or fifty plates; in the latter the area is covered with several hundred very small plates.

Horizon and Locality. — Specimen found by E. C. Case. Trenton limestone, Escanaba River, three miles north of Escanaba, Mich., Loc. 8, Zone 2. Expedition of 1927. No. 10346, Museum of Paleontology, University of Michigan.

PLEUROCYSTITES SQUAMOSUS Billings

(Plate I, Figs. 2, 6)

Description. — This specimen shows the great size of the anal area which was originally covered by a great number of very small plates; these plates have been crushed inward and a number of them still adhere to the inner surface of the plates on the antanal side.

Horizon and Locality. — Specimen found by the author. Trenton limestone, Escanaba River, three miles north of Escanaba, Mich., Loc. 8, Zone 2. Expedition of 1927. No. 10348, Museum of Paleontology, University of Michigan.

Genus EDRIOASTER Billings

EDRIOASTER BIGSBYI Billings

(Plate I, Fig. 4)

Description. — The specimen is partly embedded in the hard limestone matrix and has been somewhat distorted by crushing. Diameter of body as exposed 35 mm. Ambulacural areas covered by quadrangular plates which alternate with each other in the center of the area. Interambulacural plates covered with blunt tubercles.

Horizon and Locality. — Specimen found by E. C. Case. Trenton limestone. Escanaba River, three miles north of Escanaba, Mich., Loc. 8, Zone 2. Expedition of 1927. No. 10347, Museum of Paleontology, University of Michigan.

PLATE I

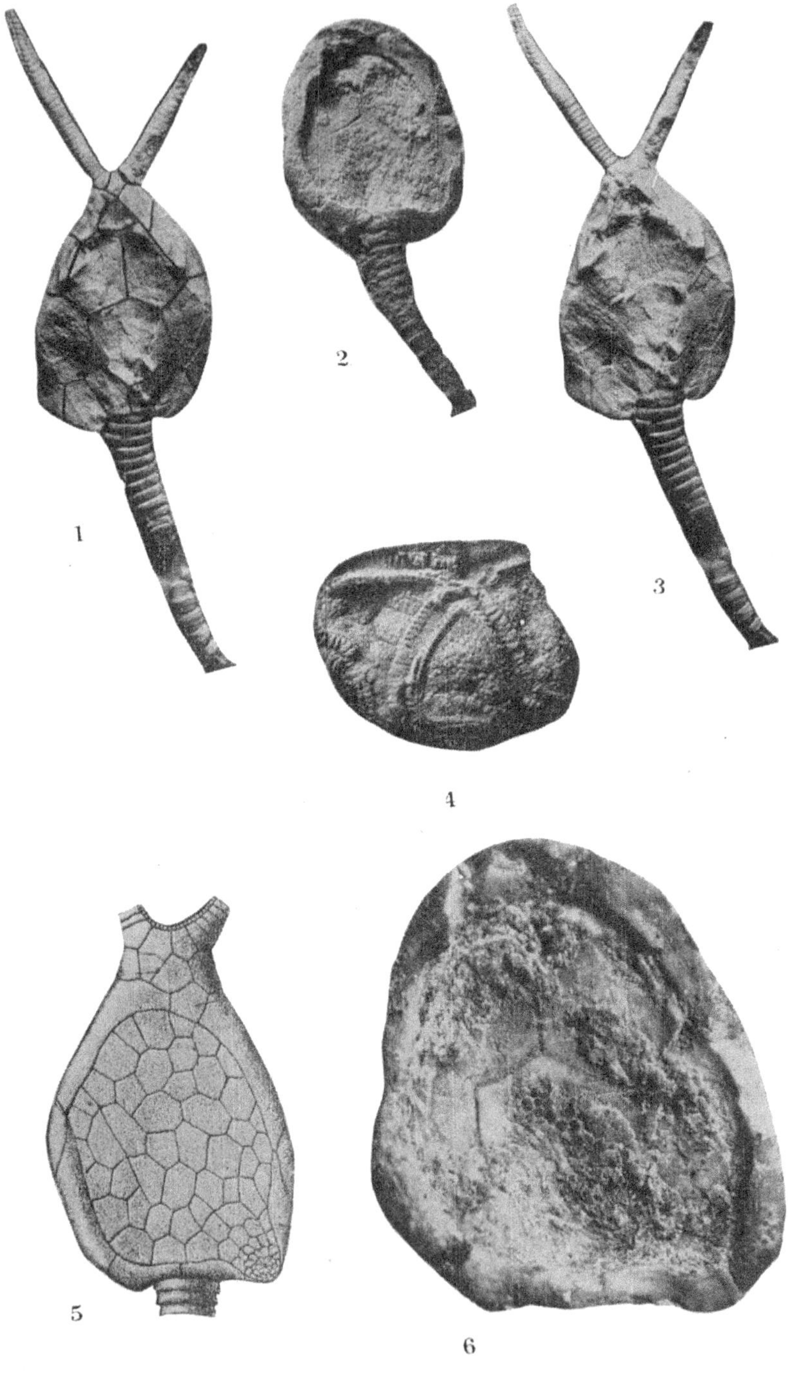

EXPLANATION OF PLATE I

FIGURES

1, 3. *Pleurocystites squamosus* Billings
  1. Antanal side, with sutures inked. Natural size
  3. The same specimen shown in Figure 1, but without the sutures inked. Natural size
  5. *Pleurocystites filitextus* Billings. After Jaekel.

2, 6. *Pleurocystites squamosus* Billings
  2. Anal side showing the great size of the anal area. Natural size
  6. The same specimen shown in Figure 2 enlarged a little over two times, showing some of the very small plates originally covering the anal area
  4. *Edrioaster bigsbyi* (Billings). Natural size

Vol. III, No. 5, pp. 81–107 (3 pls., 12 text figs.) July 15, 1929

DESCRIPTION OF A NEARLY COMPLETE SKELETON OF *OSTODOLEPIS BREVISPINATUS* WILLISTON

By E. C. CASE

IN 1913 Williston[1] published an account of a series of seven articulated vertebrae with the ribs in position and an associated mass of dermal scutes, contained in a block of red sandstone from the Clear Fork, Permian, beds of Coffee Creek, Willbarger County, Texas. To this specimen he gave the name *Ostodolepis brevispinatus.*

The vertebrae were described and figured as being as broad as long, with short, stout diapophyses located near the anterior part of the neural arch, directed forward and a little backward and with the free oval end a little cupped; the centra notochordal with the lower surface gently concave longitudinally and transversely; the neural arches low and flattened, broad from side to side and with a small, almost vestigial tuberculiform neural spine; the zygapophyses broad and flat with the articular faces looking almost directly up and down; the ribs with rather prominent and short tuberculum and capitulum, the latter attached intervertebrally and near the ends of a very slender intercentrum.

The scutes were described as about 8 mm. long and nearly as broad, about as thick as writing paper, the outer surface showing under the hand lens a very distinct series of raised lines quite concentric with the free margin and quite like the "cycloid scales of a fish."

Later Williston[2] suggested "that the name *Ostodolepis* may

[1] Williston, S. W., "*Ostodolepis brevispinatus*, a New Reptile from the Permian of Texas," *Journal of Geology*, Vol. XXI, No. 4, 1913.

[2] Williston, S. W., "The Osteology of Some American Permian Vertebrates," *Contributions from Walker Museum*, Vol. I, No. 9, p. 175, 1916.

be retained as a probable though as yet uncertain synonym of *Pantylus.*"

As will be seen from the following account the vertebrae and scutes of the specimen here described, number 11,156 of the Museum of Paleontology of the University of Michigan, correspond very closely with those of the type specimen of *Ostodolepis brevispinatus*, and the specimen is referred to that genus and species. It was obtained from the University of Chicago, about 1912, in exchange, and in all probability came from the same locality and geological horizon as the type specimen. It was enclosed in a nodule of hard calcareous red sandstone and seemed so impossible of development that it was long allowed to lie untouched. A fortunate exploration with a fine chisel revealed that the specimen could be cleaned and, though it has cost the author more than a month's time chiseling with a needle under a binocular microscope, the results seem worth the effort.

As shown in Plate I, Figure 1, the animal collapsed in a bent position after death, with the head, the right forearm and the anterior vertebrae in the natural relations. The posterior part of the vertebral column and the posterior limbs were disturbed and some portions were lost. The skull was fortunately fractured longitudinally almost in the median line, and transversely just anterior to the orbits. There has been but very slight distortion of the skull or other parts of the skeleton.

Seen from above (Plate I, Fig. 2), the outline of the skull is a regular triangle strongly suggestive of that of *Captorhinus;* the greatest peculiarity is the lack of a pineal foramen. Seen from below (Plate I, Fig. 3) the short, powerful jaws, the long conical teeth, the long and broad parasphenoid rostrum, the relatively very large interpterygoid space and the broad articular process of the pterygoid bones are distinctly amphibian. Seen from the side (Plate II, Fig. I), the elevated posterior region of the skull, the cutting away of the lower border of the temporal region, the flat and projecting rostrum of the nose, and the large orbits are evidence of a systematic position remote from *Captorhinus* or any other known cotylosaur. This assemblage of seemingly irreconcilable characters is increased by those revealed in the fractured sections of the skull. As explained below, it seems certain that the upturning of the anterior end of the nose is due to a slight distortion, but the flattening and extension of the rostrum are not explainable by any theory of distortion or compression. Many apparent anomalies, including the absence of the pineal foramen, the peculiar proportions and relations of the bones of the cranial region and the presence of a large cranial cavity directly above the true brain cavity, may be reconciled with more normal structures on the assumption that they are the result of the elevation of the posterior portion of the skull.

The surface of the bones shows a very fine pitting or sculpture not resembling the linear sculpture of *Captorhinus* and not approaching in degree of coarseness the sculpture of many other cotylosaurs.

As with every other large group in which many forms have been placed, there is a tendency to split up the Cotylosauria, but for the present this form is retained in that Order, necessarily in a new family, the Ostodolepidae, containing the single aberrant genus *Ostodolepis*.

THE CRANIAL REGION

Basioccipital. — The occipital condyle is apparently formed largely by the basioccipital. The condition of the specimen is such that the amount played by the exoccipitals in the condyle cannot be exactly determined, but it is certain that it was a relatively small part, as the basioccipital forms the floor of the foramen magnum. The basioccipital portion of the articular face of the condyle is peculiar in that the upper half is convex and the lower half concave (Text Fig. 1). The concave portion may be interpreted as the rudimentary continuation of the notochord, but, as shown in the longitudinal section (Text Fig. 2), is very shallow and relatively lower down than in most of the Permian and Triassic reptiles. The basioccipital and the exoccipital are separated by a considerable space on one side and, apparently, by an equally large space on the other side. This separation has evidently taken place along the line of the suture.

The anterior half of the lower surface of the basioccipital is underlain by the parasphenoid which terminates in an abrupt shelf, the tubera basioccipitalia. Posterior to this shelf the lower surface is nearly flat except for a shallow groove, and is sharply deflected at the posterior end to form the lower edge of the condyle. The upper surface has a process rising to the lower edge of the very short canal of the foramen magnum; indeed, it may be said that there is no canal, the foramen being formed by thin edges of bone. The floor of the brain cavity is decidedly below the lower edge of the foramen magnum; it is smooth and the basioccipital portion terminates

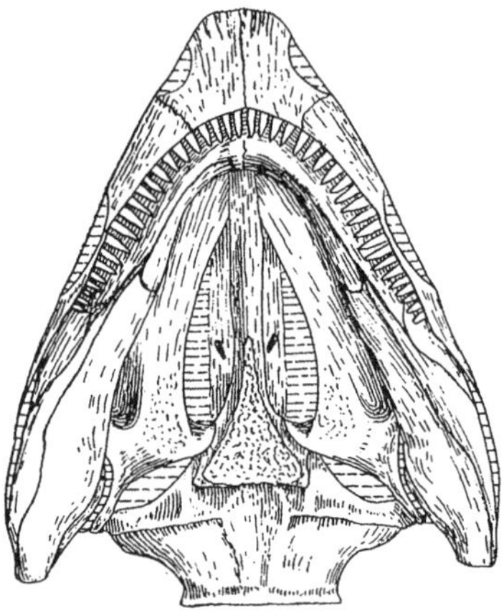

Fig. 1. Lower surface of the skull. × 1

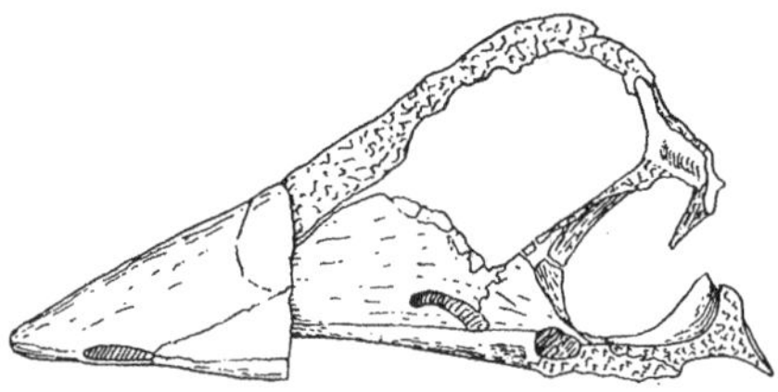

Fig. 2. Longitudinal section of the skull. × 1

anteriorly in a strong suture with coarse interdigitations where it unites with the basisphenoid. In the longitudinal fracture this suture can be traced between the basioccipital and the parasphenoid to the tubera.

The median portion of the floor of the brain cavity is somewhat higher than the sides, but the outer edges turn upward slightly, forming a shallow channel on either side. This upturning is most marked at the posterior end where the basioccipital joins the exoccipital and the otic bones. There are apparently two pairs of foramina in the floor of the brain cavity, a pair just anterior to the basioccipital-basisphenoid suture, which probably transmitted the sixth pair of cranial nerves, and a pair near the posterior end.

Basisphenoid and parasphenoid. — These bones are indistinguishably fused in the specimen, but the large development of the parasphenoid rostrum shows the persistence of that bone. The combined bones reach from the most anterior point visible, the anterior edge of the mandibular symphasis, to the middle of the lower surface of the basioccipital. The rostrum of the parasphenoid is a broad bar, hardly diminishing in breadth in its anterior portion. The anterior half is flattened below, but the posterior half is drawn down into a thin ridge. Posterior to the basipterygoid processes the bone widens and sends flanges outward as far as the middle of the length of the paroccipital process, much farther than the basioccipital. The outer edges of the flanges are closely adjacent to the lower edges of the opisthotics, but not in contact with them. The basipterygoid processes, which rise from the basisphenoidal portion of the bone, are strong and stand out at right angles, terminating in articular faces for the basisphenoidal processes of the pterygoids.

On the lower surface there is a triangular, elevated patch of rugose bone beginning anterior to the basipterygoid processes and extending back as far as the tubera. This is exactly the position occupied by a cluster of teeth in some of the amphibians, but there are no teeth present. Examination at a magnification of somewhat over 100 diameters reveals only a coarse, almost spongy surface, with the reticulations approaching a vermiculate character in some places. A similar appearance, but less pronounced, occurs on the adjacent portions of the pterygoids. Anterior to this patch the parasphenoid supports the ethmoidal elements.

The longitudinal fracture of the skull passes through the basioccipital and through the basisphenoid-parasphenoid as far forward as the middle of the rugose patch and a little to the left of the midline; anterior to this it passes entirely to the left of the parasphenoid rostrum so that the ethmoidal element of the left side is preserved partly on one of the pieces and partly on the other. This fortunate fracture permits an examination of the cranial cavities.

Just anterior to the basioccipital-basisphenoidal suture there is a strong rising process on either side of the basisphenoid, inclined slightly forward and outward. The bone is perforated at the base of each process by a large foramen which extends outward into the basipterygoid process and finds exit in the normal reptilian position just posterior to the basipterygoid process. The canals from either side meet and open into the brain cavity by a common foramen between the rising processes. These are evidently the canals and openings for the external carotid arteries. The floor of the brain cavity is slightly depressed at this point, forming a feeble sella turcia. The anterior edges of the rising processes are in articulation by coarse sutures with the posterior edges of the ethmoidal elements. The posterior portion of the upper edge of each process is in contact, or nearly so, with the plates described below as extending forward from the supraoccipital.

Epipterygoid. — Just outside the rising process of the basisphenoid there is a second rising element, based on both the basisphenoid and the pterygoid, which is evidently the epipterygoid. It is very thin, with nearly parallel anterior and posterior edges and in the specimen there are no notches at the upper end indicating the passage of nerves or vessels.

THE WALLS OF THE BRAIN CAVITY

The make-up of the walls of the brain cavity is at first very puzzling, but if, as suggested above, the elevation of the posterior portion of the skull is taken into account, the various elements may be brought into normal relationships. Beginning at the anterior end of the cavity, the rising process of the basisphenoid which articulates with the ethmoidal element is normal and understandable. The epipterygoid is normal in position and relations. There remain to be accounted for the proötic, epiotic and laterosphenoid (alisphenoid + orbitosphenoid). This complex should be continuous with the opisthotic, reaching upward to touch the parietal and perhaps, the supraoccipital, with the laterosphenoid portion touching the rising process of the basisphenoid, enclosing between them the large foramen for the fifth cranial nerve.

In the specimen there is a pair of plates, convex upward, extending forward from the supraoccipital and dividing the brain cavity from a larger one above it (see Plate II, Figs. 3 a and 3 b, and Text Fig. 2). The plates are united posteriorly, but are separated anteriorly, each lateral portion extending forward and downward until it comes in contact, or very nearly so, with the process from the basisphenoid. On the left side of the fracture the separation is by a very small space and on the right side they seem to be continuous by small fractured bits of bone in the matrix. On the right side the lower portion of the plate is marked by a sharp, deep, relatively large notch, evidently an imperfect foramen. The most careful examination shows these plates to be in undisturbed position and fused, no suture detectable, with the supraoccipital.

The assumption that the posterior portion of these plates is the epiotics and proötics is the most plausible. In most reptiles the epiotics and laterosphenoids extend upward to the under surface of the roof of the skull; the epiotic is attached to, or very near to, the supraoccipital. The elevation of the posterior portion of the roof of the skull in *Ostodolepis* would conceivably leave these elements in their natural relation to the brain, with the only possible attachment to the supraoccipital. By careful checking on the specimen it can be seen that the lower edge of the plate, epiotic and proötic portions, is separated by a very narrow space from the upper edge of the opisthotic, a space which could have been filled with cartilage or could be accounted for by the slight spreading of the skull indicated in the separation of the exoccipital from the basioccipital. The anterior portion of the plate, the laterosphenoid portion, comes into close association with the basisphenoid process. The notch described above as possibly an incomplete opening is on the posterior edge of the plate and cannot, therefore be the foramen for the fifth cranial nerve. Although the bone seems unbroken and the edges of the notch complete, it is possible that the notch is a partial section of the anterior semicircular canal of the inner ear.

THE LOWER SURFACE OF THE SKULL

(Plate I, Fig. 3, and Text Fig. 1)

Pterygoids. — These have the usual tripartite form with palatine, quadrate and external processes. The interpterygoid space is large and this, with the strong parasphenoid rostrum, is one of the strikingly amphibian characters of *Ostodolepis*. The general form of the bone is shown in Fig. 1. The palatine process extends forward as a broad, nearly flat plate with a decided rugose surface opposite the basisphenoid process. The inner edge is smooth and regular, slanting inward to touch the parasphenoid rostrum about opposite the anterior edge of the orbit. Posteriorly this edge is continued upon the lower surface of the bone leaving a notch between it and the basisphenoid process. The outer edge joins the palatine by a rather coarse suture. Because of the condition of the specimen, the presence of the hyoids (?), and the closed lower jaw, it is impossible to observe a distinct ectopterygoid, but the outer edge of the pterygoid rises and then turns sharply downward, again leaving a deep groove anterior to a point opposite the basisphenoid process. It is possible, even probable, that the outer, descending, wall of the groove is the ectopterygoid.

The process for the basisphenoid is broad with a smooth articular surface. It is distinctly articulated to the basisphenoid rather than the parasphenoid, a reptilian character. Immediately posterior to the process the quadrate process turns to the vertical position and rises nearly to the roof of the skull in the temporal region. The posterior end of this process reaches to and wraps around the posterior face of the quadrate. The lower edge is separated from the outer flange of the parasphenoid by an oval space. The posterior portion of the lower edge is slightly convex and concentric with the inner edge of the lower jaw just anterior to the articular region.

Parasphenoid. — It is impossible in the condition of the specimen to determine the relation of the anterior end of the parasphenoid to the adjacent bones. The inner edges of the palatine processes of the pterygoids touch the sides of the rostrum at points on a line with the anterior edges of the orbits.

Ethmoidal plates. — The paired ethmoidal plates rise from the outer edges of the parasphenoid rostrum, separated from it by distinct sutures, and curve outward enclosing a large cavity in the anterior portion of the skull. The lower portion of the posterior edge of each plate is joined by suture to the lower part of the anterior edge of the rising process of the basisphenoid, but the upper part curves forward leaving a space between the upper parts of the two elements. The plates reach to the roof of the skull a little posterior to the posterior edge of the orbit. The outline of the cavity enclosed and the relation of the plates to the parasphenoid and other bones are shown in the transverse fracture of the skull just anterior to the orbits (see Fig. 3). The anterior termination cannot be determined. Each plate is perforated obliquely by a foramen which originates on the inner side at a point just opposite the apex of the rugose patch on the lower surface of the parasphenoid and passes outward and forward; these foramina evidently transmitted the second pair of cranial nerves, and other nerves and vessels, to the eyeball.

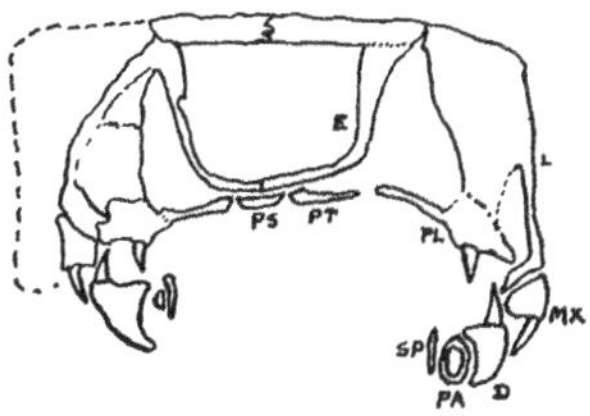

FIG. 3. Transverse section of the skull. The left side a little anterior to the right side, as the section was taken from a fracture surface. E, ethmoidal plate; PS, parasphenoid; PT, pterygoid; PL, palatine; L, lachrymal; MX, maxillary; SP, splenial; PA, prearticular; D, dentary. × 1.3

Prevomers (?). — These cannot be made out as the region is covered by the hyoid and the symphasis of the mandible.

Palatines. — The pterygoids occupy the greater visible portion of the palatal surface, but some portion of the palatines can be made out from this surface and from the transverse fracture. Excavation on the right side revealed a row of strong teeth on the outer edge of the palatine. These begin at the posterior end of the bone and five are shown in series behind the transverse fracture. All are stout and conical, with a narrow pulp cavity extending almost to the tip, as in the maxillary teeth. A tooth of nearly equal size is shown on the left side in the piece of the skull anterior to the fracture; this is so far anterior to the most anterior tooth shown on the right side as to account for three more teeth, making at least eight in all. The relation of the palatine to the other bones is shown in Fig. 3, as drawn from the surface of the transverse fracture. The palatine joins the maxillary and lachrymal on the outer side. Its inner edge is in contact with the pterygoid. Its outer edge has a strong buttress which affords attachment for the row of teeth.

Ectopterygoid. — This bone is not visible and its presence cannot be demonstrated, but its probable presence is suggested in the description of the pterygoid.

Hyoids (?). — As shown in Plate I, Figure 3, there are two pairs of long, slender bones lying on the palatine surface somewhat symmetrically. These are thin transversely and stand on edge in the specimen. Their ends show no structure. It was at first supposed that they were displaced ribs or elements, splenial or otherwise, loosened from the inner surface of the jaw. Complete examination revealed the impossibility of these assumptions, except that the anterior one on the left side may be the splenial. They resemble in form and position the long hyoids of some reptiles and can be identified only as such elements. There was no trace of a median element unless it be a small fragment of unrecognizable form found at the anterior end of the first pair. This small bit of bone suggested by its form that it was part of some crushed phalanges; it was necessarily removed.

THE POSTERIOR FACE OF THE SKULL

(Plate II, Fig. 2 and Text Fig. 4)

The outline of the bones of the posterior face of the skull seems at first impossible, but when considered in relation to the other peculiarities of the skull they may be brought into normal adjustment. The elevation of the posterior portion of the skull above the basis cranium resulted in certain changes in the proportions of the various bones that obscure their essentially normal relations. As yet unexplained are the absence of the pineal foramen and the large cavity above the true brain cavity. These are, in all probability, closely related. As the brain

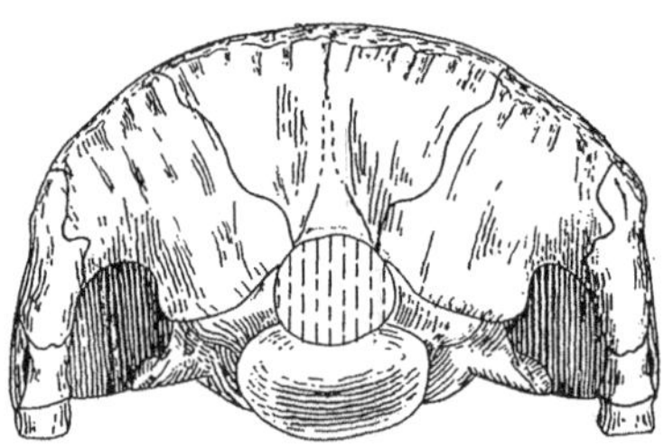

FIG. 4. Posterior surface of the skull. × 1

could not have extended above the otic and sphenoidal elements, the space above must have been in part empty or occupied by a large amount of tissue of some sort. The epiphysis and related structures probably passed upward in the space formed by the separation of the anterior ends of the otic and laterosphenoid plates, and were lost in the space or tissue above, never reaching the surface of the skull.

The elevation of the posterior end of the skull is accomplished by a uniform rise in the upper surface and a large development of the bones of the posterior surface, which lies at an angle of 100° to 105° to the plane of the roof. A close examination of the region shows that the portion immediately above the foramen magnum has been pushed slightly inward and is overlapped by the bones on either side. The depressed area is the supraoccipital. As shown in Figure 4, the dermsupraoccipitals are very long, coming down to contact with the exoccipitals. No suture or possible separation of the tabulare from the supratemporals can be detected; if such a separation was present, it must have been in the groove which marks the change from the posterior face of the skull to the upper surface. On the left side there is a fracture in the groove which might be interpreted as a fracture along the line of a suture; on the right side, where the other sutures are very evident, the bone is smooth and absolutely continuous. If the double character of the bone is assumed, the tabular is in contact with the exoccipital either by merely overlapping the suture between the basioccipital and the exoccipital, or by a direct overlapping suture with the exoccipital. Impossible as this may seem at first, it is not essentially different from the relations which occur in the stegocephalians and the most primitive reptiles where the exoccipital sends a process upward and touches the tabular. The tabular occupies its usual position on the mesial side of the otic notch. The post-temporal fenestra is not at first apparent because it is nearly closed by the great quadrate process of the pterygoid, but slight excavation reveals the opening into the space beneath the temporal roof.

If the cavity above the brain cavity is a separate thing, as seems certain, the question arises whether it is simply the continuation of the space beneath the temporal roof, or whether it is a distinct cavity. If the latter, what would form its lateral walls? It seems impossible that it can be a continuum of the temporal cavities; there must have been some septum, either single or double, extending from the surface of the epiotics to the roof of the skull. The cavity was excavated on the right side with the greatest care to a point well beyond the median line, but no trace of a bony septum nor any fragments that might be the remnants of such a septum were detected. If there was a septum, it must have been membranous or cartilaginous. The closest analogy to the condition is found in the turtles, in which the supraoccipital extends far forward and sends a median process to the roof. This median process joins the descending plates of the parietals in front, which are in the position of the epipterygoids, and part of the ethmoidal plates of this specimen. The supraoccipital of the turtles forms a good portion of the posterior part of the roof of the brain case and unites by complete anchylosis with the epiotic. Anterior to the supraoccipital the cranial cavity rises to the roof formed by the parietals. So far the analogy is good, but in this specimen there was no median bony septum dividing the space above the brain cavity. A cartilaginous or membranous septum in this position is unknown to the author, in any recent or fossil form.

The detailed description of the bones of the posterior face of the skull makes the apparently abnormal, but essentially normal, proportions and relations even more obvious.

Exoccipitals. — On the sides of the foramen magnum there is a pair of collar-like bones which touch the tabulare (+ supratemporals) and the dermsupraoccipitals; they are separated above by the supraoccipital. Traced laterally, the bone of the right side continues into a prominent condylar portion with a concave posterior face; this portion is apparently separated by a considerable space from the basioccipital below. The separation leads to the supposition that the whole exoccipital has been rotated inward on the upper surface, bringing the collar-like portion nearer to the median line and increasing the apparent width of the occipital condyle. As the collar-like portions of the exoccipitals are in similar positions on the two sides of foramen magnum, it is probable that the bones have been separated on the lower surface only. In the restoration the exoccipitals have been brought closer together on the lower side to reduce the size of the condyle and to bring it into consonance with the suggestion of the atlas. On the side of the condylar porton there is a large foramen for the exit of the last pair of cranial nerves. Just anterior to the foramen is the suture for the opisthotic.

Supraoccipitals. — The supraoccipital appears on the posterior surface in a small triangular area with the lower edge clearly showing the upper edge of the foramen magnum. The extent of the supraoccipital visible on the surface is somewhat doubtful. As stated above, it is overlapped on either side by the dermsupraoccipitals, but whether it originally extended to the top of the skull or only part way, is uncertain. On the fractured surface it can be traced to the roof of the skull beneath the dermsupraoccipitals. From the anterior face of the exposed portion rise the horizontal plates which divide the cavity of the skull; the fracture is through the place of origin of the plates and shows definitely that they are fused with the supraoccipital without any trace of a suture.

Dermsupraoccipitals. — As shown in Figure 4, the dermsupraoccipitals are large plates standing vertically in the posterior surface of the skull. In the specimen they apparently meet in the median line above the supraoccipital and join the tabulare laterally by suture. They meet the parietals above and the exoccipitals below. As in all the bones of the posterior surface, the surface is smooth with no indication of sculpture.

Tabulare and supratemporals. — As mentioned above, these two bones are indistinguishably united; the portion on the posterior face must be very largely, if not entirely, the tabular part, for this portion forms the inner edge of the otic notch.

Squamosals. — The posterior surface of the squamosal wraps around the quadrate covering all of it except the articular end. The lower part of the bone becomes very thin and terminates as a mere scale on the quadrate.

Quadrates. — Only the lower portion of the quadrate is visible.

It is covered mesially by the quadrate process of the pterygoid, posteriorly by the squamosal and externally by the quadratojugal. The quadrate foramen is about opposite the lower end of the squamosal. Unfortunately the quadrate is so far covered that little more than the broad articular end can be made out.

Quadratojugals. — The quadratojugal is visible only as a thin line from the posterior surface; its position is indicated by the quadrate foramen. Anteriorly it terminated very quickly as a sharp point against the lower edge of the squamosal, leaving quite a space between it and the posterior end of the jugal.

Opisthotics. — The opisthotic is in the normal reptilian position with the paroccipital process extending outward at a right angle to the basis cranium and terminating against the quadrate with a beveled facet on the anterior face of the extremity. The suture for union with the exoccipital is clearly apparent, being accentuated by a slight separation. The articulation with the epiotic and proötic is obscured by the deep implantation in the skull. Just below the origin of the paroccipital process is the narrow opening of the foramen ovale and just above the process is a pit of considerable size with an exceedingly thin scale of bone forming its bottom. Exploration of the inner side resulted in the perforation of the scale by the needle revealing its tenuous character. The pit may be entirely accidental owing to the thinness of the bone.

Stapes (?). — No stapes was found in the specimen, but considering the development of the opisthotic and its position, the stapes must have been an extremely delicate rod, easily displaced or destroyed.

THE UPPER AND LATERAL SURFACES OF THE SKULL

(Plate I, Fig. 2, Plate II, Fig. 1, and Text Figs. 5–6)

The outline of the various bones, as definitely as they can be determined, is shown in Figures 5–6. The probable course of such sutures as cannot be made out from one side or the other is shown in dotted lines. The skull has been but slightly distorted and no parts have been displaced, but there are a multitude of minute fractures which in places render it difficult to follow the sutures exactly. The slight distortion of the skull and the undistorted condition of the bones of the rest of the skeleton are very significant in the interpretation of the anterior end of the nose.

As shown in Plate II, Figure 1, the anterior end of the nose is turned slightly upward. That this is due to distortion is evident from the angular break shown on the surface, but the distortion is not sufficient to account for the flattening of the rostrum and the consequent sturgeon-like appearance from the side and from below. The general profile is much like that of the insectivore, *Necrolestes*, from the Santa Cruz beds of Patagonia. It was naturally assumed at first that the nose was rounded and elevated, as in *Captorhinus*, and had been reduced to its present form by crushing, but, as just stated, there is no evidence of such action in any other part of the skeleton. There is no lateral bulging of the rostrum and the lower surface is perfectly flat. The assumed original form was modeled in plastic clay and compressed to the form of the specimen. It was found that it could be reproduced only by a combined pressure and forward pull; such an application of forces was impossible in the specimen. The flat lower face of the rostrum could not have been formed by compression against any other element because it lay free in the matrix. Around the edge of the rostrum there is a break filled by matrix which might have been formed by compression, but it is not continuous and the edges of the bone on either side do not overlap. The elongate oval form of the external nares might be taken as evidence of compression, but their edges are complete and rounded, with little or no evidence of fracture or displacement. This portion of the skull has been discussed with one skilled in mechanics who was unable to see how compression and fracture could have produced the condition found. It has been discussed with a skilled preparateur who is convinced that the condition is natural, and with several friends familiar with osteology, all of whom consider the condition natural. However it may be, the specimen is described as it is. Accordingly, as the peripheral break is interpreted as a separation along a suture or an accidental fracture, the make-up of the rostrum may be interpreted in one or the other of two ways. Either the nasals form the entire upper surface of the rostrum and the premaxillaries the entire lower surface, or the nasals terminate at the point of upturning and the premaxillaries form the tip of the nose and the lower surface. In support of the second possibility there is a fracture at the line of upbending on the left side which might be along the line of a suture and there are certain indications on the right side which might indicate that the upbending took place along the nasal-premaxillary suture. In opposition to this idea the upper edge of the external narial opening of the right side is complete with the exception of two very straight lines which are evidently the lines of fracture without displacement. The first supposition seems the most plausible and is supported by the complete edges of the narial openings and the nature of the peripheral parting which, as shown above, is much more like a suture than a fracture. According to this interpretation, the rounded upper surface is formed entirely by the nasals and the flat lower surface entirely by the premaxillaries. Both surfaces show the same type of sculpture. The oval nares are almost entirely enclosed by the nasals and premaxillaries, the lachrymals and maxillaries taking only a small part in the posterior edge.

Premaxillaries. — The alveolar edge of the left premaxillary is injured, but that of the right premaxillary carries six imperfect teeth of the same form as the maxillary teeth and not larger than the largest of the maxillary series. These teeth lie almost horizontally in the specimen and give the strongest evidence for the crushing of the rostrum, for if they were as long as the longest of the maxillary series they would have interfered with the motion of the lower jaw. It seems certain that they have been bent from an oblique to a horizontal position.

Nasals. — The posterior and lateral outlines of the nasals are clearly shown. Each bone joins the frontal, prefrontal and lachrymal of its side and the nasal of the opposite side, and forms the upper border of the narial opening.

Septomaxillaries (?). — Within the orifice of the right narial opening there is a small bit of bone which might be regarded as a septomaxillary, but might equally well be regarded as a bit of the premaxillary detached from position. The point is in doubt.

Lachrymals. — The lachrymals are long and rather narrow,

extending in the very primitive position from the orbit to the nares. Each bone forms the greater part of the anterior border of the corresponding orbit and, as seen in the transverse fracture, sends a wall inward toward the median line which partially cuts off the orbit from the cavity of the skull anterior to it. No trace of a lachrymal foramen can be found, though the condition of the specimen is such that the author would expect such a foramen to be visible if present.

Prefrontals, postfrontals and frontals. — The shape and relations of these bones are sufficiently well indicated in Figure 5. The noticeable irregularity of the median suture and the asymmetry of the bones are common amphibian characters.

Maxillaries. — The maxillary just touches the posterior lower corner of the narial opening and forms the lower border of the orbit; it joins the jugal by a horizontal suture behind the orbit and just on a level with its lower edge. The suture between it and the premaxillary, visible only on the lower surface, runs inward and backward from the posterior edge of the narial opening. The maxillary is narrow vertically and carries long, simple, conical teeth set in very shallow sockets practically throughout its length. Seventeen teeth can be counted on the left side and sixteen on the right. Most of the teeth are seen in longitudinal section, for their substance was softer than the matrix and in cleaning the surface came off with the scale. They all show a narrow pulp cavity reaching nearly to the apex and of nearly equal diameter throughout. The teeth are longest directly under the orbit and decrease both anteriorly and posteriorly. They are all straight with no posterior curvature and with no cutting edge. Examination of the broken surface under a

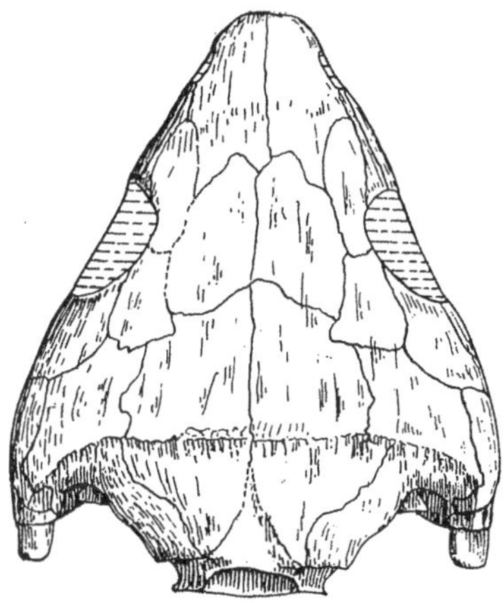

FIG. 5. Upper surface of the skull. × 1

magnification of more than a hundred diameters shows no radial or other infolding of the substance. All the teeth, on both sides, are inclined inward, but not sufficiently to interfere with the movement of the lower jaw. The inclination is normal and passes into the horizontal position of the premaxillary without sudden change, a character which argues for the natural position of the latter.

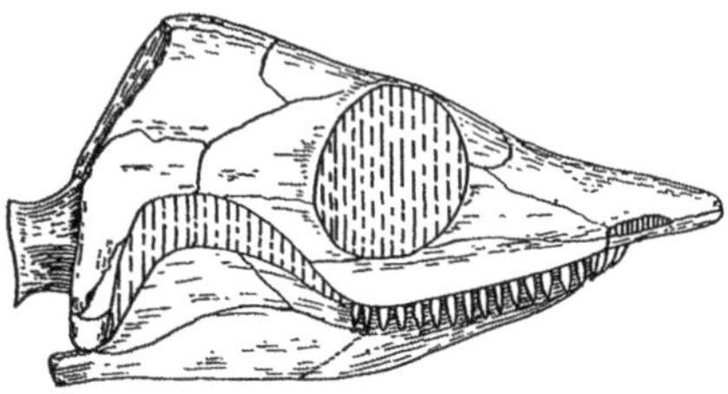

FIG. 6. Right side of the skull. The upturning of the nose corrected. × 1.

Jugals. — The jugal is hardly visible from the upper surface, but if the orbits were empty it would appear as a thin bar forming the posterior half of the lower border. From the side it is more evident, articulating with the maxillary as described and with the postorbital above. It does not reach back as far as the quadratojugal, leaving a space in which the postorbital and the squamosal form a part of the lower edge of the temporal region.

THE LOWER JAW

(Text Figs. 7a–7b)

The lower jaw is hidden partly by the complete closure and partly by the matrix at the posterior end, but most of the elements can be made out. The teeth are completely hidden, but the transverse fracture reveals that there was but a single row, similar in form to the maxillary teeth.

Surangulars. — The surangular rises sharply from the posterior end and forms a plate fitting into the space left by the cutting out of the lower edge of the temporal region.

Angulars. — The angular forms the lower part of the posterior end of the jaw and laps over slightly on the inner side below. Anteriorly it is wedged in between the coronoid, splenial and dentary.

Coronoids. — The coronoid is a sickle-shaped bone which forms the anterior half of the elevation of the posterior portion of the jaw and joins the angular, surangular and dentary. Because of the condition of the specimen it is impossible to say whether there is more than one coronoid, but apparently there is but one and it bears no teeth.

Articulars. — The articulars are largely concealed by the quadrates with which they are in close articulation and nothing can be made out except the presence of a distinct postarticular process.

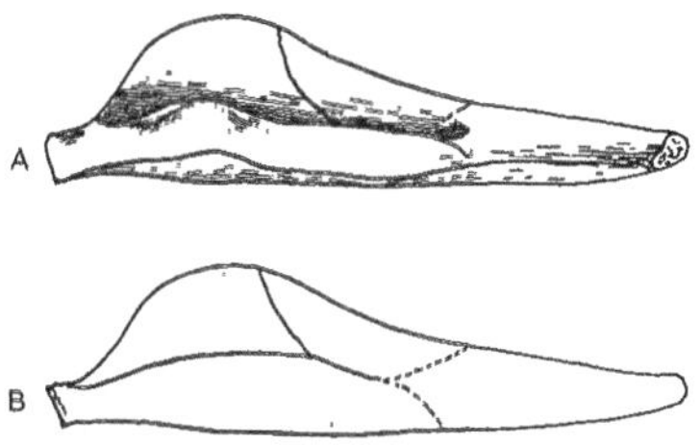

FIG. 7. Lower jaw. A, inner surface; B, outer surface. × 1

Prearticulars. — The prearticular extends forward on the inner side of the jaw. Just anterior to the articular it swells outward and the upper edge is separated, in the specimen, from the edge of the pterygoid by a thin line of matrix. The anterior end is overlapped for a short distance by the splenial.

Splenials. — The splenial is a thin, slender element which reaches almost to the symphasis, but does not take part in it.

Dentaries. — The dentary is the sole tooth-bearing element of the lower jaw. As shown in the transverse fracture, Figure 3, the teeth are confined to the outer edge. The palatal teeth pass inside the inner edge of the lower jaw. The fracture shows a considerable space for the Meckel's cartilage. The symphasis is strong.

THE AXIAL SKELETON

(Plate III, Fig. 4, and Text Figs. 8–12)

The first twelve vertebrae are in place with the ribs attached. The clavicles, probably the interclavicle, the scapula, humerus, radius and ulna, and part of the foot of the right side are in position; the coracoid, if preserved, is hidden by a mass of dermal scutes. The left humerus lay across the under side of the lower jaw and was removed to reveal more clearly the lower surface of the skull; the ulna lay above the humerus and the radius along the left maxillary just anterior to the orbit; portions of the left foot lay just upon the nose and were removed.

Posterior to the twelfth vertebra the column was folded upon itself and can be traced forward and backward through the nodule. The column was broken and disturbed in the sacral region; the sacral vertebrae are preserved, but no trace of the pelvis can be found except one pubis. The femur, tibia and fibula of the left side lie near their proper position in the folded vertebral column, on the under side of the nodule. No traces of the bones of the posterior foot were found unless some isolated podial elements may belong to them.

THE VERTEBRAL COLUMN

Atlas. — The atlas is represented by its disturbed elements which cannot be made out with entire certainty. Just anterior to the axis and lying upon it is the neural arch of the atlas showing

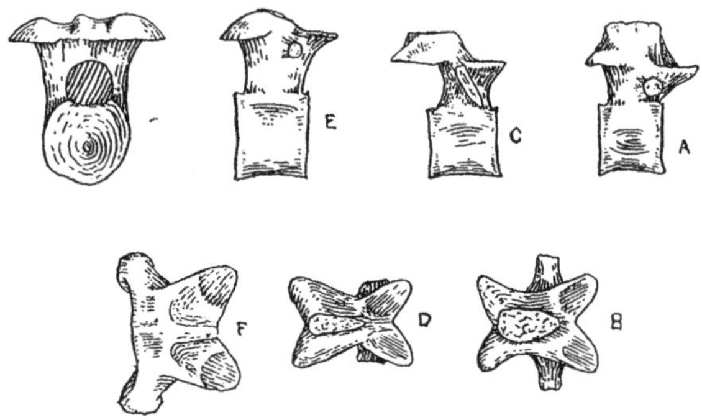

FIG. 8. Vertebrae. A, lateral view of axis; B, upper view of same; C, lateral view of twenty-third vertebra; D, upper view of same; E, lateral view of sacral vertebra; F, upper view of same; G, posterior view of same. All × 1

a low, almost rudimentary spine. On the right side is a small separate plate lying against the neural arch which may be a part of the atlas or may be a part of some other bone, possibly a part of two displaced caudal vertebrae which lie just below the axis-atlas. A rib of considerable size with distinct capitulum and tuberculum was attached to the atlas, but was necessarily destroyed in separating the pieces of the nodule to study the interior of the skull. Unfortunately the anterior face of the atlas cannot be made out, but there is nothing that could warrant the extremely broad face of the occipital condyle suggested by the position of the basioccipital and the exoccipitals; partly for this reason the exoccipitals have been rotated from their position in the specimen and the width of the condyle reduced in the restoration of the skull.

Axis (Text Figs. 8a–8b). — The axis is rather more elongate than the succeeding vertebrae with a neural spine of corresponding length but no greater height. The anterior and posterior zygapophyses are large and their faces, as in all the vertebrae, look directly up and down. The transverse process is well developed, rising from the anterior part of the neural arch and the centrum, entirely anterior to the neural spine. The attached rib is fairly long with well-developed capitulum and tuberculum and the distal part is expanded almost as much as in the succeeding vertebrae (see Fig. 9a).

Third to twelfth vertebrae. — These vertebrae are all very similar in form, so far as may be ascertained; the centra are covered by matrix. The neural spines are low and formed like that of the axis, but are shorter antero-posteriorly. The transverse processes occupy the same anterior position. The ribs are attached in position and show a gradual increase in length. The eighth rib still shows some expansion of the distal end, but beyond it the ribs are more slender and of equal diameter throughout. No trace of any intercentra can be found in this or any other part of the column.

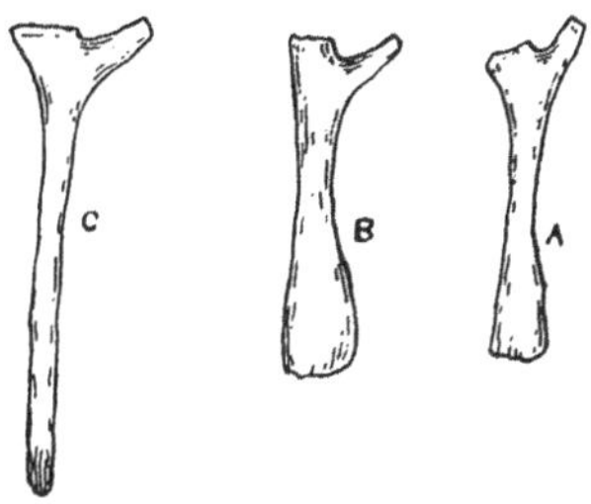

FIG. 9. Ribs. A, rib of the axis; B, rib of the fifth vertebra; C, rib of the twenty-third vertebra. All × 1

Thirteenth to twenty-third vertebrae.— At the thirteenth vertebra the column is sharply flexed upon itself and a series of eleven vertebrae, the thirteenth to the twenty-third, runs uninterruptedly back toward the head. This series differs little from the preceding series, but some vertebrae are revealed in section and some show the form of the centra. It is here that the characters described in the type specimen of *Ostodolepis* are apparent (see Figs. 8a–8g).

Twenty-fourth to twenty-eighth vertebrae. — Posterior to the twenty-third vertebra there is a break and then come three associated vertebrae with lower, broader arches much more closely resembling the typical vertebrae of the cotylosaurs and especially those of *Seymouria.* A second break is followed by two vertebrae in disturbed position; these have very broad and heavy neural arches and short, stout transverse processes; they can be only the sacral vertebrae, (see Figs. 8e, f and g). The number, twenty-eight, of sacral and presacral vertebrae corresponds exactly with the number assigned by Williston to *Seymouria.*

Caudal vertebrae.—Beyond the sacrum in the sequence of vertebrae, but between it and the two previously described series of vertebrae in the nodule, there is a series of eight vertebrae sharply folded upon itself; these diminish gradually in size; the neural spines are thin and low, scarcely more than rudimentary ridges on the neural arches. The transverse processes are still in the anterior position, but are smaller and more slender, and the ribs (only one in position and visible) appear to be shorter and more sharply curved. This series is considered to be entirely caudal. Besides this series there are two caudal vertebrae immediately below the axis-atlas.

Clavicles. — The clavicles lie nearly in the normal position, slightly displaced to the right side, with the proximal ends slightly overlapping. The shaft of the bone is slender and bent near the center of the length, dividing it into a horizontal portion and a nearly vertical portion. The proximal ends are but little expanded and show slight or no rugosity.

Interclavicle. — The interclavicle is nearly covered by the matrix in such a position that it is impossible to clear it. The form cannot be made out exactly, but it can be seen that there is a broad proximal and a narrow posterior extension. If the bone is of the form suggested by the small portion exposed it is extraordinarily broad at the anterior end.

Cleithrum (?). — There is a small, irregular bit of bone in close association with the distal end of the right clavicle, but it is so imperfectly preserved that it cannot be made out clearly. This fragment is in the proper position for the cleithrum with relation to the clavicle and the scapula. If this determination is correct the lower end of the cleithrum was rather broad.

Scapula (Fig. 10). — The scapula of the right side is preserved complete or nearly so, and in its normal position with relation to the vertebrae, ribs and limb bones. The blade is rather narrow and elongate and the cotylar portion heavy.

Coracoid (?). — A mass of dermal scutes covers the spot where the coracoid should lie, if present, and could not be sacrificed.

Humerus (Plate I, Fig. 1, and Figs. 11a–11b). — The head of the humerus is broad with a prominent deltoid process. The

Fig. 10. Right scapula. × 1

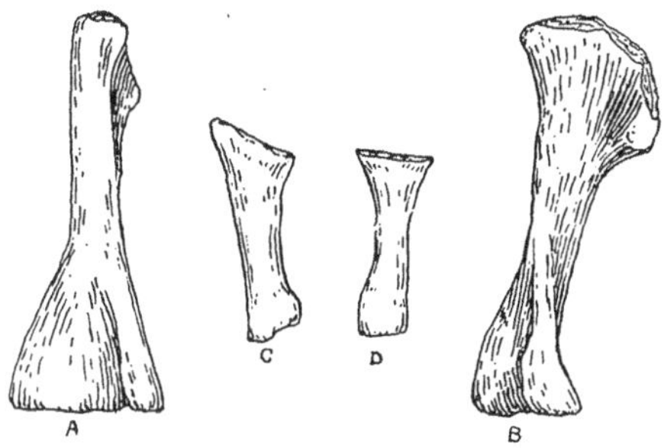

Fig. 11. A, posterior view of the humerus; B, anterior view of the same; C, ulna; D, radius. All of the right side. All × 1

shaft is contracted and angular in section rather than rounded. The lower end is turned almost at right angles to the upper end and there is no entepicondylar foramen. The articular surfaces are not well modeled, indicating the presence of considerable cartilage in the joints. The absence of the entepicondylar foramen and the presence of a considerable amount of cartilage in the articulations are indicative of amphibian affinities in the Permian vertebrates of North America.

Radius (Fig. 11d). — The radius is short, not quite half the length of the humerus, and stout.

Ulna (Fig. 11c). — The ulna is longer than the radius, stout and slightly curved. The olecranon process is poorly developed showing a poor articulation of the elbow joint.

Front foot. — A few small carpal bones, some metacarpals and phalanges, were found near the distal end of the right ulna and radius, but they are so disturbed in position that nothing of the form, number or arrangement can be made out, beyond the fact that the short and heavy metacarpals and phalanges indicate a short, stout foot such as would be expected from the character of the limb bones.

Pelvis. — Nothing of the pelvis was found except a single pubis in the proper position. The pubis is flat and oval with a large foramen. The pubis and ischium were, in all probability, flat and plate-like, as in all the more primitive reptiles.

Femur (Plate III, Fig. 4, and Text Fig. 12a). — The femur of the left side is preserved in close association with the tibia and fibula. The anterior face of the head is deeply concave with a distinct trochanter. The adductor ridge runs down the middle of the face instead of crossing diagonally to the outer edge of the lower end. This is amphibian rather than reptilian in character. The upper articular face is hidden by a mass of scutes, but the lower face shows the same lack of modeling as do the faces of the humerus.

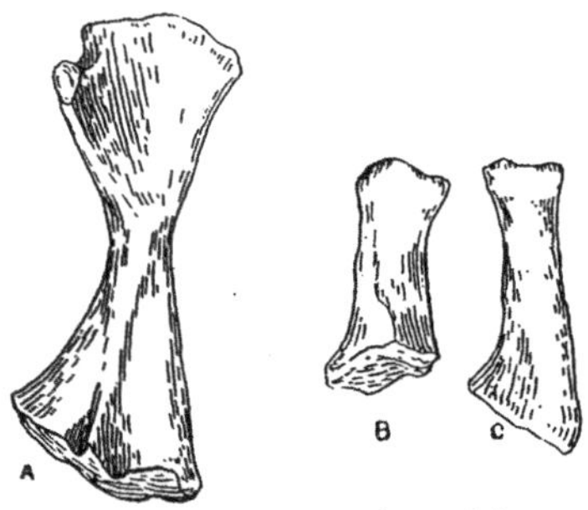

FIG. 12. A, Anterior view of femur; B, tibia; C, fibula. All of the left side. All × 1

Tibia and *fibula* (Text Figs. 12b–12c). — These are very short and stout. There are no certain indications of the posterior foot; some isolated foot bones may belong to the posterior foot.

THE DERMAL SCUTES

(Plate II, Figs. 4a–4b and 5, and Plate III, Figs. 1–3)

In the specimen the dermal scutes form a fairly solid mass on the ventral side of the thoracic and abdominal regions. It is evident that the abdominal surface was protected by closely set scutes which in the process of preservation have been packed one upon another (see Plate III, Fig. 1). As in the type specimen of *Ostodolepis* described by Williston, no trace of scutes was found on the dorsal surface. In no place has the complete form of a scute been made out, but it is evident that they were oval. Williston described them in the type specimen as approximately equal in longitudinal and transverse diameters; in this specimen they are distinctly elongate oval (see Plate II, Fig. 4a). Williston described the scutes as marked with concentric raised lines; the figures showing enlarged views of the scutes (Plate II, Fig. 5, and Plate III, Figs. 2–3) demonstrate that they are in no way like the "cycloid scales of fishes," but that they are made up of concentric rods or lamellae. This type of scute has not yet been recorded, so far as the author is aware. Credner in his description of *Discosaurus permianus*[3] notes that the discoid scutes are made of separate rods with cross connections and and cites their similarity to the scutes of living Gymnophiona, *Epicrum*, and to the Permian form, *Sparodus crassidens* of Fritsch. The scutes of *Ostodolepis* resemble most closely the scutes figured by Fritsch from *Sparodus*, in his *Fauna d. Gaskohle*, Band I, Plate 10, Figure 4.

CONCLUSION

Williston in his original description of *Ostodolepis* says: "The very short spines, the broad and flat zygapophyses and the strong rib attachments suggest a slender, lizard-like form for the living animal, one probably with long limbs and prehensile feet, cursorial or climbing in habit." It is evident from the character of the limbs and feet that Dr. Williston was very excusably misled in his interpretation of the habits and form of the animal by the

[3] *Zeitschrift d. Deut. Geol. Gesell.*, Jahrgang 1883.

slight amount of material at his disposal. So peculiar and aberrant is *Ostodolepis* that the author is unwilling to hazard more than a conjecture as to its habits. The strong limbs and feet, the low neural spines, the peculiar projecting rostrum and the large orbits are suggestive of fossorial habits and the sharp, conical teeth would be eminently fitted for the capture of worms and other soft-bodied forms.

The systematic position is even more puzzling. At present it can be safely placed only as an aberrant form of the Cotylosauria, though it could easily be placed in a separate new group, dependent entirely on individual opinion as to the value of its characters. Below is a list that indicates its peculiar assemblage of characters:

Characters indicating amphibian relationships
- Strong parasphenoid
- Wide interpterygoid space
- Course of adductor ridge on the femur
- Absence of entepicondylar foramen in the humerus
- Asymmetry of the skull bones
- Patch of spongy bone on the parasphenoid

Characters indicating reptilian relationships
- Single occipital condyle
- Well-developed basioccipital
- Pterygoids articulating with basisphenoid
- Intervertebral position of the capitulum of the rib
- Position of the paroccipital process
- Presence of a quadrate foramen
- Character of the teeth

Characters indicating specialization
- Elevation of the posterior portion of the skull
- Absence of a pineal foramen
- Large orbits
- Row of teeth on the palatine bone
- Flattened rostrum of the nose

PLATE I

FIG. 1

FIG. 2

FIG. 3

PLATE II

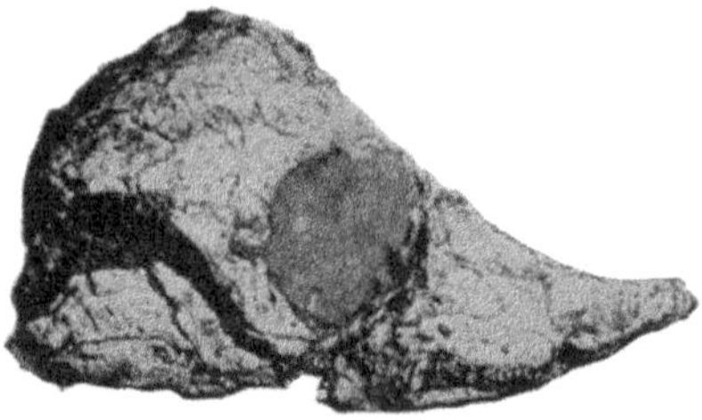

FIG. 1

FIG. 2

FIG. 3 a

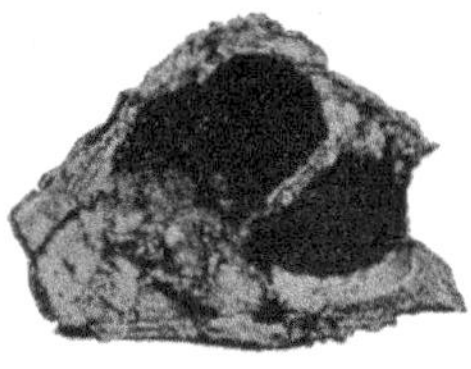

FIG. 3 b

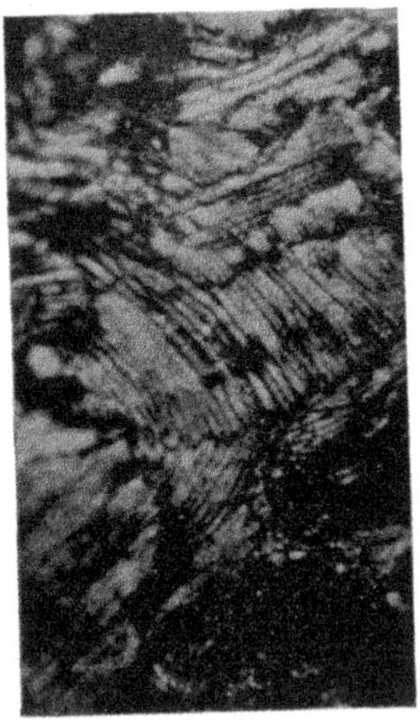

FIG. 5

FIG. 4 a

FIG. 4 b

PLATE III

Fig. 1

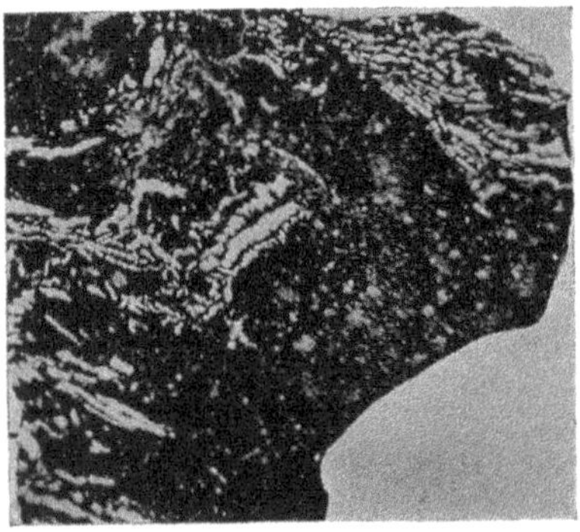

Fig. 2

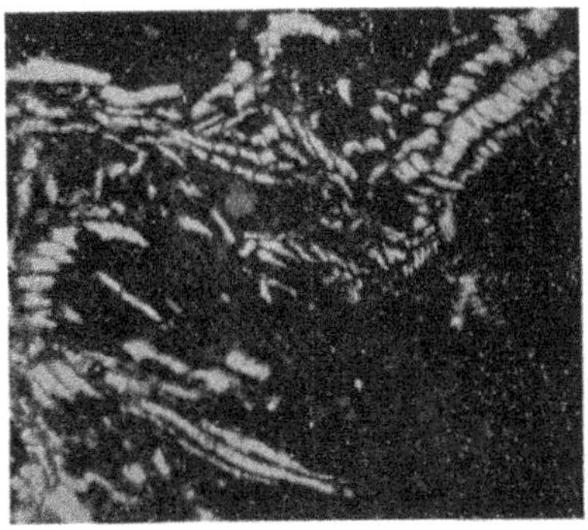

Fig. 3

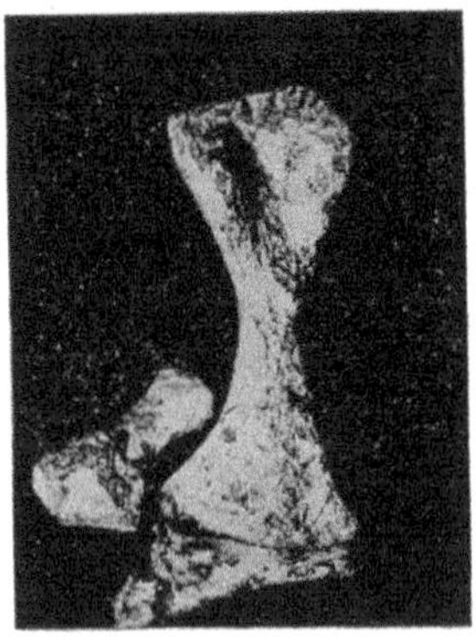

Fig. 4

DESCRIPTION OF PLATES

PLATE I

FIG. 1. Upper surface of the nodule showing the position of the specimen. × 1

FIG. 2. Upper surface of the skull. × 1

FIG. 3. Lower surface of the skull. × 1

PLATE II

FIG. 1. Right side of the skull. × 1

FIG. 2. Posterior surface of the skull. × 1

FIG. 3. a, median section of the skull, left side. × 1

b, median section of the skull, right side. × 1

FIG. 4. a, a group of dermal scutes. × 1.8

b, a second group of dermal scutes. × 1.8

FIG. 5. A portion of Figure 4 a, showing the parallel rods. × 14.5

PLATE III

FIG. 1. A surface showing edges of the dermal scutes. × 6

FIG. 2. A polished surface with edges of the dermal scutes, showing the parallel rods. × 6

FIG. 3. A portion of the same scutes shown in Figure 2. × 14.5

FIG. 4. Right femur, tibia and fibula. × 1

VOL. III, No. 6, pp. 109–150 (5 pls.) FEBRUARY 20, 1930

THE COLOR PATTERNS OF FOSSIL CEPHALOPODS AND BRACHIOPODS, WITH NOTES ON GASTEROPODS AND PELECYPODS

By AUG. F. FOERSTE

CONTENTS

ACKNOWLEDGMENTS

THE incentive toward the studies recorded in this paper was furnished by the loan from Professor E. C. Case of the University of Michigan of several specimens showing color patterns. Two of these were specimens of *Orthoceras* showing transverse zigzag color lines and one was a fragment of limestone retaining parts of two valves of a terebratuloid brachiopod, both of which showed concentric color lines. These were supplemented by a loan from Professor Carl Dunbar of Yale University of several specimens of *Orthoceras*, belonging to two species, both of which showed color markings, one of these having zigzag lines, and the other, relatively straight transverse ones. Finally, Dr. H. Hedström of the Geological Survey of Sweden loaned me two specimens of a small erect cyrtoceroid in which the steeply inclined color lines formed a zigzag pattern only on the ventral side of the conch. I had previously been permitted by Dr. Jan Koliha to study the unrivaled collection of cephalopods described by Barrande, a considerable number of which show color markings; similar markings occur also on specimens formerly borrowed from Dr. Rudolf Ruedemann and Professor T. E. Savage. To all of these I here express my appreciation of the favors so generously granted.

The following studies are concerned chiefly with color markings in cephalopods and brachiopods, but a few notes on gasteropods and pelecypods are added, enough to give some notion of the relative number of color-marked species known at present and their stratigraphic distribution. The bibliography at the end includes all the more important papers bearing on this subject, especially those listing species from the widely scattered literature. Among these the papers by Deecke, Leidhold, Newton, Oppenheim and Richter will be found very useful.

I. Introduction

Fossils retaining evidences of former coloration are always of interest, especially when they belong to the remote past. In fossils of more recent age the original colors are frequently preserved, but in earlier fossils, especially those of Paleozoic age, the former presence of colors is usually indicated only by varying shades of light and dark. Sometimes these traces of color are in the form of dots or blotches, irregularly scattered, or arranged according to some definite pattern. At other times they may take the form of radiating or concentric lines or bands, or of combinations of the two. Occasionally the lines are of a zigzag form, or the patterns may be even more intricate.

In the more recent fossils all the colors at present known from animals of a similar type are to be expected, but in fossils from

Paleozoic strata the colors usually found are confined to various shades of brown or are nearly black. Occasionally the browns verge on red.

In some cases there is evidence that the color shown by a particular fossil is determined in some manner by the mineral matter present in the inclosing matrix. Occasionally this may be indicated by a single individual. For instance, specimens have been found which possessed brownish color markings where preserved in whitish limestone, but which had almost black markings where extending into a blackish matrix.

All traces of color tend to disappear with the increasing age of a deposit of fossils, especially during the recrystallization of the calcareous components of the shell. Hence color patterns are retained more frequently by the more recent species. This is shown especially by the gasteropods. A total of 293 species of gasteropods has been recorded as retaining color markings. Of these, 224 species occur in the Tertiary, 25 in the Mesozoic, and 44 in the Paleozoic. Even among Paleozoic strata great differences are shown, the Carboniferous retaining 33 species of gasteropods with color markings; the Devonian, 5 species; the Silurian, 3 species; and the Ordovician, also 3 species.

In cephalopods, on the contrary, almost all color-marked species belong to the Paleozoic. This is readily understood when it is remembered that color markings are extremely rare among all the strongly coiled cephalopods, both nautiloid and ammonoid. They are not numerous among straight conchs, and they appear to occur most frequently among the breviconic cyrtoceroids, which are known only from Paleozoic strata.

In a similar manner, only two brachiopods preserving color patterns have been recorded from the Tertiary. But this type of life was much more abundant in Paleozoic and Mesozoic times, from which a larger number of species with color patterns is reported.

The general vertical distribution of the different types of shell-bearing animals is indicated on the following page.

The very imperfect character of the data secured so far is shown by the fact that of the 33 color-marked species of gasteropods recorded at present from the Carboniferous of the entire world, 15 were discovered by J. Brookes Knight only recently, and at a single horizon, in the Pennsylvanian division of the Carboniferous, near St. Louis, Missouri. In a similar manner, of the 25 color-marked species of cephalopods known so far from the Silurian of the entire world, 22 were found in a relatively small area north of Prague, formerly in Bohemia, now a part of Czechoslovakia. Similar rich but isolated localities probably await discovery elsewhere.

Number of Species Recorded as Color-marked

	Tertiary	Mesozoic	Paleozoic	Total
Brachiopods ..	2	10	11	23
Pelecypods ...	6	12	7	25
Gasteropods ..	224	25	44	293
Cephalopods ..	1	1	39	41
Total	233	48	101	382

The imperfection of our knowledge regarding color markings in fossil shells is shown also by the irregular vertical distribution of the gasteropods in the Tertiary and Mesozoic. Of the 293 species recorded as showing color markings, 16 occur in the Pliocene, 75 in the Miocene, 10 in the Oligocene, and 72 in the Eocene; 2 additional ones are recorded as from the Tertiary, without any statement of their more exact horizon. The abundance of color-marked species of gasteropods in the Eocene should be contrasted with the small number of 5 species recorded from the Cretaceous, 11 from the Jurassic, and 9 from the Triassic. It is evident that the sudden increase of recorded color-marked species from 5 in the Cretaceous to 72 in the Eocene does not indicate that color-marked species were relatively scarce during Cretaceous times, but either that the conditions for the preservation of the colors of Cretaceous species, for some unknown reason, was less favorable, or that by some chance certain localities unusually favorable to the preservation of color markings have been found in Eocene strata, while similar good localities of Cretaceous await future discovery.

II. Brachiopods

Recently a terebratuloid brachiopod shell showing black concentric lines and bands on an otherwise light gray field was found by Professor E. C. Case in the Alpena member of the Traverse group, in the Middle Devonian of eastern Michigan. Its exact horizon was about 90 feet below the top of the quarry of the Michigan Alkali Company, at Alpena, Michigan. This horizon is between 22 and 30 feet above the present base of the quarry, where the limestone is dark and fine-grained, irregularly bedded, and with streaks of black. At one point within this zone were found the cephalopods *Acleistoceras casei*, *Acleistoceras nummulatum*, *Alpenoceras ulrichi* and *Nephriticerina alpenensis*, described in *Contributions from the Museum of Geology, University of Michigan*, in 1927. In addition, this horizon contains a species of *Acervularia*, a long form of *Favosites*, various cyathophylloids, and large, flat, fan-shaped stromatoporoids. The *Acervularia* occurs also in abundance in a zone 10 feet thick at a horizon 16 feet above this cephalopod horizon.

The terebratuloid species with color markings is referred provisionally to *Cranaena*, but its exact generic relationship cannot be determined until specimens are secured permitting the discovery of the structure of the interior of their valves in the vicinity of their beaks, according to the methods recently used by Bielanski on Iowan Devonian material. The Alpena species is described here as new. At least no terebratuloid species with concentric color markings has yet been described from Middle Devonian strata of America. The name **Cranaena casei** is proposed. To judge from fragments of *Cranaena* in the matrix of other specimens from the same locality and horizon, further search will no doubt result in securing better material for specific description, though probably lacking the color markings. At least, to judge from occurrences elsewhere, even where color-marked species are unusually frequent, the number of specimens of any species actually retaining color markings forms only a small part of the total number of specimens of that species occurring at the same horizon. The following description gives a more detailed account of the color-marked specimens of *Cranaena* from Alpena.

Cranaena casei, sp. nov. (Plate II, Fig. 5). — Two valves, their interiors facing each other, evidently not belonging to the same individual, since the color patterns on the two valves do not match. Their lateral displacement is about 5 mm. The width of the valve best preserved is about 25 mm., and the other is of about the same size. Their lateral outlines appear to have been similar to those of *Cranaena calvini* (Hall & Whitfield), but the valves of the latter are spotted, instead of marked with concentric lines and bands. The posterior half of the better preserved valve is almost smooth, as seen without a lens, but its anterior half is striated distinctly, though not strongly, in a concentric direction, and very faint radiating lines can be detected if the shell is illuminated transversely. The surface of the shell is minutely punctate, with the punctae arranged in diagonal rows which cross each other at an angle of approximately 45 or 50 degrees. There are about 15 punctae within a length of 1 mm., counted along one of the diagonal rows. The punctae are elongated distinctly in a radial direction from the beak. Concentric black lines cross the otherwise gray surface of the shell. The different black lines vary in width from about 0.66 mm. to fully 0.7 mm. They occur at irregular intervals and in an irregular order of succession, broader and finer bands and lines being intermingled. In fact, under a lens some of the broader black bands may be seen to be made up of a series of very narrow but closely crowded black lines, separated by still finer intermediate lighter-colored lines. In other specimens the broader black bands are not interrupted by lighter-colored lines, but appear continuous even under a lens. In the two valves here under consideration the black lines and bands are more distant from each other posteriorly, but more closely crowded anteriorly.

The relative scarcity of concentric lines in fossil brachiopods is shown by the following list, which includes all known fossil brachiopods retaining color markings; also a brief indication of the character of the color markings presented.

LIST OF SPECIES

Pliocene

Terebratula ampulla; spotted

Eocene

Terebratula sp. (from Säntis); spotted

Cretaceous

Rhynchonella octoplicata; spotted
Terebratula biplicata; with radial lines

Jurassic

Terebratula eudesi; with radial lines
Terebratula maxillata; with radial lines
Terebratula subovoides; with radial lines
Terebratula umbonella; spotted
Waldheimia perforata; with concentric lines

Triassic

Terebratula falsani; with radial lines
Terebratula potieri; with radial lines
Terebratula vulgaris; with radial lines

Carboniferous

Dielasma elongatum; with radial lines
Dielasma hastatum; with radial lines
Martinia glaber; with radial lines
Schizophoria resupinata; with radial lines

Devonian

Cranaena calvini; spotted
Cranaena navicella; spotted
Cranaena morsii; with radiating rows of dots and spots
Cranaena (?) *sp.;* a species from Arctic America, evidently incorrectly referred to *Terebratula;* spotted
Cranaena casei (from Alpena, Michigan); with concentric lines and bands
Newberria cimex; with concentric bands
Pugnax pugnus (often still cited under *Rhynchonella*); spotted

This list includes 23 species of brachiopods retaining color markings. Of these, 18 belong to the terebratuloids (*Terebratula, Dielasma, Cranaena, Newberria*); 1 is a terebratelloid (*Waldheimia*); 2 are rhynchonelloids (*Rhynchonella, Pugnax*); 1 is a spiriferoid (*Martinia*), and only 1 is an orthoid (*Schizophoria*). Among these species the forms with punctate shells predominate greatly; they include the 18 terebratuloids and 1 terebratelloid. Rhynchonelloids, spiriferoids and orthoids are relatively few. Eleven of these species have radiating lines. In 8 species there are dots or spots irregularly arranged; in 1 species the dots are arranged in radiating series. In 3 species the lines or bands are concentric.

The Devonian brachiopod brought back from Arctic America by Sir John Richardson, and now preserved in the British Museum of Natural History, was cited in 1854 by Forbes [1] under the generic name *Terebratula.* That genus does not occur, however, in Paleozoic strata, and it is more likely to be a *Cranaena* or to belong to some genus closely related to *Cranaena.* Forbes described this specimen as being beautifully spotted. According to Suess,[2] this spotting was irregular. The specimen is mentioned also by Kayser.[3] It is of interest especially on account of the presence of *Cranaena calvini* and *Cranaena navicella* in the Hackberry stage of the Upper Devonian in Iowa. This Hackberry fauna is regarded as of northern origin, a view which finds support in the occurrence of an apparently closely related species in the American Arctic. The general color of *Cranaena calvini* is described by the Fentons [4] as brownish to olive-brown, with touches of purple and chocolate, one specimen being olive-brown with dots and longitudinal splotches of purple. The color of *Cranaena navicella* is described by them as brown or olive-brown, with purplish-brown mottlings.

The Alpena limestone fauna also may be of northern origin. The terebratuloid *Cranaena casei* from this limestone differs, however, from any other known species of this genus in having concentric color lines and bands instead of irregularly arranged dots and splotches of color.

The presence of concentrically arranged lines and bands in the Alpena limestone is of interest, moreover, because concentric lines

[1] Forbes, E., "Note on an Indication of Depth of Primaeval Seas, Afforded by the Remains of Colour in Fossil Testacea," *Proc. Royal Soc. London,* Vol. 7, pp. 21–23, 1854.

[2] Suess, Ed., "Über die Wohnsitze der Brachiopoden," *Sitzungsberichte der k. Ak. Wiss., Math.-nat. Kl.,* Vol. 37, pp. 185–248, 1859.

[3] Kayser, E., "Notiz über Rhynchonella pugnus mit Farbenspuren aus dem Eifler Kalk," *Zeitschrift der deutsch. geol. Gesell.* (Berlin), Vol. 23, pp. 257–265, 1871.

[4] Fenton, C. L. and M. A., *The Stratigraphy and Fauna of the Hackberry Stage of the Upper Devonian* (Macmillan Co., New York), pp. 132–133, 1924.

and bands are extremely rare in fossil species, and are relatively rare even among living forms. Among fossil forms, they occur in *Waldheimia perforata* (Piette), from the Lower Liassic of Gloucestershire, in England; the transverse or concentric bands deviate but slightly from the course of the lines of growth. They occur also in *Newberria cimex*, in which, however, the concentric bands deviate conspicuously from the lines of growth. Among living brachiopods the following species are listed by Richter,[5] on the authority of Professor Blochmann of Tübingen, as having concentric color lines: *Terebratula coreanica* Ad. & Reeve, *Terebratula cruenta* Dillw., *Terebratula rubicunda* Sow., *Krausina rubra* Pallas, *Laqueus rubellus* Sow. and, with some deviations, also *Waldheimia grayi* Davidson. In these living forms, however, the concentric lines follow the lines of growth closely. In this respect, the Alpena specimens of *Cranaena* correspond more closely to the living terebratuloids than to *Newberria cimex* from the Devonian of western Europe, its concentric lines and bands being strictly parallel to the lines of growth.

III. Pelecypods

Fossil pelecypods with color markings are known at present only in small numbers, with the following stratigraphical distribution: Miocene, 3 species; Oligocene, 1; Eocene, 2; Cretaceous, 9; Jurassic, 1; Triassic, 2; Carboniferous, 8; a total of 26 species.

Of the Carboniferous species, Forbes cited 4 as occurring in the collections of the Geological Survey of Great Britain; namely, *Aviculopecten elongatus*, *Aviculopecten intercostatus*, *Aviculopecten sublobatus*, and an unnamed species of *Aviculopecten* with spotty markings on the ribs. The species called *Aviculopecten sublobatus* by Forbes is referred at present to the genus *Streblopteria*. To this list given by Forbes, Oppenheim added the three species *Aviculopecten rugulosus* McCoy, *Syncyclonema colorata* Koninck and *Syncyclonema sowerbyi* McCoy. Recently J. Brookes Knight found a pelecypod with color markings in the Pennsylvanian part of the Carboniferous, in the vicinity of St. Louis, Missouri, thus

[5] Richter, Rudolf, "Zur Färbung fossiler Brachiopoden," *Senckenbergiana*, Vol. 1, no. 3, p. 84, 1919.

raising to 8 the total of Carboniferous pelecypods known to have color markings. The discovery of 15 color-marked species of gasteropods and of the single peleycpod in the St. Louis area was announced to the Paleontological Society of America at its meeting in New York city in December, 1929 (*Bull. Geol. Soc. Am.*, 40: 212. 1929), but the details have not yet been published.

IV. Gasteropods

As already noted on a preceding page, the total number of species of fossil gasteropods recorded as retaining color markings equals 293, of which 224 occur in the Tertiary, 25 in the Mesozoic, and 44 in the Paleozoic. Of the Paleozoic species, 33 occur in the Carboniferous, 5 in the Devonian, 3 in the Silurian, and 3 in the Ordovician. These Paleozoic species, as far as published, are recorded in the following list:

LIST OF PALEOZOIC SPECIES

Carboniferous

- *Capulus margarita* Boulanger
- *Euomphalus (Solarium) pentangulatum* Sowerby
- *Glyptobasis marshalli* Roundy
- *Holopea proutana* Hall
- *Macrocheilus maculatus* Koninck
- *Mourlonia (Pleurotomaria) carinata* Sowerby (= *flammigera* Phillips)
- *Natica (Naticopsis?) lirata* Phillips
- *Naticopsis* cf. *altonensis* McChesney
- *Naticopsis picta* Girty
- *Naticopsis plicistria* Phillips
- *Patella retrorsa* Phillips
- *Patella solaris* Koninck
- *Pileopsis (Capulus) pileus* Phillips
- *Platyostomella (Littorina) scotoburdigalensis* Etheridge
- *Pleurotomaria conica* Phillips
- *Pleurotomaria rotundata* Sowerby
- *Trochus sp.*, in the collection of the Geological Survey of Great Britain
- *Turbo cryptogrammus* Koninck
- Fifteen additional species found by J. Brookes Knight near St. Louis, but not yet published.

Devonian

- *Platyceras repletum* Barrande
- *Pleurotomaria beaumonti* Archiac & Verneuil
- *Pleurotomaria limbata* Phillips
- *Pleurotomaria orbignana* Archiac & Verneuil

Turbonitella (*Naticopsis*) *subcostata* Archiac & Verneuil (= *harpula* Sowerby)

Silurian

Cyrtolites Pharetra Lindström
Platyceras nobile Barrande
Polytropina (*Oriostoma*) *helicina* Lindström.

Ordovician

Cyclonema (*Turbo*) *rupestris* Eichwald, from the Bala of Chair of Kildare, near Dublin, Ireland
Holopea symmetrica Hall, from the Trenton of New York and New Jersey
Holopea (*Straparollina*) *harpa* Hudson, from the Valcour member of the Middle Chazyan of New York

All the Silurian gasteropods in the preceding list belong to the Gotlandian division of the Middle Silurian, as exposed in Europe. Among the few Ordovician species here listed *Holopea harpa* [6] is the oldest. The next oldest probably is the *Holopea symmetrica.*[7] The age of the *Cyclonema rupestris* [8] from the Chair of Kildare is not known definitely. The exposures here consist of the Middle and Upper Bala. I have been unable to determine from the literature from which of these two horizons the Kildare specimen was obtained, but it was from the equivalents of either the Trenton or the Cincinnatian.

V. Cephalopods

Among cephalopods retaining color markings a total of 41 species has been recorded, all of which are discussed on the following pages. Among these, 1 ranges from the Eocene into the Miocene, 1 occurs in the Middle Liassic, 1 is found in the Pennsylvanian part of the Carboniferous in America, 3 occur in the Carboniferous limestone division of the Carboniferous of western Europe, 3 occur in the Middle Devonian, 25 are found in the Middle Silurian, and 7 are known from the Ordovician.

Among these 41 species the following 9 occur in America:

[6] Raymond, Percy, "An Ordovician Gasteropod Retaining Color Markings," *Nautilus*, Vol. 19, p. 101, 1906; also *Annals Carnegie Mus.*, Vol. 4, p. 212, 1908.

[7] White, Theodore G., "The Faunas of the Upper Ordovician Strata at Trenton Falls, Oneida Co., N. Y.," *Trans. New York Acad. Sci.*, Vol. 15, p. 85, 1896.

[8] Forbes, E., *op. cit.*, pp. 21–23.

LIST OF AMERICAN SPECIES

Orthoceras dunbari; from the Ochelata formation in the Pennsylvanian of Oklahoma
Orthoceras anguliferum alpenense; Alpena limestone, Middle Devonian of Michigan
Orthoceras trusitum; Guelph, Middle Silurian of New York
Sactoceras manitoulinense; Waynesville member of Richmond, Manitoulin island in southwestern Ontario
Rizoceras coronatum; Shamattawa member of Richmond; west of Hudson Bay
Sactoceras westoni; Lorraine, northwest of Toronto, in southwestern Ontario
Geisonoceras tenuitextum; Trenton, New York
Orthoceras tenuistriatum; Trenton of New York and New Jersey.
Orthoceras sp.; Plattin, in northeastern Missouri

1. BREVICONIC ORDOVICIAN, SILURIAN AND DEVONIAN CEPHALOPODS

From the Silurian of Bohemia, now included in Czechoslovakia, Barrande[9] figured the following eighteen species as preserving color markings. All these he referred to the genus *Cyrtoceras*, though none of them show the internal structure of the siphuncle characteristic of the genotype, *Cyrtoceras depressum.* In this list the name of the species is followed by the number of the plate or plates on which this species is illustrated, and by the page in the text where it is described. Those numbers which are followed by the letter *S* refer to pages in the supplement to the text published in 1877. The remaining page numbers refer to the first part of the text, published in 1867. The words "black" and "white" indicate merely whether the color designs on the shell are lighter or darker than the general body color of the latter. The word "erect" is used for those conchs which are straight, or nearly straight; usually their curvature is very small, and in some species it is confined to the apical part of the conch. The word "curved" is used for those conchs in which the lengthwise curvature is more conspicuous. Conchs in which the siphuncle is located on the convexly curved side of the conch are called "exogastric"; those in which it is located on the concavely curved side are called

[9] Barrande, Joachim, *Systeme silurien du centre de la Boheme,* Text: pt. I (1867); pt. II (1870); pt. III (1874); pt. IV (1877); pt. V, including Supplement (1877).

"endogastric." In some of these conchs, as already stated, the lengthwise curvature may be confined to the apical part of the conch. In those erect conchs in which both the siphonal and the antisiphonal outlines of the conch are slightly convex, one side is usually more convex than the other. In those cases the conch is called endogastric if the siphuncle is located near the less convex side. Strictly speaking, the ventral side of a conch is known definitely only when the location of its hyponome can be determined. In the fossil shell this is shown by the location of the hyponomic sinus. In the ellipochoanoidal nautiloids this is usually on the convexly curved side of the conch, at least during early stages of its growth. The cross-sections of most of the conchs here listed are circular, but in two species they are slightly compressed laterally; in one this lateral compression is more conspicuous.

LIST OF BOHEMIAN SILURIAN CYRTOCEROIDS

1. *Cyrtoceras cyathus;* pls. 153, 481, 507; pp. 542, 22–S, 134–S. Black, erect, exogastric, with circular cross-section
2. *Cyrtoceras intricans;* pl. 514; p. 149–S. White, curved, exogastric, with circular cross-section
3. *Cyrtoceras sinuatulum;* pl. 153; p. 566. White, erect, exogastric, with circular cross-section
4. *Cyrtoceras zebra;* pls. 168, 481; pp. 632, 49–S. Black, erect, exogastric, with circular cross-section
5. *Cyrtoceras bonum;* pl. 167; p. 651. White, erect, endogastric, slightly compressed laterally
6. *Cyrtoceras gentile;* pl. 505; p. 144–S. White, erect, endogastric, with circular cross-section
7. *Cyrtoceras haesitans;* pl. 505; p. 147–S. White, erect, endogastric, with circular cross-section.
8. *Cyrtoceras iridis;* pl. 153; p. 554. White, curved, exogastric, with circular cross-section
9. *Cyrtoceras parvulum;* pls. 177, 481, 504, 505; pp. 689, 39–S, 168–S. White, curved, endogastric, with circular cross-section
10. *Cyrtoceras veteranum;* pls. 208, 514; pp. 571, 190–S. White, moderately curved, exogastric, with circular cross-section
11. *Cyrtoceras chrysalis;* pl. 506, p. 121–S. White, erect, endogastric, with nearly circular cross-section
12. *Cyrtoceras pseudomorphum;* pl. 506; p. 174–S. White, erect, probably endogastric, with slightly compressed cross-section
13. *Cyrtoceras eurus;* pl. 506; p. 142–S. White, erect, endogastric, with circular cross-section
14. *Cyrtoceras jubatum;* pl. 506; p. 151–S. White, erect, exogastric, with circular cross-section

15. *Cyrtoceras vittatum;* pl. 153; p. 572. White, erect, exogastric, with circular cross-section
16. *Cyrtoceras decurio;* pl. 240; p. 545. Black, curved, exogastric, with laterally compressed cross-section
17. *Cyrtoceras jugale;* pl. 506; p. 152–S. White, erect, exogastric, with circular cross-section
18. *Cyrtoceras maculosum;* pl. 155; p. 556. Black, moderately curved, probably exogastric, with circular cross-section

All the species here listed have short conchs. These enlarge rapidly and have relatively large living chambers. They differ from the typical *Rizoceras* of Hyatt,[10] of which the genotype is *Orthoceras indocile* Barrande (his Plate 185), in having more or less distinctly curved conchs, instead of conchs which are almost strictly orthoconic, though usually showing a slight lengthwise curvature, at least at the apical end of the conch.

The color pattern shown by these Bohemian forms varies in the different species.

In *Cyrtoceras cyathus*, *Cyrtoceras intricans*, *Cyrtoceras sinuatulum* and *Cyrtoceras zebra* the color lines are transverse and wavy, with the sinuations numerous within the circumference of the conch.

In *Cyrtoceras bonum* the transverse color lines show a chevron pattern, zigzagging up and down so as to produce 6 or 8 chevrons within the circumference of the conch. In *Cyrtoceras gentile*, *Cyrtoceras haesitans*, *Cyrtoceras iridis* and *Cyrtoceras parvulum* the chevrons arch evenly upward and meet laterally at acute, downward pointing angles. In these species the number of chevrons within the circumference of the conch appears to be 6 in some species and 8 in others. In *Cyrtoceras veteranum* the transverse color lines zigzag up and down, but curve at the turning points. The number of chevrons within the circumference of the conch is 6. In all these 6 species one of the chevrons occupies the median part of the siphonal side of the conch, and this chevron arches upward. This curvature of the median chevron is shown in *Cyrtoceras bonum* by Figure 20 on Barrande's Plate 167; in *Cyrtoceras gentile* by Figures 23 and 30 on Plate 505; in *Cyrtoceras*

[10] Hyatt, Alpheus, "Genera of Fossil Cephalopods," *Proc. Boston Soc. Nat. Hist.*, Vol. 22, p. 276, 1884.

haesitans by Figure 19 on Plate 505; in *Cyrtoceras iridis* by Figure 21 on Plate 153; in *Cyrtoceras parvulum* by Figures 4, 5, 11 and 12 on Plate 481, and also by various figures on Plate 504; and in *Cyrtoceras veteranum* by Figure 21 on Plate 208. In the type of the species last named the chevron along the median part of the convexly curved siphonal side of the conch rises. A similar rising chevron is shown along the median part of the opposite or antisiphonal concavely curved side in Figure 15 on Plate 504.

The five species, *Cyrtoceras chrysalis*, *Cyrtoceras pseudomorphum*, *Cyrtoceras eurus*, *Cyrtoceras jubatum* and *Cyrtoceras vittatum*, show a similar upward arching of the chevrons, but the upper side of each chevron is fringed by short branches having the same color as the chevron itself. In three of these species one of the chevrons occupies the median line of the siphonal side of the conch. This is well shown in *Cyrtoceras pseudomorphum* by Figure 8 on Barrande's Plate 506; and in *Cyrtoceras jubatum* by Figure 23 on Plate 506. In *Cyrtoceras eurus* it is indicated very vaguely by Figure 18 on Plate 506. It may be present on *Cyrtoceras chrysalis*, but nothing is known of the color lines on the ventral side of *Cyrtoceras vittatum*.

In *Cyrtoceras decurio* the transverse color lines are replaced by broad color bands. On the lateral side of the type only a single chevron is preserved, which rises to an angular point, two of its sides meeting along the median part of the conch at a point which is directed downward. According to Barrande there is a possibility that an additional series of chevrons once occupied the convexly curved side of the conch, but no trace of such a series remains.

In *Cyrtoceras jugale* there appear to be several vertical series of short horizontal color lines, the individual members of which are somewhat yoke-shaped.

In *Cyrtoceras maculosum* the shell appears ornamented with irregularly arranged black blotches irregular in shape.

Regarding a specimen of *Cyrtoceras parvulum* Barrande in the British Museum of Natural History, Foord [11] stated that about

[11] Foord, Arthur H., *Catalogue of the Fossil Cephalopoda in the British Mus.*, pt. 1, pp. 292, 297, 1888.

seven upward arching chevrons occur within the circumference of the conch. On the septate part of one of the specimens of *Cyrtoceras expandens* (Barrande, Plate 167, p. 659), in the same museum, Foord observed light-colored zigzag bands resembling those of *Cyrtoceras iridis*, as illustrated by Figure 21 of Barrande's Plate 153.

Zigzag types of color patterns have been observed also in areas other than Bohemia. For instance, in the Shamattawa limestone, on the Shamattawa river, west of Hudson Bay, Savage and Van-Tuyl found a small, erect, rapidly enlarging rizoceroid conch, later described by Foerste and Savage[12] as *Rizoceras* (?) *coronatum* (Plate II, Fig. 3). The number of waves within the circumference of this conch is estimated at about fifteen. The pattern of these color lines differs, however, from that shown by the Bohemian cyrtoceroids described above in the direction of curvature of their chevrons; the latter curve moderately downward instead of arching strongly upward. The siphuncle is close to the ventral wall of the conch and consists of narrow, fusiform segments. The Shamattawa limestone in which this specimen was found is regarded as of Richmond age.

Of *Cyrtoceras depressum* Bronn, Foord[13] stated that "zigzag bands of colour are seen upon the cast in one of the specimens in the Museum Collection" (No. 66380, British Museum of Natural History). From this brief description it cannot be determined to what extent these zigzag colors are likely to have been visible on the exterior of the shell.

The cyrtoceroid described by Barrande under the name *Cyrtoceras bolli* (Barrande's Plate 119, pp. 485, 20–S) differs strongly from the rizoceroids described in the earlier paragraphs of this paper in the much slower rate of enlargement of the conch and in its dorso-ventral depression. The siphuncle is located on the convexly curved side of the conch and its segments enlarge slightly within the camerae. The transverse color lines are narrow and black. Their course is wavy, and that wave which occupies

[12] Foerste, Aug. F., and Savage, T. E., "Ordovician and Silurian Cephalopods of the Hudson Bay Area," *Denison Univ. Bull., Journ. Sci. Lab.*, Vol. 22, p. 50, pl. 5, fig. 7, 1927.

[13] Foord, Arthur H., *op. cit.*, p. 266.

the median part of the ventral side of the conch curves more strongly downward than do the waves ventro-laterally. These waves are preserved only on the ventral side of the conch, but it is estimated that, if they extended in the same manner also around the dorsal side of the conch, their number would be six or seven within the circumference of the conch. This Bohemian species is of Devonian age, occurring in Barrande's horizon Gg3.

Another breviconic cyrtoceroid showing color markings is the species here named *Hedstroemoceras haelluddenense* (Plate I, Figs. 3 A, B; Plate V, Figs. 1, 2 A, B; Plate III, Figs. 3 A, B). In this species there are 14 almost vertical color bands, of which 12 slope downward in a ventral direction at an angle of about 12 degrees with the vertical axis of the conch. The remaining 2 rise in a ventral direction at about the same angle, thus forming a zigzag pattern across the median part of the ventral side of the conch. The rising chevron of the zigzag pattern has its angle along the median line of this side. It is assumed that a similar rising angle is formed by the 2 bands nearest the median line of the dorsal side, but, if so, this angle is not preserved in the two specimens at hand. Since these specimens are types of both a new species and a new genus, a more detailed description of the conch here follows.

Hedstroemoceras haelluddenense, gen. et sp. nov.

Conch small, erect, dorsal outline faintly convex, ventral outline more distinctly convex. Cross-section nearly circular, slightly compressed laterally. Living chamber enlarging but slightly. Siphuncle located near the ventral side of the conch, but not in contact with it; its segments fusiform in outline.

The length of the larger specimen (Plate I, Fig. 3 A; Plate V, Figs. 2 A, B; Plate III, Figs. 3 A, B) is 46 mm., of which 22 or 23 mm. belongs to the living chamber. The phragmacone enlarges dorso-ventrally from a diameter of 10 mm. at its base to 15 mm. at the base of the living chamber 23 mm. farther up, the diameter at its aperture being 16 mm. The lateral diameter at the base of the living chamber is 14.3 mm. The cast of the interior of the living chamber contracts to within 5 or 6 mm. of the aperture and then expands again. This is probably due to a thickening of the interior of the walls of this chamber; no corresponding contraction of the exterior of the shell is noted. The number of camerae in a length equal to the dorso-ventral diameter of the conch at the base of the living chamber is a little more than 6; the uppermost camera is a little shorter than those immediately beneath. The sutures of the upper septa rise distinctly ventro-laterally and curve downward both ventrally and dorsally. The ventral lobe is distinctly narrower and slightly deeper, but the dorsal lobe is broad and shallow and tends to be straight and directly transverse along its median part. Within the third camera beneath the living chamber, where the dorso-ventral diameter is 14.2 mm., the center of the siphuncle is 2 mm. from the ventral wall of the conch. The segment of the siphuncle is fusiform in outline, with a maximum diameter of 1.3 mm., narrowing to 0.8 mm. at its passage through the septum. The dorsal vertical outline of this segment is faintly convex, but the ventral outline is distinctly convex, with the maximum convexity above its mid-height.

The length of the second specimen (Plate I, Fig. 3 B; Plate V, Fig. 1) is 32 mm., 18 mm. of which belongs to the living chamber. Its dorso-ventral diameter at the base of the living chamber is 14.5 mm.; its lateral diameter is about 13 mm. The number of camerae within a length equal to the dorso-ventral diameter of the conch is eight. The sutures of the septa are almost straight, with very faint undulations that cannot be asserted to be strictly conformable to those of the preceding specimen. This suggests that the undulations of the septa in the preceding specimen are not to be considered diagnostic. Both specimens are from Haelludden, on Oeland, southeast of Sweden, in the gray *Endoceras vaginatum* limestone.

Hedstroemoceras is distinguishable from the following genera.

In *Clinoceras* Mascke the siphuncle is located dorsad of the center of the conch and its segments are almost cylindrical, though distinctly contracted at the passage of the siphuncle through the septa. Moreover, the conch is distinctly depressed dorso-ventrally.

In *Cyrtactinoceras* Hyatt the conch also is depressed dorso-ventrally. Its siphuncle, however, is ventrad to the center of the conch. At earlier stages of growth its segments are obliquely globular or slightly nummuloidal in form, but later the siphuncle moves nearer the center of the conch though remaining ventrad to the latter, and its segments become narrowly elongated, enlarging only slightly within the camerae.

In *Wetherbyoceras* Foerste the siphuncle is near the ventral wall of the conch, but not in actual contact with it. Its segments are obliquely globular, and are strongly constricted at the septal necks. The interior of the siphuncle is occupied by ten to twelve converging vertical lamellae which do not reach its center.

In *Chicagooceras* Foerste the conch is only slightly depressed dorso-ventrally, and the siphuncle is located slightly ventrad of the center of the conch.

2. PHRAGMOCEROIDS

One of the phragmoceroids described by Barrande under the name *Phragmoceras panderi* (Barrande's Plate 429, p. 236–S) is marked with numerous vertical and transverse black lines, of which the vertical ones tend to be more distinct, more regularly spaced, and more widely separated, with about fifty occurring within the circumference of the conch. The transverse lines, on the contrary, are fainter, more irregularly spaced, and more numerous. These specimens, however, differ from typical *Phragmoceras panderi* (Plates 48, 50, p. 232) in being more strongly curved lengthwise, and in the stronger dorso-lateral divergence of the median dorsal lobes of the aperture. These lobes are longer and the included median dorsal crest is much broader than in the type of the species. It belongs, however, to Hyatt's genus *Hexameroceras,* of which typical *Phragmoceras panderi* (Barrande's Plate 48, p. 278) is the genotype. For the color-marked specimens figured on Barrande's Plate 429 the term **Hexameroceras barrandei** is proposed. They were found at Grosskuchel, Bohemia, in the Middle Silurian.

Hedström [14] described four species of Phragmoceras with black

[14] Hedström, Herman O., "Über die Gattung Phragmoceras in der Obersilurformation Gotlands," *Sveriges Geologiska Undersökning, Uppsatser och avhandlingar* i 4: o, Ser. C a, n: o 15, 1917.

shells. This black color is ascribed to the retention of the chitinous layer which once covered the entire surface of the conch. It is evenly distributed, and therefore did not make a color pattern as in the species forming the subject of this paper. Hedström's species include *Phragmoceras costatum*, *Phragmoceras munthei*, *Phragmoceras fasciatum* and *Phragmoceras convolutum*. All were obtained in the Middle Silurian of Gotland.

3. ORDOVICIAN AND SILURIAN ORTHOCEROID CEPHALOPODS

Barrande figured five specimens under the name *Orthoceras pellucidum* and one under the name *Orthoceras pellucidum* var. *contrahens*, all of which have white vertical color bands or lines on a general brown field. On his Plate 261 the bands are broad but irregular in width and arrangement. They appear to occur on all sides of the conch. This conch is gently curved lengthwise, with the siphuncle slightly ventrad of the center of the conch. On his Plate 400 the whitish bands are relatively few and far separated. In one specimen on Plate 420 the bands appear to be few, but on the other, they are relatively numerous, approximately of the same size and equally spaced. Apparently they become narrower toward the right side of the second specimen. In the specimen figured on Plate 452 the color bands consist of paired white lines, the pairs being equally spaced around the circumference of the conch. The specimen figured on Plate 266 under the name *Orthoceras pellucidum* var. *contrahens* differs from the five specimens figured under the name *Orthoceras pellucidum* in having numerous vertical white lines alternating with similar narrow brown lines. In *Orthoceras pellucidum* the segments of the siphuncle are globular (Plate 420, Fig. 3), as in *Orthoceras richteri* (Plates 318, 349), the genotype of Hyatt's genus *Sactoceras*,[15] to which genus it should be referred.

Several conchs with vertical color bands similar in width, number and arrangement to those of *Orthoceras pellucidum* were described from the Waynesville member of the Richmond at Clay Cliff, on the eastern margin of Manitoulin Island, in the northern part of Lake Huron, under the name *Sactoceras manitoulinense*

[15] Hyatt, Alpheus, *op. cit.*, p. 273.

Foerste.[16] In these conchs (Plate II, Fig. 2) the black color bands appear to belong to a layer immediately beneath that forming the surface of the shell. A second species, with narrower and more closely arranged vertical black and white lines, similar in appearance to *Orthoceras pellucidum* var. *contrahens*, was described in the same publication from the Lorraine of the area northwest of Toronto, in southern Ontario, under the name *Sactoceras westonense* Foerste (Plate II, Fig. 1). In this second species the color banding is confined to one side of the conch, assumed to be ventral. In *Sactoceras manitoulinense* the color banding also appears to be confined to one side of the conch, but whether this side is ventral or dorsal was not determined at the time this specimen was studied.

In a specimen of an unknown species of *Orthoceras* found in the top of the Plattin limestone in Ralls County, in northeastern Missouri,[17] vertical color banding is present. These colors are merely lighter and darker shades of the light brownish tint characterizing the inclosing rock. The specimen is 22 mm. wide and the color bands equal or slightly exceed 1 mm. in width; the intermediate lighter-colored intervals are 1 mm. or slightly less in width.

Angelin and Lindström [18] figured from the Silurian of Gotland, under the name *Orthoceras columnare* Marklin, a part of the surface of a conch showing vertical color lines. These lines are half a millimeter or less in width and 1 mm., or moderately more, apart. They are distinctly darker than the remainder of the shell, which is whitish.

Blake [19] figured, under the name *Orthoceras annulatum*, a specimen from the Middle Silurian of Great Britain, from 28 to 29 mm. in diameter, showing on one side vertical color bands, alternately

[16] Foerste, Aug. F., "Upper Ordovician Faunas of Ontario and Quebec," *Geol. Surv. Canada, Memoir 138*, pp. 224, 226, 1924.

[17] Foerste, Aug. F., "The Kimmswick and Plattin Limestones of Northeastern Missouri," *Denison Univ. Bull., Journ. Sci. Lab.*, Vol. 19, p. 212, 1920.

[18] Angelin, N. P., and Lindström, G., *Fragmenta Silurica*, p. 7, pl. 10, fig. 1, 1880.

[19] Blake, J. F., *A Monograph of the British Fossil Cephalopoda*, pt. 1, p. 91, pl. 4, fig. 4, 1882.

light and dark. According to Blake, these bands are one eighth of an inch (3 mm.) wide, but his figure indicates only five bands on the side of the conch illustrated. The lateral outlines of these bands do not appear to be sharply defined, as in the orthoceroids described in the preceding pages. *Orthoceras annulatum* is the genotype of *Dawsonoceras*. For a note on *Orthoceras pictum* Blake, see Addendum, p. 146.

Clarke [20] figured, under the name *Orthoceras tenuistriatum* Hall, a specimen from the Platteville member of the Black River formation in Minnesota which is ornamented by narrow, vertical, opaque or dull lines crossing the otherwise shining surface of the shell. These vertical lines are assumed to correspond to original color lines. According to the figures published by Clarke, these color lines are about one tenth of a millimeter wide, and at least fifty occur within the circumference of the conch. It is possible that these lines represent differences in the texture of the shell rather than in color.

Vertical color lines are shown also by *Orthoceras trusitum* Clarke and Ruedemann,[21] from the Guelph member of the Middle Silurian of New York. In the specimen represented by Figure 2 on their Plate 13, there are nine or ten vertical light brown lines in a width of 3 mm. These lines are confined to one side of the specimen. In the specimen represented by Figure 9 on the same plate, the structure usually accompanying color banding is present, but no distinctive coloring remains. Those vertical lines which correspond to the color bands consist of denser material and weather less rapidly than the intermediate parts of the shell. In a more recent paper Doctor Ruedemann has pointed out that the specimen illustrated by Figure 4 on Plate 12 of the publication by Clarke and Ruedemann also belongs to the species *Orthoceras trusitum*, and not to *Kionoceras darwini*.[22] In this specimen the

[20] Clarke, John M., "The Lower Silurian Cephalopoda of Minnesota," *Geol. Surv. Minnesota, Final Report*, Vol. 3, pt. 2, p. 788, 1897.

[21] Clarke, John M., and Ruedemann, Rudolf, "Guelph Fauna in the State of New York," *New York State Museum, Mem. 5*, p. 77, pl. 13, and also pl. 12, fig. 4, 1903.

[22] Foerste, Aug. F., "American Arctic and Related Cephalopods," *Denison Univ. Bull., Journ. Sci. Lab.*, Vol. 23, p. 99, pl. 14, figs. 3 A, B, 1928.

surface of the shell is smooth, but the vertical brown color lines are preserved distinctly, though here again on only one side of the shell. Dr. Ruedemann suggests the possibility that this restriction of the vertical color lines to one side of the conch may have been a part of the original color scheme, but he cautiously abstains from asserting definitely that this is the fact.

In the case of *Geisonoceras tenuitextum* (Hall), however, from the Trenton of New York, Dr. Ruedemann [23] was convinced not only that the vertical color lines were restricted to one side of the conch in the present state of preservation of the specimens, but that they had never been present on that side where now they are absent. This is indicated by the following facts. In the specimen illustrated by Ruedemann's text Figure 21 there are seventeen vertical brown color lines 0.5 mm. wide, separated by intervals of 0.4 mm. along the median part of the ornamented area, but by a much wider interval, 0.9 mm., at the lateral margin of this area. Moreover, the intervals between the vertical brown lines are decidedly darker (light brown) than the general color on the opposite side of the conch (whitish), where no vertical color lines appear. In the specimen represented by his Figure 22 the vertical color lines again are confined to one side of the conch. This is true also of the specimen represented by his Figure 23. In that case the color of the vertical lines is light brown where the interior of the conch is filled with crystalline calcite, and black where it is filled with black calcareous mud. In the specimen represented by his Figure 24 there are only ten vertical dark color bands. From the restriction of these vertical color lines and bands to one side of the conch Dr. Ruedemann concluded that in the species in question the conch usually occupied, during the life of the animal, an approximately horizontal position, with the side ornamented with color lines facing upward and the lighter-colored side facing downward. The species *Geisonoceras tenuitextum*, here under consideration, is that originally described by Hall under the name *Endoceras proteiforme* var. *tenuitextum*. According to Dr. Ruedemann, the

[23] Ruedemann, Rudolf, "On Color Bands in Orthoceras," *New York State Mus. Bull.*, Nos. 227, 228, p. 79, 1921. (*Orthoceras trusitum*, pp. 81, 86. *Geisonoceras tenuitextum*, pp. 81–84.)

species *Orthoceras strigatum* Hall belongs to the same species as *Geisonoceras tenuitextum;* the types happen to be specimens in which the surface of the shell was weathered so as to leave the color bands more or less distinctly elevated, the intermediate parts being reduced to furrows.

It is customary in studies of the color markings of fossil shells to cite the statement by Marsh (*Proc. Am. Assn. Adv. Sci.*, 17th meeting, p. 325, 1869), that he noted in a specimen of *Endoceras proteiforme,* from the Trenton of New York, distinct traces of color arranged in a delicate cancellated pattern. Unfortunately the specimen so described apparently can no longer be located, and therefore its generic reference cannot be confirmed. This generic reference, however, must be very doubtful, in view of the fact that very heterogeneous material was included by Hall himself under this name, so that the orthoceroids later separated as *Orthoceras lineolatum*, *Orthoceras tenuistriatum* and *Orthoceras tenuitextum* were originally described by Hall as varieties of his species *Endoceras proteiforme.* Under these circumstances it is evident that, in the absence of any knowledge of the siphuncle of the specimen so identified, it is unwise to accept the earlier identifications of orthoconic shells as *Endoceras proteiforme.*

4. DEVONIAN AND CARBONIFEROUS ORTHOCEROID CEPHALOPODS

The first known record of color design in a fossil cephalopod is that by D'Archiac and De Verneuil,[24] who in 1843 figured *Orthoceras anguliferum* from the Middle Devonian of Paffrath, in the Rhenish part of Germany, accompanied by the following description:

> Shell of an elongate cone shape, smooth, but covered with *chevrons* or zigzag coloured marks, of which the acute angles correspond regularly from the upper to the lower end of the cone. These angles, about 18 in number, are alternately turned toward the summit and the base. The little brown bands which form them are half a millimeter in width, and are separated by an uncoloured space of nearly equal width. If a line be passed through the successive summits of the upper and lower angles formed by the fracture of the same band an

[24] Archiac, d', and Verneuil, de, "On the Fossils of the Older Deposits in the Rhenish Provinces," *Trans. Geol. Soc. London*, Second Series, Vol. 6, p. 303, 1842.

oblique ellipse is obtained whose axis corresponds with a line passed through the siphon. Septa nearly flat and somewhat oblique. Siphon small, nearer to the centre than the edge. Paffrath: very rare.

Marsh[25] in 1865 secured another specimen of *Orthoceras anguliferum* from the Paffrath locality, where it was associated with *Natica subcostata,* also retaining color markings.

At present there are several specimens of *Orthoceras anguliferum* in the Peabody Museum, at Yale University, one of which was figured recently by Dunbar.[26] This specimen is illustrated as viewed from the side, and shows that the color lines are confined to the upper or ventral side of the conch, with the lower or dorsal side free from markings. Dunbar recognized the similarity between the restriction of the color markings to one side of the orthoconic shell in *Orthoceras anguliferum* and the restrictions in the species *Geisonoceras intertextum* and *Orthoceras trusitum,* as described by Ruedemann in the publication cited on page 131, and from this he drew the same conclusion: namely, that during the life of the animal the conch of *Orthoceras anguliferum* had been carried in a more or less horizontal position.

This opinion was further corroborated by the distribution of transverse color markings in a species of *Orthoceras* from a locality about 18 miles north of Tulsa, in Oklahoma, where it was found in the Ochelata member of the Middle Pennsylvanian. These transverse bands are confined to one side of the conch, assumed to be ventral; the dorsal side is free of color markings, thus again suggesting a horizontal position for the conch during the life of the animal. This specimen also belongs to the Peabody Museum, and the name **Orthoceras dunbari** is here proposed in honor of Professor Dunbar, a distinguished member of the force there in charge.

Since these specimens of *Orthoceras anguliferum* and *Orthoceras dunbari* offer conclusive evidence that at least some of the orthoceroids carried their conchs in an approximately horizontal posi-

[25] Marsh, O. C., "On the Preservation of Color in Fossils from Palaeozoic Formations," *Amer. Assn. Adv. Sci., Proc. of 17th Meeting,* p. 325, 1869.

[26] Dunbar, Carl, "Phases of Cephalopod Adaptation," in Thorpe, M. R., *Organic Adaptation to Environment,* p. 192, 1924.

tion, a more detailed description of their color markings is offered on page 139.

Of *Orthoceras anguliferum* the Peabody Museum contains three specimens, numbered 4132, 4129 and 4131, the order in which they are discussed.

Specimen No. 4132 (Plate I, Fig. 1; Plate III, Fig. 1) is 90 mm. long and enlarges from a diameter of 5.8 mm. at its base to 17 mm. at its top, the apical angle being 7.5 degrees. The ventral side is fairly well preserved, but the dorsal side is well preserved only along the upper third of the specimen; its middle and lower parts are crushed in. In its present condition the ventral outline of the conch has a convexity of 1 mm., but its original course, before the specimen was crushed, may have been nearly straight. Where the lateral diameter of the conch is 15.5 mm., its dorso-ventral diameter is 14.7 mm., again possibly due to crushing. The ventral side of the conch is marked by acutely angular zigzag brownish transverse lines, which are interpreted as consisting of three rising chevrons, as indicated at *A*, *B* and *C* of Figure 1 on Plate III. There are four descending angles, lettered *D*, *E*, *F* and *G*. The ascending chevron *A* and the descending angle *G* are interpreted as doubled. Of the ascending angles corresponding to *D*, seven consecutive ones can be recognized with confidence along the right side of the ornamented area; their lower points are located 5, 10, 15.5, 22, 29, 37 and 46 mm. above the base of the specimen, and the right arms of these angles (marked *y* in the figure) have a length of 1.5, 2, 2.5, 3, 3.5, 3.7, and 4 mm., respectively, each arm terminating abruptly, as seen in Figure 1 on Plate I, and as shown also by Figure 17 in the chapter on "Phases of Cephalopod Adaptation," published by Dunbar. The two chevrons *A* 1 and *A* 2 are preserved only by the uppermost two transverse color lines; the chevron *A* 1 is not distinguishable lower down, owing to cracking and crushing of the conch previous to fossilization. It is assumed, by analogy with several of the other specimens here illustrated, that the left end of each transverse line should terminate with a short ascending arm, such as is indicated at *x* in Figure 1 on Plate III, but the coloring of the specimen at this point is so indistinct that nothing definite can be learned regarding the character of the terminations of the transverse lines here. Where the circumference of the conch is 47.5 mm., the width of the ornamented area is apparently 19 mm., or approximately four tenths of the circumference of the conch. Along the remaining six tenths of the circumference there is no trace of color lines. These lines are brown. They weather less readily than intermediate parts of the shell. The colored part evidently is very thin, and is measured only in hundredths of a millimeter. The shell itself is estimated to be 0.4 mm. thick at the upper end of the specimen. Its surface, where not weathered, is smooth. The dorsal side of the conch is slightly lighter in color than that part of the shell which is intermediate between the color lines. The color lines are merely of a darker tint of brown than the remainder of the shell.

Specimen No. 4129 (Plate III, Fig. 2) is 28 mm. long, and enlarges in this length from a diameter of 6.5 mm. at its base to 10 mm. at its top. The area marked by transverse zigzag color lines is 10.75 mm. wide where the circumference of the conch is 25.5 mm., the ratio of the former to the latter being 42 per cent. The ascending chevrons and descending angles are more symmetrical in form and arrangement than in any other specimens studied and have been adopted here as a standard. Of the right arms corresponding to y in Figure 2 on Plate III, three are defined clearly on the specimen, their lengths in ascending order being 3, 4 and 5 mm. The two left arms, corresponding to x, are respectively 3 and 4.5 mm. long; the uppermost one of this series is not defined distinctly. Here, again, the dorsal side of the conch has a lighter shade of brown than the general ground color of the ventral side. The septum at the top of the specimen has a concavity of 1 mm. The siphuncle is 3.2 mm. from the dorsal wall of the conch, where the diameter of the latter is 10 mm. The passage of the siphuncle through the septum is 1 mm. in diameter. Five camerae occupy a length of 8 mm., corresponding to six in a length equal to the diameter of the conch.

Specimen No. 4131 is 43 mm. long and enlarges from 3 mm. at its base to 7.25 mm. at its top. The area marked by transverse zigzag lines is estimated to be 7.4 mm. wide where the circumference of the conch is 18.3 mm.; the ratio of the former to the latter is 40 per cent. There are three series of upward directed chevrons corresponding to *A*, *B* and *C* of the other diagrams, and the two downward directed angles corresponding to *E* and *F* are preserved distinctly. Along the lower part of the right side of the color-marked area there are also traces of the angles corresponding to *D*, with the right arms *y* scarcely one third of a millimeter in length. This specimen is of interest because its lower end cannot be far removed from the original apical end of the conch.

These three specimens of *Orthoceras anguliferum* are from Paffrath, in the Rhenish part of Germany, and came from the *Stringocephalus* or Givetian division of the Middle Devonian. Evidently they were obtained from some soft argillaceous layer which permitted their being cleaned without the slightest injury to the surface of the shell. The color of the body of the shell shades very gradually from the darker tint between the zigzag color lines to the slightly lighter tint on the opposite or dorsal side of the shell. The right and left arms corresponding to the parts marked *x* and *y* on the accompanying diagrams terminate so abruptly, and yet at such exact agreement with the rate of enlargement of the conch, that, in the absence of a corresponding abrupt change in color on the general surface of the conch, it seems impossible to regard this absence of color lines on the dorsal side of the conch as due merely to the exposure of this side to weathering previous to fossilization. If, however, the color lines were already confined to the ventral side of the conch during the life of the animal, this is in favor of the more or less horizontal position of the conch during this period of its existence. Further evidence is presented by the two specimens of *Orthoceras anguliferum alpenense* next described.

Recently two specimens of *Orthoceras* showing zigzag transverse color lines almost identical in pattern with those of *Orthoceras anguliferum* were found in the Alpena limestone of the Traverse group in the quarry of the Michigan Alkali Company at Alpena in eastern Michigan. Only Alpena limestone occurs in this quarry and the specimens were found about 90 feet below the top of the quarry. According to Irwin Pohl, the entire Traverse group, of which the Alpena limestone is a member, is of later age than the Onondaga and of earlier age than the Hamilton. Its fauna occupied northwest Ohio, southwest Ontario, and all of the southern peninsula of Michigan, connecting by way of James Bay and Hudson Bay with the Arctic. The group thins out southward and overlaps the Onondaga eastward.

Specimen No. 10775 (Plate I, Fig. 2; Plate IV, Fig. 1), from the Alpena limestone, is 82 mm. long, but of this length only 58 mm. belongs to the conch; the remainder belongs to that part of the matrix which extends beyond the upper end of the conch. The specimen enlarges from a diameter of 10 mm. at its base to 17 mm. at a point 45 mm. farther up; the apical angle is 9 degrees. One side of the specimen is ornamented by conspicuous transverse zigzag lines, tinted medium brown, which are defined sharply from the much lighter colored intervening spaces, which are light brown. Where the circumference of the conch is 50.3 mm., the width of the ornamented area is almost 24 mm., or 45 per cent of the circumference. The zigzag lines vary from 0.25 to 0.6 mm. in width; the intervening spaces vary from 0.5 to 1.0 mm. in width. The longest straight courses of the zigzag lines are 14.5 mm. long; the included angles vary from 28 to 32 degrees. The middle part of the color pattern is well preserved. The left arms x can be recognized only with difficulty. The right chevrons $C1$ and $C2$ and the right arm y are distinctly shown by the two uppermost transverse lines; the corresponding parts of the lower transverse lines are not exposed.

Specimen No. 10776 (Plate IV, Fig. 2) is 57 mm. long, enlarging from a diameter of 11.3 mm. at its base to 17 mm. at a point 40 mm. farther up, the apical angle being 8.3 degrees. Where the circumference of the conch is 50.3 mm., the width of the ornamented area is 22 mm., which is equivalent to 41 per cent of the circumference. Within this area the zigzag lines are not colored as strongly as in the preceding specimen, but their course can be traced distinctly, especially along the upper part of the specimen. Nothing is known of the structure of the interior of the conch beyond the fact that the concavity of the septum is 4 mm. where the diameter of the conch is 17 mm. No trace of the siphuncle remains.

The only difference that can be detected between the Michigan specimens and those of typical *Orthoceras anguliferum* from Paffrath, Germany, is that the ascending and descending angles of the former tend to be slightly broader, with the sides forming these angles correspondingly shorter. Moreover, the concavity of the septa in the former appears to be greater. On this account the Michigan specimens are here designated as **Orthoceras anguliferum alpenense**; the differences are evidently of a minor character. It is possible, however, that, when the structure of the interior of the Michigan specimens becomes better known, other differences may be noted.

Special attention is called to the doubling of the rising chevron *C* on the right side of the two specimens from Alpena, Michigan, and the corresponding doubling of the rising chevron *A* in one of the specimens from Paffrath. This doubling is known at present only from lateral parts of the color pattern, and is opposed to the opinion that the absence of color lines on the dorsal side of the conch is due to their removal by exposure to weathering previous to fossilization, rather than to the confinement of the color pattern to the ventral side of the conch during the lifetime of the animal. Attention is called also to the relative constancy of the ratio between the width of the color-marked area and the entire circumference of the conch in the different specimens here described from Paffrath and Alpena. In the three specimens from Paffrath this ratio is 40, 42 and 40 per cent, respectively. In the two Alpena specimens it is 45 and 41 per cent. This comparative constancy in the ratio between the width of the color-marked area and the entire circumference of the conch can hardly be a matter of chance, but favors the view that the restriction of the color-marked area to the ventral side of the conch is original, and not secondary, resulting from the weathering away of former color markings on the dorsal side of the conch.

Another specimen (Plate II, Fig. 4; Plate V, Fig. 3) bearing on the problem of the horizontal position of some orthoceroid conchs is the one from the Pennsylvanian of Oklahoma, briefly mentioned by Dunbar in his chapter on "Phases of Cephalopod Adaptation." This specimen, already referred to in an earlier paragraph, is from the Ochelata formation, near the middle of the Pennsylvanian as exposed in Oklahoma. It was found 15 miles north of Tulsa, and is numbered 10549 in the Peabody Museum, Yale University. For this specimen the name **Orthoceras dunbari** is here proposed, although further information is needed about the location and structure of its siphuncle. This specimen is 74 mm. long, and is strongly crushed in an obliquely dorso-ventral direction. In its present state the conch enlarges from a maximum diameter of 15 mm. at its base to 22 mm. at a point 67 mm. farther up, suggesting an apical angle of 6 degrees. The middle of the ventral area of the shell is crossed at regular intervals by dark transverse lines. The area crossed by these lines varies in width from 6 mm. at the lower end of the specimen to 15.5 mm. at its upper end. The lines consist merely of a deeper shade of brown than the general color of the shell. In a length equal to the present maximum diameter of the conch their number is sixteen. Along the upper part of the conch these darker lines vary in width from 0.6 to 0.8 mm.; the lighter brown intervening spaces vary from 0.4 to 0.7 mm. in width. Along the upper and lower parts of the specimen the lines are almost directly transverse, but near midlength there appears to have been a break, above which they at first slope strongly downward from left to right, gradually becoming less oblique along a vertical interval of 10 to 15 mm. Both sides of the transversely lined vertical area are margined laterally by vertical areas in which a blackish color predominates. Along the right side of this area the blackish vertical band is quite constant in width, increasing from 4 mm. at its base to 5 mm. at its top. The vertical black band on the left side of the transversely lined area is about 12 mm. wide along the upper part of the specimen, but its preservation is very irregular farther down, since only a small part of its original width is preserved at the base of the specimen. It is estimated that the transversely lined area occupied about 30 per cent of the circumference of the conch, and that, including the two vertical black bands, the entire more or less ornamented area formed about 52 per cent of the entire circumference. The black vertical color bands forming the lateral margins of the transversely lined area are confined to a very thin film on the surface of the shell, possibly less than one hundredth of a millimeter in thickness. Apparently this film was rubbed off or weathered off readily, or was not deposited evenly over the vertical band. The shell itself is half a millimeter thick and its surface is smooth. The color of the dorsal side of the conch is medium brown, faintly lighter than the general color of its ventral side.

In the preceding specimen the area covered by transverse color lines is definitely limited laterally. The transverse banding does not continue beneath the black film forming the lateral border of the transversely banded area, as may be seen readily where this film is removed. This fact, again, suggests that the color-lined area was already restricted to the ventral side of the conch during the lifetime of the animal, and that it was never present on its dorsal side. Apparently the living animal carried its conch in an approximately horizontal position, as already suggested by Dunbar.

There is no evidence, however, that all orthoconic cephalopods, or even a great number of them, ever carried their conchs in this position. Such a conclusion is premature and based on too little evidence. In the specimen of *Orthoceras pellucidum*, for instance, figured by Barrande on his Plate 261, the conch is distinctly curved lengthwise, and yet the vertical color bands occur on both the ventral and dorsal sides of the conch. The evidence is not as conclusive in the specimens illustrated on Plates 420 and 452, since it is not known how the figures of them were oriented, but in the variety *contrahens*, figured on Plate 266, the slight lengthwise curvature of the conch again makes it possible to determine which side is ventral and which is dorsal, so that here, again, the even distribution of the vertical color bands both ventrally and dorsally can be confirmed. This even distribution certainly does not suggest a horizontal position for the conch.

Possibly the horizontal position of the conch was adopted by only a few of the smaller species of orthoconic cephalopods. At least the restriction of the color patterns to the ventral side of the conch is known so far only in the case of a few species, none of which attained a large size.

5. CARBONIFEROUS CYRTOCEROID CEPHALOPODS

Transverse color lines are shown also by a conch found in the upper part of the Carboniferous limestone at Vise, Belgium, where it was identified by De Koninck [27] as *Cyrtoceras unguis* Phillips. By Phillips [28] this species was referred to *Orthoceras*, and this procedure was followed by Foord,[29] probably owing to the central location of its siphuncle. The specimen figured by De Koninck is about 105 mm. long. Its lower part, for a length of 20 mm., is curved distinctly on both sides of the conch, the convex side being assumed as ventral. Above this lower part the dorsal side is straight and the convexity of the ventral side becomes very faint. The cross-section is circular near the base of the specimen, but becomes oval farther up. The siphuncle is located slightly ventrad of the center of the conch and is narrowly cylindrical. The surface of the shell is smooth, but is marked by transverse color bands. These bands are alternately light and dark. Those near the lower part of the conch are relatively broad, but along the upper half of the shell they are much narrower and more numerous. They slope downward from the dorsal toward the ventral side of the conch, with several sinuations along their course.

An entirely different type of ornamentation is shown by *Cyrtoceras acus*, from the Carboniferous limestone at Tournai, Belgium. This species originally was identified by De Koninck [30] as *Cyrtoceras unguis* Phillips, but later was recognized as distinct from it. The form of this species is very similar to that of *Cyrtoceras unguis*, having a similar moderately curved apical end, the conch becoming straight farther up. The siphuncle, however, is located much nearer the ventral side of the conch, which is somewhat depressed dorso-ventrally; and the transverse color lines, alternately light and dark, are strongly zigzag, resembling those of *Orthoceras anguli-*

[27] Koninck, L.-G. de, *Faune du calcaire carbonifère de Belgique*, p. 170, pl. 34, fig. 2, 1880.

[28] Phillips, John, *Geology of Yorkshire*, pt. 2, p. 238, pl. 21, fig. 2, 1836.

[29] Foord, Arthur H., *op. cit.*, p. 116, 1888.

[30] Koninck, L.-G. de, *Descr. des anim. fossiles de Belgique*, p. 524, pl. 47, fig. 8a, 1844; also *Faune du calcaire carbonifère de Belgique*, p. 180, pl. 35, figs. 6, 6a, 1880.

ferum, but they occur on the dorsal as well as on the ventral side of the conch. It is not known how many upward pointed chevrons occur within the circumference of this conch, but apparently their number is five, the odd one being located along the median part of the ventral side of the conch.

It is not likely that either *Cyrtoceras unguis* or *Cyrtoceras acus* carried their conchs in a horizontal position. It has already been noted by Ruedemann [31] that none of the breviconic cyrtoceroids are known to have carried their conchs in this position.

6. SILURIAN TROCHOCEROID CEPHALOPOD

The trochoceroid described by Barrande (Plate 28, p. 113, 1867), under the name *Trochoceras pulchrum*, from the Silurian of Bohemia, differs from the genotype *Trochoceras davidsoni* (Plate 27, p. 89) in having the siphuncle a short distance ventrad of the center of the conch, instead of in contact with its ventral wall; there are no lateral wing-like protuberances; the surface of the shell is ribbed transversely; these ribs curve much more strongly downward and they are crossed by relatively fine striae which are parallel to the longitudinal axis of the conch. In addition to these markings there are longitudinal brick-red color bands which appear equally distributed on all sides of the conch. In the areas between these bands the general color of the shell is whitish.

7. COILED CEPHALOPODS FROM CARBONIFEROUS, LIASSIC AND TERTIARY STRATA

Among the nautiloid cephalopods Foord cited *Coelonautilus paucicarinatus* Foord,[32] from the Carboniferous limestone of Cork, Ireland, as having, in one specimen, very distinct color markings, arranged in spots or irregular bands of a blackish hue.

A much more recent representative of the Nautiloidea is *Aturia aturi* Basterot,[32] from the Eocene and Miocene of southwestern France and Italy. Regarding this species Foord stated that the surface of its shell is ornamented with semicircular,

[31] Ruedemann, Rudolf, *op. cit.*, p. 87.
[32] Foord, Arthur H., *op. cit.*, pt. 2, pp. 116, 361, 1891.

transverse, approximate zones or bands of a reddish color, beginning in the umbilical depression and extending to the edge of the periphery, upon which they break up into little patches which follow the direction of the striae.

Among the goniatite members of the Ammonoidea, *Goniatites striolatus* Phillips [33] was cited by Oppenheim as being represented in his collection by a specimen having brown, and rather thick, markings along the suture lines, as seen through the translucent shell. This coloring, however, probably belongs to the interior of the conch, but the color patterns here under consideration occur close to the exterior surface of the shell, where they are confined to an extremely thin layer just beneath the surface. Moreover, Dr. L. F. Spath kindly called to my attention that the specimen described by Oppenheim could not have been a typical specimen of *Goniatites striolatus,* since that species, now known as *Homoceras striolatum,* came from the Sabden shales of Yorkshire, which belong to the Middle Carboniferous, but Oppenheim's own specimen came from the Carboniferous limestone, a different horizon, and therefore likely to contain species distinct from those known from the Sabden shales.

A more typical representative of the Ammonoidea is *Amaltheus (Paltopleuroceras) spinatus,* from the Middle Liassic division of the Lower Jurassic. In this species Schindewolf [34] recently noted the presence of dark, transverse lines, approximately parallel to the lines of growth of the conch.

VI. Crinoid and Trilobite

Remains of color markings are confined chiefly to the various types of shells. Bather [35] noted, however, the presence of dark spots also on the arms of a crinoid, *Cyathocrinus acutitubus,* from the Middle Silurian at Tividale, in England, stating that these

[33] Phillips, *op. cit.*, pt. 2, p. 234, pl. 19, figs. 14–18, 1836.

[34] Schindewolf, O. H., "Über Farbstreifen bei Amaltheus (Paltopleuroceras) spinatus (Brug)," *Palaeontologische Zeitschrift,* Vol. 10, pp. 136–143, 1928.

[35] Bather, F. A., British Fossil Crinoids, VIII, p. 215. Annals and Mag. Nat. Hist., Ser. 6, Vol. 9, p. 215, 1892. See also *The Crinoidea of Gotland,* pt. 1, p. 151; *Kongl. Svenska Vet.-Akad. Handl.*, Stockholm, Vol. 25, no. 2, 1893.

dark spots "may possibly be the relic of some original colouring, since similar ornament is found in recent forms."

Moreover, Raymond[36] recently discovered a specimen of *Anomocare vittata*, a trilobite from the Conasauga shale member of the Upper Cambrian in Cherokee County, Alabama, which retained three transverse bands on its pygidium. These bands were dark gray, almost black, and alternated with light gray. They were approximately parallel to the pleural elements of the pygidium. See also Addendum on page 146. There were also many small irregularly placed spots of a yellowish hue.

VII. Earliest Appearance of Color Markings in Different Groups of Animals

The oldest known fossil retaining color markings is the Upper Cambrian trilobite *Anomocare vittata*, just discussed. Furthermore, it appears to be the only trilobite in which color markings are known to occur.

Apparently the next oldest is the cephalopod described in this paper as *Hedstroemoceras haelluddenense*, from Haelludden, on the island of Oeland, off the southeastern coast of Sweden. Here it was found in the gray *Endoceras vaginatum* limestone, which belongs to the Kunda formation of Raymond.[37] Nothing strictly comparable to this formation is known on the American continent, and it is regarded either as belonging very low in the Chazyan or as intermediate between the Chazyan and the Canadian.

The oldest gasteropod retaining color markings is the *Holopea harpa* discovered by Raymond in the Middle Chazyan of Valcour Island, in Lake Champlain, New York.

No brachiopods retaining color markings are known from strata older than the Middle Devonian; and no pelecypods retaining color markings are known from strata older than the Carboniferous.

[36] Raymond, Percy, "A Trilobite Retaining Color Markings," *Am. Journ. Sci.*, Vol. 4, pp. 461–464, 1 fig., 1922.

[37] Raymond, Percy, *Expedition to the Baltic Provinces of Russia and Scandinavia*, 1914, pt. 1: *The Correlation of the Ordovician Strata of the Baltic Basin with Those of Eastern North America*, pp. 192, 216, 1916.

Of course, this does not mean that the order of appearance of color markings in different types of animals is known. The retention of color markings in the older strata is so much a matter of chance and the number of specimens retaining color markings is so very small, especially among the more ancient forms, that no conclusions as to the order of appearance of color markings in different groups of animals can as yet be drawn.

VIII. Color Markings as an Index of Relatively Shallow Seas

Most colors of shells are not visible during the life of the animal. In many shells, the prismatic layer, which carries the color markings at or immediately beneath its surface, is covered by a horny layer of conchiolin, which conceals these colors during the life of the animal. In those not covered by a horny layer the prismatic layer, with its color markings, is concealed by reflections of the mantle. Of course, the horny layer or the mantle itself may show colors, but they are not the colors retained by shells in their fossil form, and therefore have nothing to do with the present study.

Notwithstanding this fact there evidently is an intimate connection between the presence of color markings on shells and their access to light during the life of the animal. Forbes [38] has called attention to the fact that in shallow waters shells present more varied colors and more distinct color designs. At greater depths the colors tend to become uniform over the entire surface of the shell, so that color patterns disappear. Moreover, at these depths the the variety of colors is restricted more and more to various shades of dull red or of reddish brown. Newton [39] stated that in the Mediterranean only 1 out of 18 shells showed colors below the 100-fathom line; between 35 and 55 fathoms, 1 out of 3; and at depths of only 2 fathoms or less, more than half.

[38] Forbes, E., "Report on the Mollusca and Radiata of the Aegean Sea, and on Their Distribution, Considered as Bearing on Geology," *Rept. of 13th Meeting of the British Assn. for Adv. of Science*, London, pp. xxiii, 130, 1844; also pp. 21–23 of article cited in note 1.

[39] Newton, R. B., "Relics of Coloration in Fossil Shells," *Proc. Malacozool. Soc. of London*, Vol. 7, p. 280, 1907.

From the vividness of the color patterns of the Carboniferous limestone specimens studied by him, Newton concluded that the Carboniferous limestone was deposited in water less than 50 fathoms deep. The relatively numerous species of gasteropods with color patterns found by J. Brookes Knight in a single horizon of the Pennsylvanian division of the Carboniferous, in the vicinity of St. Louis, Missouri, suggests deposition in an equally shallow sea. By a similar line of reasoning, it seems probable that the portion of the Alpena limestone which retained color markings in a species of cephalopod and also in a species of brachiopod (both discussed above) was deposited in relatively shallow waters.

Addendum

Another orthoconic cephalopod with vertical color bands is described by Blake (*British Fossil Cephalopoda*, p. 144, Pl. 13, Fig. 5 *bis*) from the Bala shales at Bala in northern Wales under the name *Orthoceras pictum* Blake. It has "longitudinal bands of colour, alternately white and brown, some wider and some narrower, about 12 per diameter. The colour bands are variable in their distance and breadth, the dark colour being as broad as the light." See page 130.

In the list of papers to be presented before Section E of the American Association for the Advancement of Science, at Des Moines, Iowa, December 30–31, 1929, is one by James S. Williams, entitled: "An Indication of a Color Pattern on a Mississippian Trilobite." Hitherto only one species of trilobite retaining color markings has been known, and that is the Cambrian form found by Professor Raymond. See page 144. This trilobite is a new species of *Proetus*, which Professor Williams has named *Proetus tenuituberculus*. He informs me that it is very close to *Proetus missouriensis* Shumard.

BIBLIOGRAPHY

Alberti, Fr. von. Gefärbte Muschel-Kalk-Terebrateln. Neues Jahrb. für Min., pp. 672–673 (1845).

Angelin, N. P., and Lindström, G. Fragmenta Silurica, p. 7, pl. 10, fig. 1 (1880).

Archiac, d', and Verneuil, de. On the Fossils of the Older Deposits in the Rhenish Provinces; preceded by a general survey of the fauna of the Palaeozoic rocks, and followed by a tabular list of the organic remains of the Devonian system in Europe. Trans. Geol. Soc. of London, Ser. 2, Vol. 6, pp. 303–410 (1842).

BARRANDE, JOACHIM. Systeme silurien du centre de la Boheme. Text: pt. I (1867), pt. II (1870), pt. III (1874), pt. IV (1877), pt. V, including Supplement (1877).

BATHER, F. A. British Fossil Crinoids, VIII. Annals and Mag. Nat. Hist., Ser. 6, Vol. 9, p. 215 (1892).

BATHER, F. A. The Crinoidea of Gotland, pt. I, p. 151. Kongl. Svenska Vet.-Akad. Handl. (Stockholm), Vol. 25, no. 2 (1893).

BLAKE, J. F. A Monograph of the British Fossil Cephalopoda, pt. I, p. 91, pl. 4, fig. 4 (1882).

BLOCHMANN, F. Brachiopoda. Handwörterbuch der Naturwissenschaften, Vol. 2, p. 149, Jena (1912).

CLARKE, JOHN M. The Lower Silurian Cephalopoda of Minnesota. Geol. Surv. Minnesota, Final Report, Vol. 3, pt. 2, p. 788 (1897).

CLARKE, JOHN M., AND RUEDEMANN, RUDOLF. Guelph Fauna in the State of New York. New York State Mus., Mem. 5, p. 77, pl. 13, also pl. 12, fig. 4 (1903).

DAVIDSON, TH. Mon. British Fossil Brachiopoda. Palaeontographical Society of London (1882–84). (*Terebratula biplicata, Dielasma hastatum.*)

DEECKE, E. Über Farbenspuren an fossilen Molluskenschalen. Sitzungsb. der Heidelberger Akad. d. Wiss., Math.-naturw. Klasse, Abt. B, Jahrg. 1917, 6, Abhandl. Heidelberg (1917).

DESLONGCHAMPS, E. Brachiopodes. Paleontologie française, terrain jurassique. Paris (1862).

DUNBAR, CARL. Phases of Cephalopod Adaptation, in Thorpe, M. R., Organic Adaptation to Environment, p. 193 (1924).

FENTON, C. L. AND M. A. The Stratigraphy and Fauna of the Hackberry Stage of the Upper Devonian (Macmillan Co., New York), pp. 132, 133 (1924).

FISCHER, P. H. La Persistance des couleurs parmi les fossiles du Trias moyen. Journ. de Conchyliologie, Paris, Vol. 69, p. 5, also p. 131 (1925).

FOERSTE, AUG. F. The Kimmswick and Plattin Limestones of Northeastern Missouri. Denison Univ. Bull., Journ. Sci. Lab., Vol. 19, p. 212 (1920).

FOERSTE, AUG. F. Upper Ordovician Faunas of Ontario and Quebec. Geol. Surv. Canada, Mem. 138, pp. 224, 226 (1924).

FOERSTE, AUG. F., AND SAVAGE, T. E. Ordovician and Silurian Cephalopods of the Hudson Bay Area. Denison Univ. Bull., Journ. Sci. Lab., Vol. 22, p. 50, pl. 5, fig. 7 (1927).

FOORD, ARTHUR H. Catalogue of the Fossil *Cephalopoda* in the British Museum, pt. 1, pp. 116, 266, 292, 297, 307 (1888); pt. 2, pp. 116, 351 (1891); pt. 3, p. 292 (1897).

FORBES, E. Note on an Indication of Depth of Primaeval Seas, Afforded by the Remains of Colour in Fossil Testacea. Proc. Royal Soc. London, Vol. 7, pp. 21–23 (1854).

GIRTY, G. H. Notice of a Mississippian Gasteropod Retaining Coloration. Am. Journ. Sci., Vol. 34, p. 339 (1912).

GREGER, D. K. A New Devonian Brachiopod Retaining the Original Color Markings. Am. Journ. Sci., Vol. 25, p. 313 (1908). (*Cranaena morsii.*)

HEDSTRÖM, HERMAN O. Über die Gattung *Phragmoceras* in der Obersilurformation Gotlands. Sveriges Geolgiska Undersökning, Stockholm, i 4:o, Ser. Ca, n:o 15, pp. 24, 27, 1917.

KAYSER, E. Notiz über Rhynchonella pugnus mit Farbenspuren aus dem Eifler Kalk. Zeitschrift d. deutsch. geol. Gesellschaft, Vol. 23, pp. 257, 289 (1871).

KEYES, C. R. Preservation of Color in Fossil Shells. Nautilus, Vol. 4, pp. 30–31 (1890).

KNIGHT, J. BROOKES. Some Pennsylvanian Gasteropods and a Pelecypod Showing Color Markings. Bull. Geol. Soc. America, Vol. 40, p. 212 (1929).

KONINCK, L.-G. DE. Descr. des anim. fossiles de Belgique, p. 524, pl. 47, fig. 8a (1844).

KONINCK, L.-G. DE. Faune du calcaire carbonifère de Belgique, p. 170, pl. 34, fig. 2; p. 180, pl. 35, fig. 6, 6a (1880).

LEIDHOLD, CL. Ergänzende Besprechung von W. Deecke, Über Färbungspuren an fossilen Molluskenschalen. Naturw. Wochenschrift, N. F., Vol. 17, pp. 84–86, Jena (1918).

LINDEN, MARIA VON. Die Entwicklung der Skulptur und der Zeichnung bei den Gehäuseschnecken des Meeres. Zeitschr. für wissenschaftl. Zoologie, Vol. 61, pp. 261–317 (pp. 274, 305), Leipzig (1896).

LINDSTRÖM, G. On the Silurian Gasteropoda and Pteropoda of Gotland. Kongl. Svenska Vet. Akad. Handlingar, Stockholm, Vol. 19, no. 6, pl. 20, figs. 30–31 (1884).

MARSH, O. S. On the Preservation of Color in Fossils from Palaeozoic Formations. Proc. Am. Assn. Adv. Sci., 17th Meeting, p. 325 (1869).

NEWTON, R. B. Relics of Coloration in Fossil Shells. Proc. Malaco-zool. Soc. of London, Vol. 7, p. 280 (1907).

OPPENHEIM, PAUL. Über Erhaltung der Färbung bei fossilen Molluskenschalen. Centralblatt für Min. Geol. Pal., Jahrg. 1918, pl. 368 (1918).

ORBIGNY, A. Paleontologie française, terraine jurassique, gasteropodes, p. 56, pl. 244, figs. 4–5 (1850).

PHILLIPS, JOHN. Illustrations of Geology of Yorkshire, Vol. 2, p. 226, pl. 15, fig. 2 (1836).

RAYMOND, P. E. An Ordovician Gasteropod Retaining Color Markings. Nautilus, Vol. 19, p. 101 (1906).

RAYMOND, P. E. A Trilobite Retaining Color Markings. Am. Journ. Sci., Vol. 4, pp. 461–464, 1 fig. (1922).

RICHTER, RUD AND E. Paleontologische Beobachtungen, I, Über einzelne Arten von Acidaspis. . . . Jahrbücher d. Nassauischen Vereins für Naturkunde in Wiesbaden, 70. Jahrg., 1917, pp. 143–161, pl. 1, Wiesbaden (1918). (*Terebratula cimex*, p. 156.)

RICHTER, RUD. Zur Färbung fossiler Brachiopoden. Senckenbergiana, Vol. 1, no. 3, pp. 83–96, 172 (1919).

RICHTER RUD. Brachiopoden mit konzentrischen Farbbändern. Senckenbergiana, Vol. 6, p. 168 (1924).

ROUNDY, P. V. Original Color Markings of Two Species of Carboniferous Gasteropods. Am. Journ. Sci., Vol. 38, p. 446 (1914).

RUEDEMANN, RUDOLF. On Color Bands in Orthoceras. New York State Mus. Bull., Nos. 227–228, pp. 79–88 (1921).

SCHINDEWOLF, O. H. Über Farbstreifen bei Amaltheus (Paltopleuroceras) spinatus (Brug). Palaeontologische Zeitschrift, Vol. 10, pp. 136–143 (1928).

SOWERBY, G. B. Thesaurus Conchyliorum or Monographs of Genera of Shells, Vol. 1, London, pls. 68, 69, 70, 72, showing living terebratuloid shells with concentric color markings (1847).

SUESS, ED. Über die Wohnsitze der Brachiopoden. Sitzungsber. k. Ak. Wiss., Math.-nat. Kl., Vol. 37, Wien, pp. 185–248 (1859).

WHITE, THEODORE G. The Faunas of the Upper Ordovician Strata at Trenton Falls, Oneida Co., N. Y. Trans. New York Acad. Sci., Vol. 15, p. 85 (1896).

PLATE I

FIG. 1. *Orthoceras anguliferum* D'Archiac and De Verneuil. — Lateral view of conch with ventral outline on left, showing the abrupt termination of the zigzag color lines along the middle of this lateral side. A part of this color pattern across the entire circumference of the conch is represented by Figure 1 on Plate III. The two rising chevrons of this pattern correspond to angles *B* and *C* on the latter plate. The descending angle *F* is readily identified only along the upper left margin of the figure. The left part of the color pattern is out of view. Angle *B* of each zigzag line lies along the median line of the ventral side of the conch. This conch is slightly curved lengthwise, but has been turned so as to expose a little more than half of the color pattern. From Paffrath, in the Rhenish part of Germany; in the *Stringocephalus* or Givetian division of the Middle Devonian. Specimen No. 4132, Peabody Musuem, Yale University. Magnified 2 diameters.

FIG. 2. *Orthoceras anguliferum alpenense* Foerste. — Ventral view, with the left margin of the color pattern poorly preserved, and the right margin not exposed. A part of this color pattern across the entire circumference of the conch is represented by Figure 1 on Plate IV. The two rising chevrons of this figure correspond to angles *A* and *B* on Plate IV. Angle *B* of each zigzag line lies along the median part of the ventral side, but the specimen has been turned so as to show that part of the color pattern which is best preserved. From the quarry of the Michigan Alkali Company, at Alpena, Michigan; in the Alpena limestone. Specimen No. 10775, Museum of Paleontology, University of Michigan. Magnified 1.7 diameters.

FIG. 3. *Hedstroemoceras haelluddenense* Foerste. — *A*, ventral view of a specimen retaining the living chamber and the upper part of the phragmacone. The upper four camerae are outlined by the sutures of the septa. The lower one of this group of camerae was sectioned sufficiently to expose the siphuncle, as shown by Figure 3 on Plate III. The lower part of the specimen retains the shell with part of its color pattern. The latter, as far as preserved, is shown by Figure 2 *A* on Plate V. *B*, lateral view of another specimen, exposing the living chamber and two camerae, beneath which the specimen retains the shell with its color pattern. The latter, as far as preserved, is shown by Figure 1 on Plate V. From Haelludden, on the island Oeland, Sweden; in the gray *Endoceras vaginatum* limestone. From Museum of the Geological Survey of Sweden, in the Riksmuseum, Stockholm. Magnified 2 diameters.

PLATE I

1

2

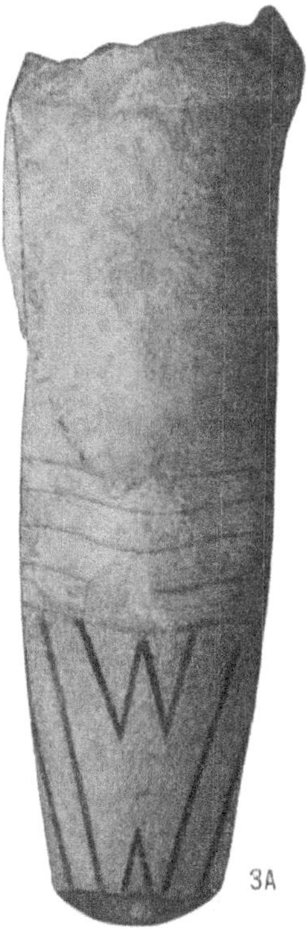

3A

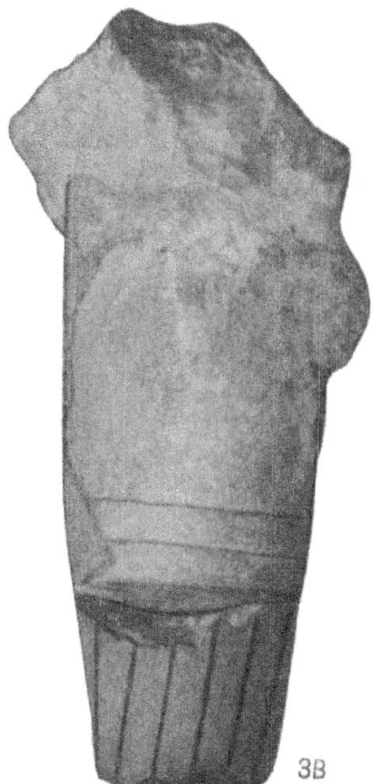

3B

PLATE II

FIG. 1. *Sactoceras westonense* Foerste. — Ventral view, showing remains of vertical color bands. On the dorsal side, although the surface of the shell is well preserved, there is no color banding. From Weston, Ontario, in the Lorraine formation. Specimen No. 2174 in the Victoria Memorial Museum. Same specimen as that figured and described in *Upper Ordovician Faunas of Ontario and Quebec*, Foerste, Memoir 138, Geological Survey of Canada, p. 226, Plate 39, Figure 5; Plate 40, Figure 2, 1924. Figure enlarged 1.5 diameters.

FIG. 2. *Sactoceras manitoulinense* Foerste. — Ventral view, showing vertical color bands. From Clay Cliff, on the northeastern shore of Manitoulin Island; in the Meaford or Waynesville member of the Richmond. Specimen No. 8539, in the Victoria Memorial Museum. Same specimen as that described and figured in *Upper Ordovician Faunas of Ontario and Quebec*, p. 224, Plate 39, Figure 4a, 1924. Figure enlarged 1.5 diameters.

FIG. 3. *Rizoceras* (?) *coronatum* Foerste and Savage. — Ventral view, showing scalloped transverse striae on surface of shell. From Shamattawa River, southwest of Hudson Bay; in the Shamattawa member of the Richmond. Specimen No. 22 HB in the Savage collection. Same specimen as that described and figured in *Ordovician and Silurian Cephalopods of the Hudson Bay Area*, Foerste. Denison Univ. Bull., Journ. Sci. Lab., Vol. 22, p. 50, Plate 5, Figure 7, 1927.

FIG. 4. *Orthoceras dunbari* Foerste. — Ventral view, showing the area crossed by transverse color lines, bordered by a vertical black area on each side. From 15 miles north of Tulsa in Oklahoma, in the Ochelata member of the Pennsylvanian. Specimen No. 10549, Peabody Museum, Yale University. Magnified 2 diameters.

FIG. 5. *Cranaena casei* Foerste. — The right anterior quarter of the surface of a valve retaining black concentric lines of different width and distribution. The upper part of this specimen exposes also a part of the interior of a second valve, filled with matrix, but retaining a remnant of part of its interior structure in the vicinity of its beak. From the quarry of the Michigan Alkali Company, at Alpena, Michigan; in the Alpena limestone. Specimen No. 10775 in the Museum of Paleontology, University of Michigan. Magnified 2 diameters.

PLATE II

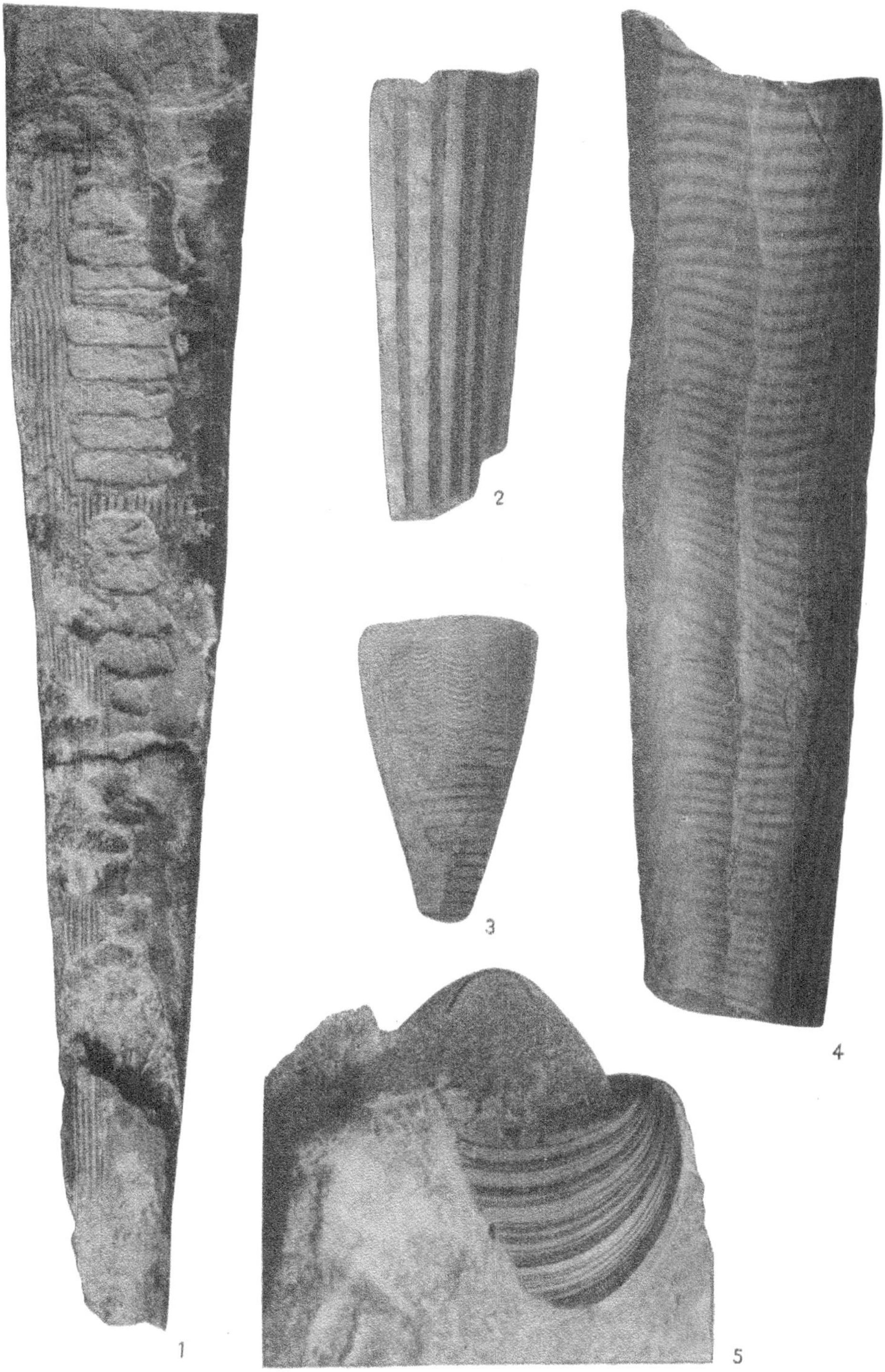

PLATE III

FIG. 1. *Orthoceras anguliferum* D'Archiac and De Verneuil. — Color pattern of the specimen represented by Figure 1 on Plate I. The figure represents the entire circumference of the conch, the blank parts represent the dorsal side of the latter where there is no color pattern. Magnified 3 diameters. *A*, *B* and *C* represent the upward directed chevrons; *D*, *E*, *F* and *G*, the downward directed angles. Angles *A* and *G* apparently are doubled. The letter *y* is the right arm of angle *D*, and *x* is the left arm of angle *G*. Specimen No. 4132, Peabody Museum, Yale University.

FIG. 2. *Orthoceras anguliferum* D'Archiac and De Verneuil. — Similar figure of the second specimen of this species described in the text. Magnified 3 diameters. Specimen No. 4129, Peabody Museum, Yale University.

FIG. 3. *Hedstroemoceras haelluddenense* Foerste. — *A*, vertical section of conch through the siphuncle, only the upper one of the two camerae figured being actually sectioned; *B*, cross-section showing location and size of siphuncle at its passage through the septum. Both figures magnified 2 diameters. Same specimen as that represented by Figure 3 A on Plate I. From Museum of Geological Survey of Sweden, in the Riksmuseum, Stockholm.

PLATE III

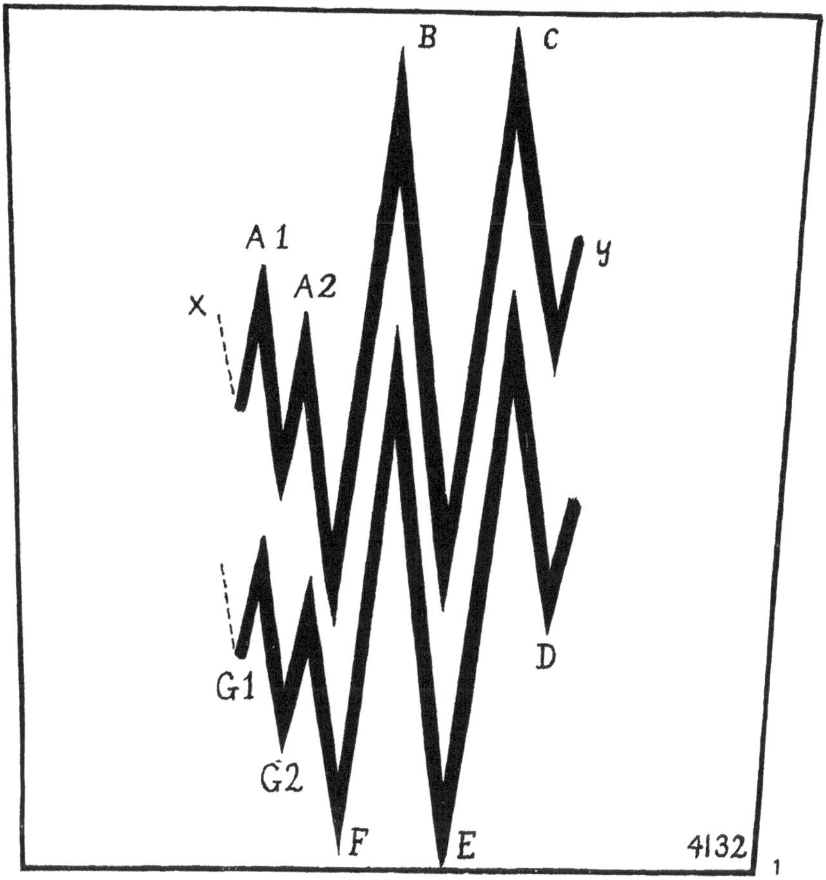

1

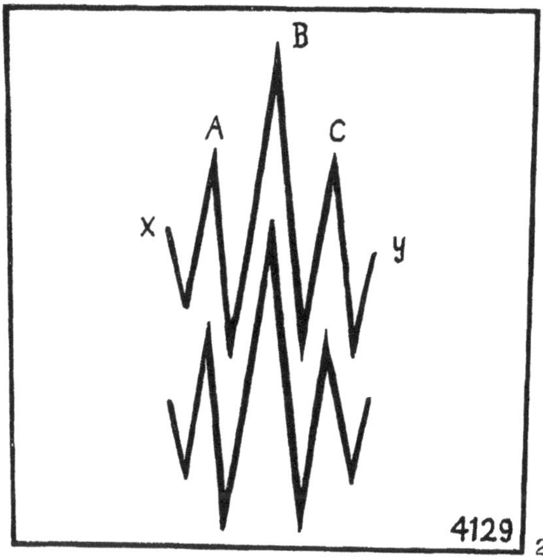

2

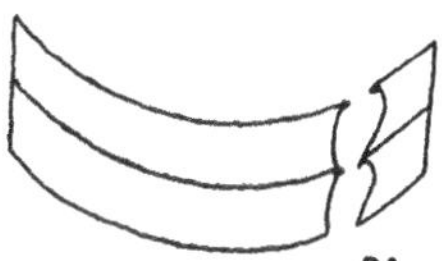
3A

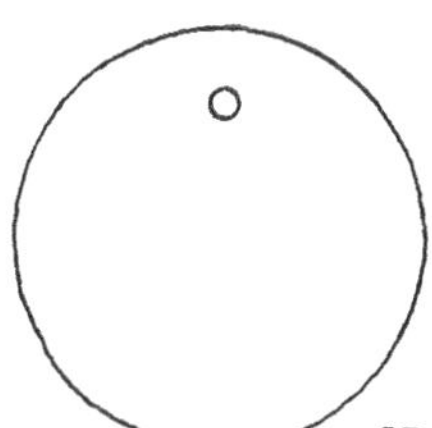
3B

PLATE IV

FIG. 1. *Orthoceras anguliferum alpenense* Foerste. — Color pattern of same specimen as that represented by Figure 2 on Plate I. In this specimen it is the angle *C* which is doubled. Magnified 2.4 diameters. Specimen No. 10775, Museum of Paleontology, University of Michigan.

FIG. 2. *Orthoceras anguliferum alpenense* Foerste. — Color pattern of second specimen described in the text. Magnified 2.4 diameters. Specimen No. 10776, Museum of Paleontology, University of Michigan.

PLATE IV

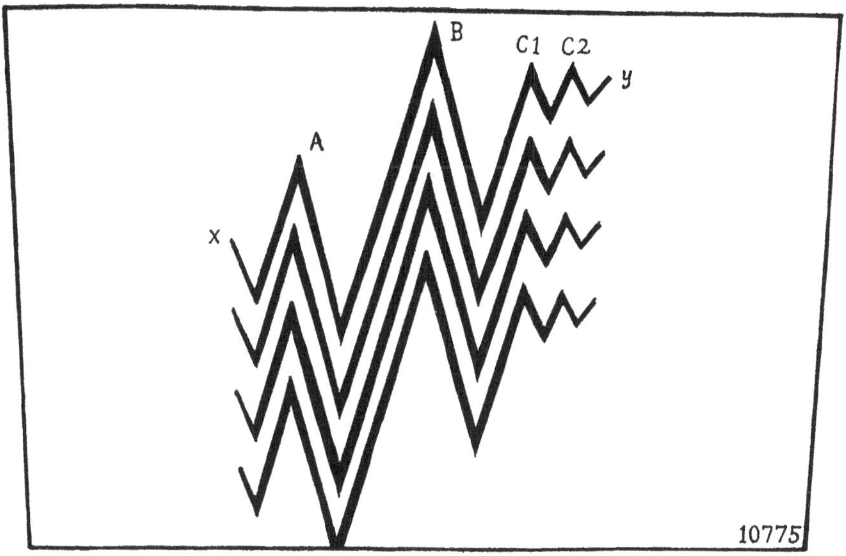

1

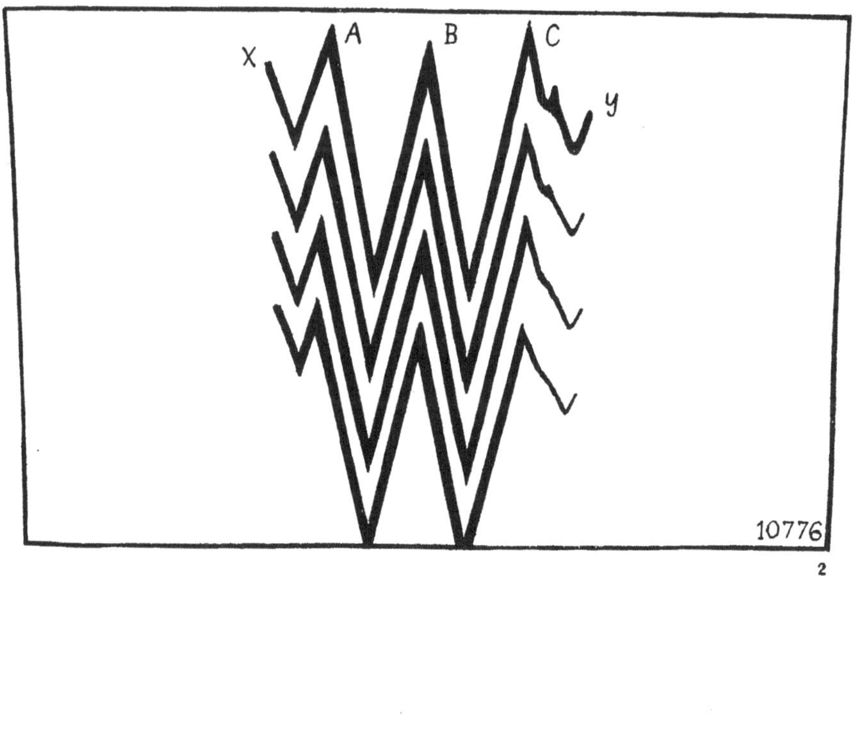

2

PLATE V

Fig. 1. *Hedstroemoceras haelluddenense* Foerste. — Color pattern of the same specimen as that represented by Figure 3B on Plate I. Magnified 3 diameters. The dotted lines represent the assumed location of color bands not preserved in the specimen. It is assumed that the two color bands nearest the median line of the dorsal side of the conch met at an upward directed angle. Along the median part of the ventral side they formed a pattern resembling the letter W, better shown in Figure 2A of this plate. Museum of the Geological Survey of Sweden, in the Riksmuseum, Stockholm.

Fig. 2. *Hedstroemoceras haelluddenense* Foerste. — *A*, color pattern of same specimen as Figure 3A on Plate I, magnified 3 diameters; *B*, diagram of suture at top of phragmacone, natural size. From *d* to *d* represents the entire circumference of the conch, *d* being the median line of the dorsal side; *v*, the median line of the ventral side; and *vl* the ventro-lateral parts of the conch. The suture rises ventro-laterally, and curves downward so as to form a ventral lobe narrower than the dorsal lobe. The median part of the dorsal lobe is flat and broader than the corresponding part of the ventral lobe. Museum of the Geological Survey of Sweden, in the Riksmuseum, Stockholm.

Fig. 3. *Orthoceras dunbari* Foerste. — Color pattern of same specimen as that represented by Figure 4 on Plate II, magnified 2 diameters. The dorsal side of the conch does not present any color pattern. Specimen No. 10549, Peabody Museum, Yale University.

PLATE V

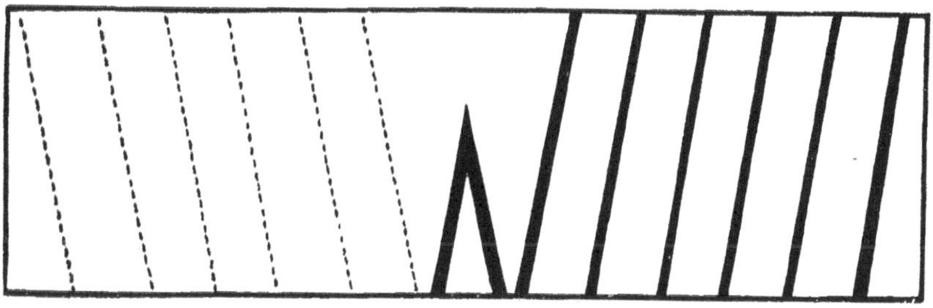

1

2A

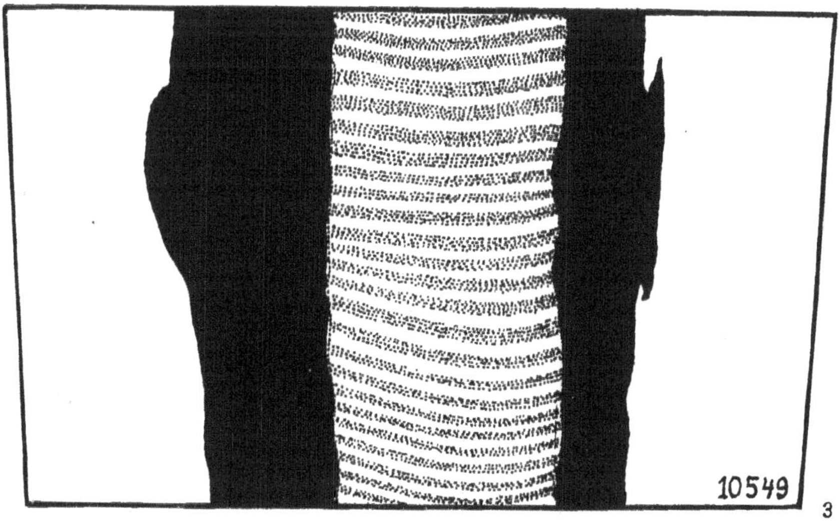

3

2B

Vol. III, No. 7, pp. 151–154 (1 pl., 1 text fig.) February 20, 1930

ADDITIONAL NOTES ON NEPHRITICERINA

By AUG. F. FOERSTE

THE type of *Nephriticerina alpenensis* showed very well the rate of enlargement of the conch, its lengthwise curvature, the ornamentation of the surface of its shell, the location of the siphuncle and the form of one of its segments. It was very misleading, however, when an attempt was made to determine the character of its aperture. A second specimen, from the same locality and horizon, presents an entirely different appearance along the upper part of its living chamber, and one much more likely to be correct. Hence it is figured and described in the following lines. It fails, however, to reveal some of the features well shown by the first specimen, such as the ornamentation of the surface of the shell and the location and character of the siphuncle.

Nephriticerina alpenensis Foerste

(Plate I, Figs. 1 A, B; Text Fig. 1 A, B)

Nephriticerina alpenensis Foerste, Contrib. Mus. Geol. Univ. Michigan, Vol. 2, no. 9, p. 195, 1927.

A second specimen of *Nephriticerina*, from the same locality and horizon as the type of *Nephriticerina alpenensis*, is assumed to belong to the same species. This specimen is a cast of the interior of the living chamber and is of interest chiefly because it gives a more accurate idea of the form of the aperture than could be obtained from the specimen described first. The margin of this aperture, at least in the second specimen, is approximately directly transverse to the curving central axis of the chamber, except on its convexly curved ventral side, where it curves downward for a distance of about 16 mm., producing a hyponomic sinus about 40 mm. wide at its top. The chamber is strongly depressed dorso-

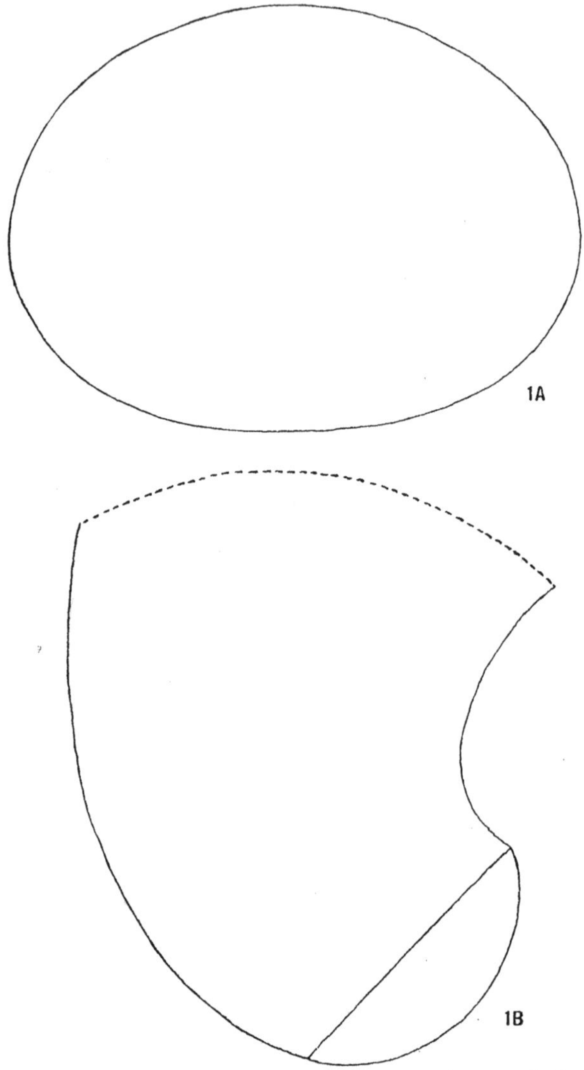

FIG. 1. *Nephriticerina alpenensis* Foerste. Same specimen as that represented on Plate I. *A*, cross-section with the ventral part of the outline at the top of the figure; *B*, lateral view, outlining more distinctly the suture of the septum at the base of the living chamber. There is no trace of the siphuncle. For the latter see Contributions from the Museum of Geology, University of Michigan, Vol. 2, Plate V, Figure 5, 1927.

ventrally. At the aperture its lateral diameter is 79 mm. and its dorso-ventral one is 59 mm. The corresponding diameters at the bottom of the specimen are estimated at 57 mm. and 36 mm. The dorsal side of the cross-section is distinctly more flattened than the ventral side. Along the convexly curved ventral outline of the chamber the level of the highest part of the aperture is 80 mm. above the suture of the septum at its base. Along the concavely curved dorsal outline this interval is about 40 mm. The concavity of the septum at its base is 15 mm. Its maximum concavity appears to be dorsad of the center of the conch. The base of the specimen is broken in such a manner that no trace of the siphuncle is retained. In the absence of any trace of the vertical ribs on the surface of the shell the reference of this specimen to *Nephriticerina* can be regarded as only tentative, but the correctness of this reference is highly probable in view of the occurrence of this specimen at the same locality and horizon as the genotype.

Locality and horizon. — From the quarry of the Michigan Alkali Company, at Alpena, Michigan, in the Alpena limestone. Specimen No. 10777 in the Museum of Paleontology, University of Michigan.

PLATE I

Fig. 1. *Nephriticerina alpenensis* Foerste. — *A*, ventral view, showing the outline of the aperture with its hyponomic sinus. The lower part of the specimen is crushed; *B*, lateral view of the same specimen, showing the rate of its lengthwise curvature. See also Figure 1*A* and *B* on page 152. Specimen No. 10777, Museum of Paleontology, University of Michigan.

PLATE I

1A

1B

Vol. III, No. 8, pp. 155–161 (5 pls., 2 text figs.) JULY 15, 1930

ON THE LOWER JAW OF *BRACHYSUCHUS MEGALODON*

By E. C. CASE

IN MAY of 1929 the author published the description of a very perfect skull (number 10336 of the Museum of Paleontology of the University of Michigan) of a large phytosaur under the name *Brachysuchus megalodon*.[1] In the summer of 1929 a collecting party from the Museum revisited the site of the discovery and Mr. Wm. H. Buettner, preparator in the Museum, recovered the nearly perfect lower jaws of a phytosaur within a hundred feet of the spot where the skull had been found two years before. The jaws accord so perfectly with the skull in characters and proportions that it may be safely assigned to the same genus and species, and with considerable confidence to the same individual; it carries the number 10336 A in the Museum (Plates I–II).

The jaw was found resting on the lower surface and is remarkable in the fact that nearly all the teeth are preserved entire and in position. This, so far as the author knows, is unique among the collected specimens of phytosaurs and permits a very satisfactory study of the dentition of the lower jaws.

The left ramus is perfect and the teeth also are perfect, with the exception of the tips of a few. The posterior part of the right ramus has suffered somewhat from decay and a few of the posterior teeth are incomplete, but are represented by their shattered bases. The anterior tusks, as described below, are in a somewhat anomalous condition.

As the individual was mature at the time of death, some of

[1] E. C. Case, *Description of the Skull of a New Form Phytosaur, with notes on the Characters of Described North America Phytosaurs*, Memoirs of the University of Michigan Museums (University of Michigan, Ann Arbor, 1929), Vol. 2.

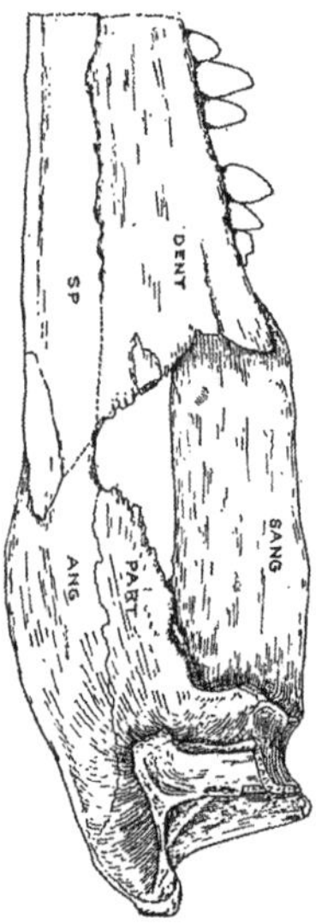

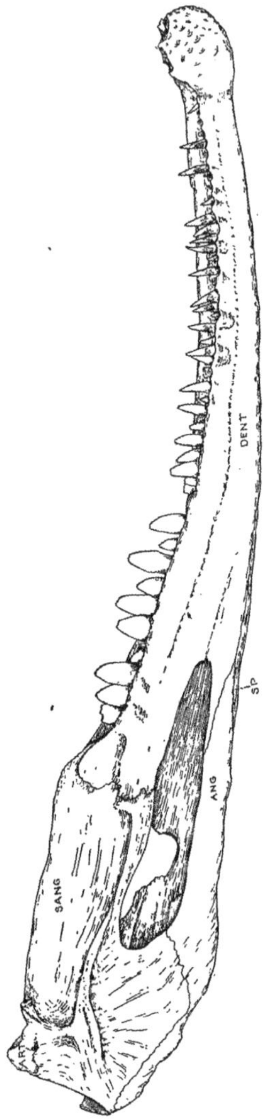

FIG. 1. Inner surface of the posterior portion of the right ramus of *Brachysuchus megalodon*

FIG. 2. Outer surface of the right ramus of *B. megalodon*

ANG, angular; ART, articular; DENT, dentary; PART, prearticular; SANG, surangular; SP, splenial the sutures in the posterior portion of the jaws have closed in part, and it is not possible to make out the complete conformation of all the bones, but enough of each suture can be traced to show the general form and relationships of all the elements (Figs. 1–2).

Dentaries. — These form the major portion of the jaws and carry the teeth. The two, with the splenials, form a broad, heavy, relatively short symphasial region corresponding in size and proportions with the rostrum of the skull. The anterior ends of the two bones are expanded, forming together a large hemispherical prominence, rugose externally, which holds the large anterior tusks. This portion was somewhat shattered because of the thin walls, but all parts were held in position by the matrix and have been carefully replaced in their proper positions. The great anterior tusks are apparently not in a normal position. In the anterior swelling of the upper jaw there are two large sockets followed by two smaller ones on each side. In the lower jaw there are three large sockets on either side. The median socket of the left side is empty, but indicates the presence of a large upright tusk. The second socket is very irregular in form and holds two large tusks in the same stage of development; they are closely crowded together, with the apices just projecting from the opening; on the lower surface of the bone at this place, near the median line, there is the apex of a third tooth pointing obliquely outward and forward toward the opening. The third socket also carries two large teeth with their apices just below the edge of the opening; one of these teeth points backward and outward.

On the right side the median socket indicates an upright tusk; it is empty, but there is a replacing tusk just appearing at the bottom. The second socket carries two large teeth in the same stage of development, both pointing almost directly forward. One of these is projecting from the socket and the tip of the other is just below the edge of the opening. The opening of this socket is very irregular; it appears to have been occluded by a thin plate of bone and the two teeth are breaking through by a process of resorption. The third socket is nearly completely occluded; the covering plate of bone is perforated by one or two small holes.

This socket is occupied by the bases of the roots of the teeth appearing in the second socket and the wall between the two sockets has disappeared.

Besides the teeth described, the upper halves of four others and the tips of four more were found in the matrix filling the cavity of the anterior expansion, a total of sixteen for the six sockets. In all probability the tips and the incomplete teeth are parts of replacing teeth not yet fully formed. The displaced teeth and fragments of others and the disturbed position of the large tusks suggest two possibilities. Either the large tusks, normal in position in life, were forced back into the sockets in the process of maceration and fossilization, a suggestion supported by a slight transverse wrinkling of the roots of some, or the pulp of the living teeth was distorted by some injury to the jaw during life and the position of the teeth is pathological. This suggestion is supported by the partly occluded sockets of the right side and by the fact that the whole anterior end is slightly distorted, the right side being a little lower than the left. It is very easy to see how such an injury could result from a vigorous snap at some heavily armored prey (Plate III, Fig. 1).

Assuming that there were normally three functional tusks in the anterior end of the jaw, there are forty-nine teeth and sockets in each ramus; there are forty-six on each side in the upper jaws. The arrangement and stage of development of the teeth are very irregular, showing that the replacement was without order and that the teeth were very loosely fixed in the jaw (Plate I, Fig. 2). As this is the first specimen which shows a complete dentition, it permits observations which clear up many hitherto uncertain points. The anterior tusks have the same form as the teeth of the median part of the whole series. They are convex on both the inner and outer faces, but the outer face is much more convex than the inner; the edge formed by the meeting of the two surfaces is crenulate; there is no fluting of the sides. The first teeth posterior to the anterior swelling are small and nearly or quite circular in section, with distinct flutings on the sides. This type of tooth when found isolated has been referred to the Stegocephalia. From the ninth tooth, backward, the section of the teeth becomes progressively more unsymmetrical until they begin to broaden into the characteristic leaf-shaped teeth of the posterior part of the series. The outer side of the tooth is much more convex than the inner and the two sides meet in distinct ridges without crenulations; the sides of the teeth are fluted (Plate IV, Fig. 1). This condition continues to the twenty-ninth tooth. The thirtieth tooth is still high and sharp, but the edges show the first trace of crenulation and the sides are not fluted. With the thirty-second tooth the shape begins to change, the teeth are thinner and begin to broaden antero-posteriorly, a character which reaches its maximum in the posterior teeth. The crenulation upon the edges and the lack of fluting upon the sides continue to the posterior end of the whole series (Plate IV, Fig. 2). It is evident that types of teeth previously assigned to several different forms of phytosaurs and even to stegocephalians occur in this one jaw. The lower edge of the jaw is slightly curved downward, antero-posteriorly, but the median elevation between the teeth is nearly horizontal. In the anterior portion of the jaws the median elevation rises so high that the smaller teeth could have had but little function other than to help hold the captured prey and work it back toward the powerful cutting teeth behind. That this median elevation was functional in capturing prey is evidenced by two areas of injured bone, one in the space opposite the thirteenth–fifteenth teeth and one a little farther back (Plate III, Fig. 2).

The dentary is divided at the posterior end of the dental series and clasps the anterior end of the surangular. Below, the dentary extends on the outer surface as far back as the anterior third of the external opening; a tongue extends some distance back of this point on the floor of the Meckelian cavity.

Angular. — This reaches from the posterior end of the jaw as far forward as the anterior end of the external opening, joining the dentary and the splenial (Fig. 2). It forms the lower border of the posterior portion of the ramus and in both rami is bent inward at the lower edge; this is somewhat emphasized by the accidents of fossilization. The separation from the articular is not distinct in all places, but a suture on the posterior face of the jaw shows that the postangular process is formed by the angular alone.

Surangular. — The surangular is marked on its outer face by a strong rugose ridge, broadening posteriorly. Above and below the ridge are rugosities arranged somewhat radially. A vertical extension of the rugose ridge rises opposite the anterior edge of the articular face for the quadrate; it is marked on either side by foramina. These foramina open into canals which unite within the bone and end in a common foramen on the inner side. The surangular forms the whole of the outer side of the jaw above the external opening and ends in a slightly thickened process between the two projections formed by the bifurcation of the posterior end of the dentary. The sutures between the surangular and the articular are mostly obliterated, but one can be followed for a short distance on the posterior side of the jaw.

Coronoid. — This is not developed in the jaw.

Prearticular. — This is fused with the articular. The suture between it and the angular is clearly seen on the inner side of the jaw, but becomes obscure at the anterior end. The bone ends at the posterior end of the notch in the middle of the edge of the inner Meckelian opening. The posterior portion of the edge of the inner opening is formed by the thin edge of the prearticular, and the anterior portion is formed by the thin edges of the dentary and, possibly, the splenial (Fig. 1).

Articular. — This has a strong concave face for the quadrate. The posterior face is triangular and concave. The inner edge is extended into a pronounced vertical process before joining the prearticular.

Splenial. — This is visible for a short distance only on the outer surface of the jaw. The bones of the two sides form the inner portion of the broad, heavy posterior portion of the symphasis, but narrow anteriorly, ending in a point opposite the fourteenth tooth, leaving the portion of the jaws anterior to this to be formed by the dentaries alone (Plate V, Figs. 1–2). On the inner surface the splenial forms the lower portion of each ramus, but the course of the posterior portion of the suture between it and the dentary is not clear. In a fracture on the left ramus the suture can be seen extending back to the anterior edge of the notch in the lower border of the inner opening and continuing around the upper

PLATE I

FIG. 1. Left side of skull of *B. megalodon.* × $\frac{1}{6}$

FIG. 2. Left ramus of *B. megalodon.* × $\frac{1}{6}$

PLATE II

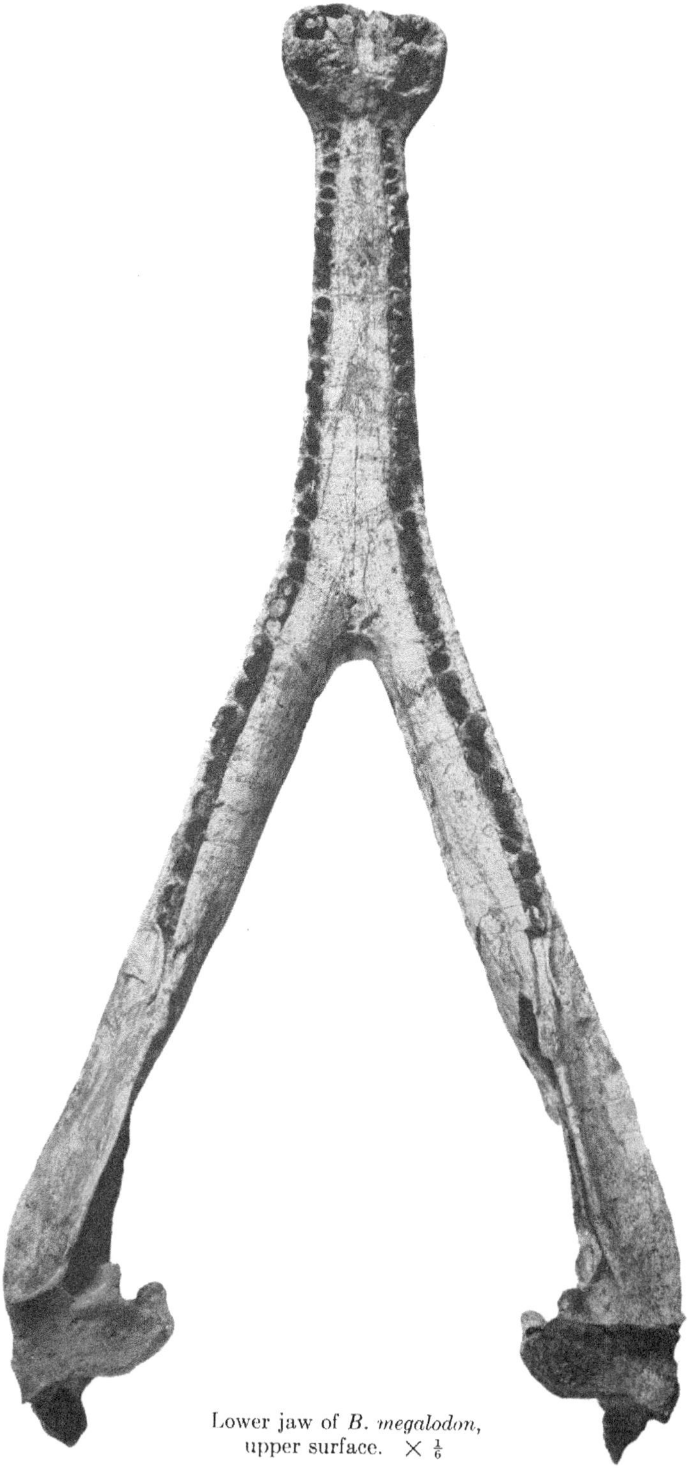

Lower jaw of *B. megalodon*,
upper surface. $\times \frac{1}{6}$

PLATE III

Fig. 1. Anterior view of the jaw, showing the distorted tusks. × ½

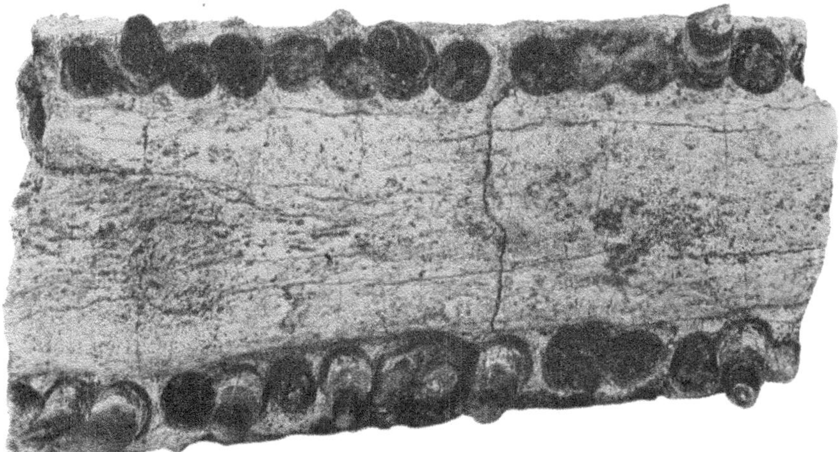

Fig. 2. Upper surface of the median elevation between the thirteenth and twenty-sixth teeth, showing injuries. Visible also in Plate IV, Figure 1. × ½

PLATE IV

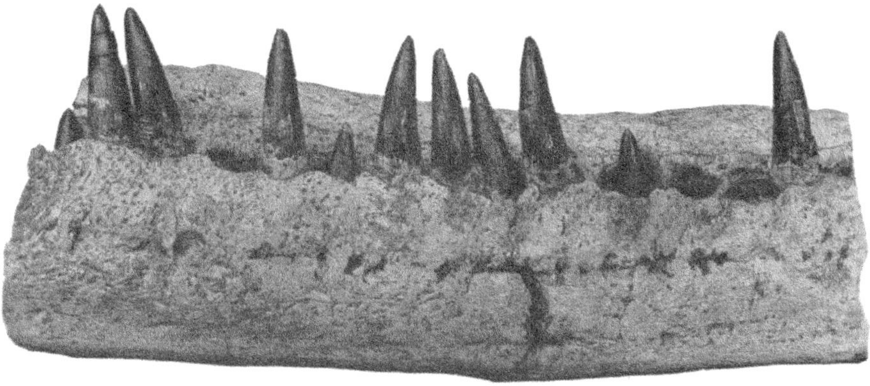

FIG. 1. Lateral surface of the right ramus of the jaw between the thirteenth and twenty-sixth teeth, showing the conical form, the lack of crenulations and the fluted sides. × ½

FIG. 2. Lateral surface of the right ramus of the jaw between the twenty-seventh and thirty-ninth teeth, showing the lack of flutings and the beginning of the broader form of tooth. The crenulations do not show in the photograph. × ½

PLATE V

FIG. 1. Section of right ramus of the jaw opposite the thirteenth tooth; anterior face of section shown in Plate IV, Figure 1. The symphasial region is formed by the dentaries only. × $\frac{2}{3}$

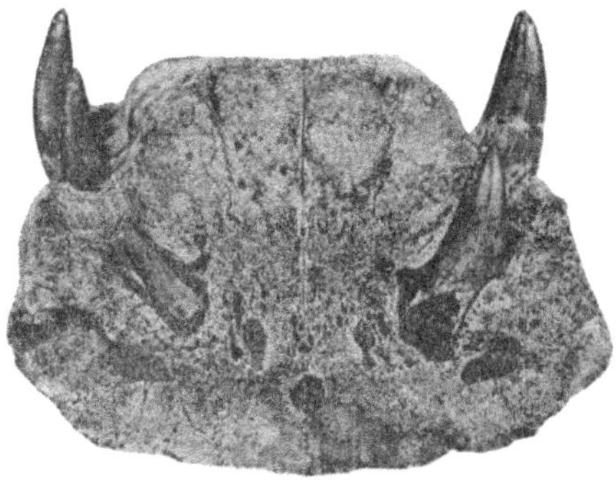

FIG. 2. Section of the right ramus of the jaw; posterior face of the section shown in Plate IV, Figure 1. The splenial takes part in the symphasis. × $\frac{2}{3}$

edge of the angular. The splenial apparently ends at this point by overlap on the inner (within the Meckelian cavity) side of the angular. Certainly the splenial does not, in this specimen, expand posteriorly and form the inner wall of the alveolar edge, as it is shown to do by Huene in his figures of the jaw of *Mystriosuchus pleiningeri*.

The inner Meckelian opening is largely formed by a lack of the inner wall of the jaw due to the wastage of the edges of the dentary and prearticular, but at the bottom there is a distinct notch with thick, rounded edges almost entirely within the angular. It appears probable that there was originally a small but complete fenestra, but that the upper edges have disappeared through the wastage of the bones as described above. The wastage was probably caused by the pressure of strongly developed biting muscles. In the much more slender jaw of *Leptosuchus crosbyensis* the inner fenestra is completed by a thin upper border formed from the dentary and prearticular bones.

It is apparent that the form described as *Brachysuchus megalodon* was of notably different feeding habits from such forms as *Leptosuchus* and *Mystriosuchus carolinensis*. The extremely heavy bones and the relatively short symphasial region in *Brachysuchus*, with the powerful anterior tusks and the large posterior cutting teeth and with the evidence of greatly developed jaw muscles, are in strong contrast to the longer, lighter jaws, the weaker teeth and the less well developed jaw muscles of the other forms. The light phytosaurs with elongate jaws which permitted a very quick snap probably fed upon fish and the more softly bodied forms, while the phytosaurs of the *Brachysuchus* type were well equipped to capture and devour the heavily armored forms, such as other phytosaurs and the giant stegocephalians. That even such powerful jaws and strong teeth as were possessed by *Brachysuchus* suffered from the force necessary to crush the defensive armor of the prey, is amply attested by the deformed anterior end of the specimen described and by the injuries to the median elevation between the teeth.

VOL. III, No. 9, pp. 163–182 (5 pls., 13 text figs.) DECEMBER 18, 1931

ARTHRODIRAN REMAINS FROM THE DEVONIAN OF MICHIGAN

By E. C. CASE

THE quarry of the Kelly Island Limestone Company at Rockport in Alpena County, Michigan, has yielded fragmentary remains of Arthrodires for some years, but nothing that could be identified or described has previously been recovered. During the summer of 1930 Professor G. M. Ehlers and Mr. Charles Campbell made a small collection from a recently opened face in the quarry which was more than usually rich in fish remains. The specimens were of such interest that Professors Case, Ehlers and Hussey made a subsequent trip to the quarry, which resulted in the recovery of more material.

A party from the Geological Survey of Michigan made a section in this quarry in 1926. At a later date Case added to this section an upper bed revealed by an extension of the quarry face.

As now known the section is as follows:

	Feet
ZONE 6	
Layer 3. Yellow weathered limestone	2
Layer 2. Bluish white, plastic clay, when present; frequently absent. No fossils	2
Layer 1. Blue clay, fossiliferous. Shading into the yellow weathered limestone above. Sometimes separated from it by the layer of blue, plastic clay	8
ZONE 5. Limestone, gray, lithographic. Base becomes dark gray at contact with the black limestone	10
ZONE 4. Limestone, black, with many corals and stromatoporoids. Fish plates	10
ZONE 3. Black shale with many heads of *Acervularia*	1
ZONE 2. Limestone, black; similar to that of Zone 4 above	12
ZONE 1. Bell shale. Exposed	5

The limestone above the Bell shale is designated by the Geological Survey of Michigan as the Rockport limestone of the Long Lake formation in the Traverse series of the Devonian. The fish remains occur in the black limestone of Zone 4 and in certain shaly and light-colored limestone phases of the same zone. The dark color is due to the infiltration of bituminous material and the stone will sometimes burn with a dark, smoky flame in the heat of a blast lamp. The bones are deep black in the rock, but weather to a light blue, having the characteristic appearance of the mineral vivianite into which they have been converted.

The most perfectly preserved Arthrodire remains are those of a small form approaching so closely one from the Columbus limestone near Delaware, Ohio, described by Eastman [1] as *Protitanichthys*, that it is placed in the same genus, but considered a new species, *Protitanichthys rockportensis*, with specimen number 12980 U.M., consisting of the complete impression of the cephalic shield and some portions of the shield itself, as the holotype, and specimen numbers 12976 U.M., 12978 U.M. and 12979 U.M., used in interpreting the holotype, as paratypes.

The specimen described by Clarke [2] as *Coccosteus* (?) *halmodeus* and regarded by Eastman and Hussakof as a *Dinichthys* is so similar to the specimens from Ohio and Michigan that it seems to the author of this paper that it should be included in the same genus. Both the New York and the Ohio specimens are from the Marcellus horizon in the Devonian.

Eastman notes three points in which his *Protitanicthys* differs from *Dinichthys* and approaches *Coccosteus:* "First of all, there is to be noted the small size, correlated with the finely tuberculate ornament of the external surface; secondly, the sinuous suture line between the pair of central plates; and thirdly, the exclusion of the pineal from contact with the latter pair." Eastman continues: "The form of the head shield, too, although this can be determined only approximately, appears to have been long and narrow, without much lateral expansion at the posterior border. One striking peculiarity, however, distinguishes it from the species

[1] Eastman, Charles R., *Devonian Fishes of the New York Formations, Memoir 10, New York State Museum*, pp. 144–149, pl. 10, fig. 30, 1907.

[2] Clarke, J. M., "New and Rare Species of Fossils from the Horizons of the Livonia Shaft," *Thirteenth Annual Report of the State Geologist* (*N.Y.*) *for 1893*, pp. 161–168, pl. 1 and figures, 1894.

of *Coccosteus* and *Dinichthys*, and points to the ancestral relationships with *Titanichthys*, in recognition of which we have given it the generic name. *** We refer to the large size and transverse elongation of the pineal plate, which, according to our view, indicates initial modification along the line of descent culminating in *Titanichthys*."

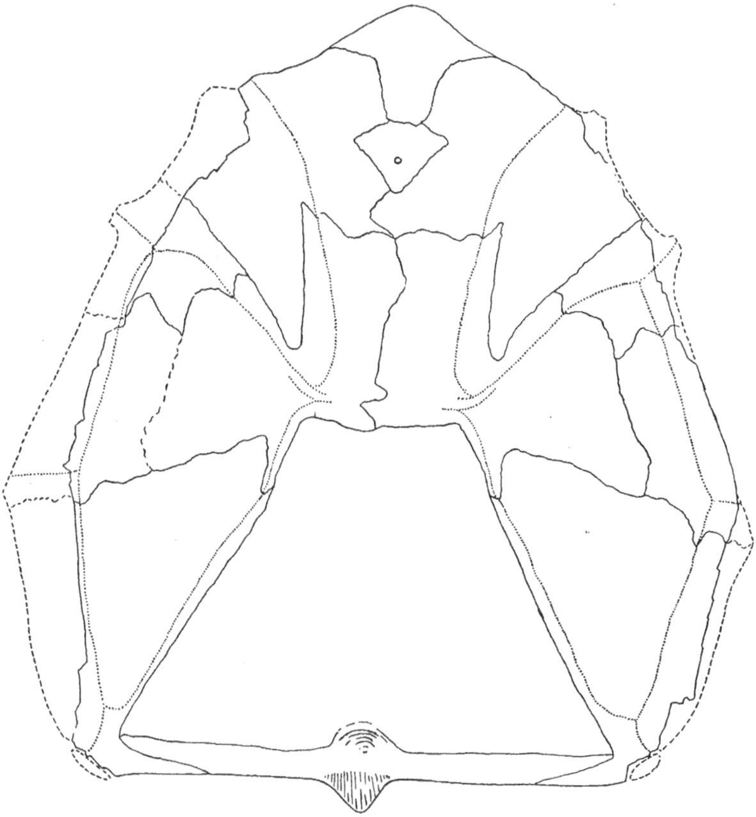

FIG. 1. Cephalic shield of *Protitanichthys rockportensis*. Number 12980 U.M. Restored portions in broken lines; sensory grooves dotted. × ½

Clarke's figure of *halmodeus* does not show the broad pineal plate, but the similarity in the rest of the cephalic shield of his specimen to those from Ohio and Michigan suggests that there may have been a misinterpretation of the New York specimen in this region.

The description and illustrations of the cephalic shield of *P. rockportensis*, Figure 1, and Plate I, Figure 1, are based on the holotype checked by the paratypes and two less perfect specimens.

As shown in the section of the quarry, Zone 4, in which the

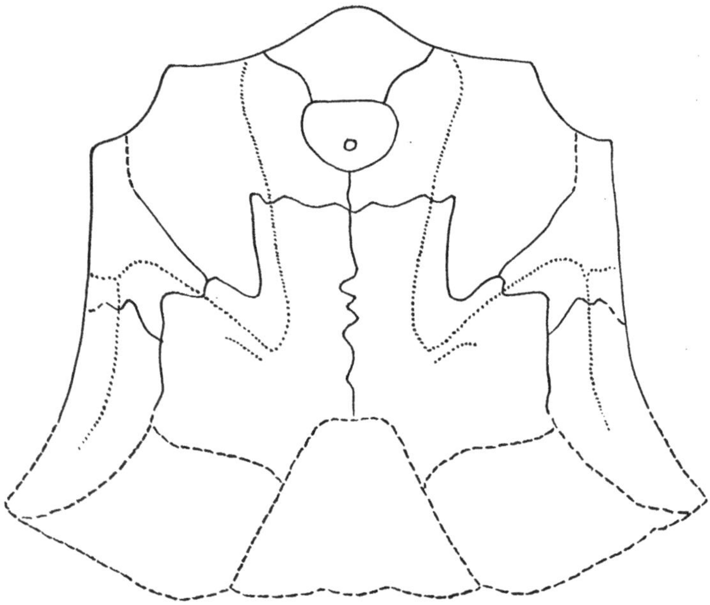

FIG. 2. Restoration of the cephalic shield of *Protitanichthys fossatus* Eastman, from Eastman

Arthrodire remains occur, carries numerous large stromatoporoid and coral heads. The cadavers came to rest upon a very irregular bottom and the cephalic shields and separate bones were somewhat distorted by being pressed down upon and between the resistant heads. This has rendered it necessary to make some readjustment in the drawings of the shield, which is represented as flat instead of convex and irregular. The obvious asymmetry of the two sides has been preserved.

The outline of the cephalic shield is oval. The exact periphery is not preserved in the holotype, but has been restored in broken line from the paratypes. This is not the shape suggested by Eastman for the genotype (see Fig. 2), but it will be noted that the whole posterior portion of his figure is restored and does not recognize the reëntrant posterior portion of the border. His restoration was evidently influenced by Clarke's restoration of *halmodeus* (see Fig. 3.)

The greatest differences between *rockportensis* and the two

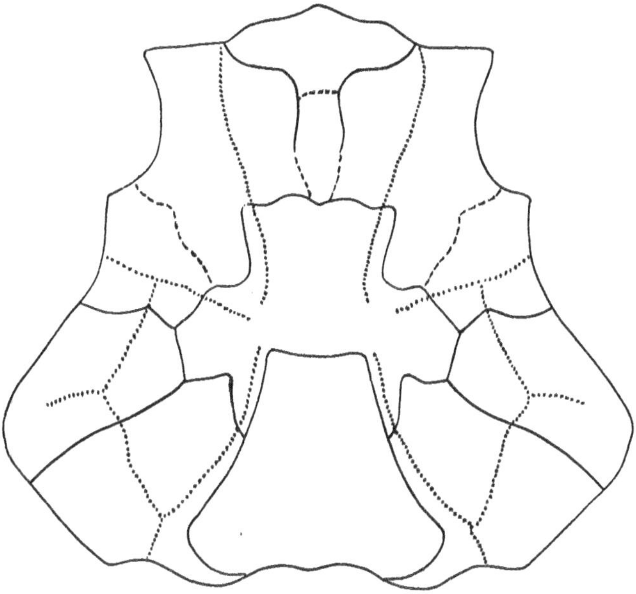

FIG. 3. Restoration of cephalic shield of *Protitanichthys* (*Coccosteus*) *halmodeus* (Clarke), from Clarke

other species, which are the definitive characters of the species, are the larger size of the lateral wings of the central plates, the correlative narrowing of the marginals, and the larger size of the median occipital plate. The latter point is somewhat conjectural in Eastman's *P. fossatus*, as the whole plate is restored in his figure.

A detailed description of the plates is unnecessary; their outline and position are shown in Figures 1 and 4. It will be noticed that the outline of the plates does not correspond exactly with that of either of the other described species, but the difference is not greater than that between the outlines in the five specimens of *P. rockportensis*. The course of the sutures is notably irregular

FIG. 4. Restoration of cephalic shield of *Protitanichthys rockportensis*. Number 12976 U.M., paratype. Restored portions shown in broken wavy lines. × ½

and the exact form of the plates was a matter of considerable individual variation. Much of the difference between Clarke's figure of *halmodeus* and the figures of the other species is due to the fact that he did not recognize the constant presence of the sharp reëntrant angle in the sutures wherever they are crossed by one of the sensory grooves.

The ornamentation is closely similar to that of *Coccosteus.* The tubercles are all small and frequently show a radiating, asteroid structure at the base. They are very uniformly distributed and become slightly smaller near the periphery of the shield. There is a decided lack of the tubercles near the center of the shield in the holotype, where the median occipital and the paired central plates meet, but in another specimen of the collection, number 13052 U.M., they are as abundant here as elsewhere. In the last specimen the sensory grooves are not so well marked as in the holotype, suggesting that with age the shield may have become thicker and more strongly marked with partial obliteration of the sutures and grooves. The median portion of the posterior edge of the median occipital is thickened and elevated and has very few and very small tubercles; the whole posterior border has a similar lack of tubercles and is marked by a series of fine transverse striations.

The system of sensory grooves is very clearly marked on the specimens and is shown in dotted lines in Figures 1 and 4. Certain additions and corrections may be made in completion of previous descriptions. The distal extremities of the prominent V-shaped grooves which cross the preorbital and postorbital plates and unite on the centrals clearly reach the edge of the shield just anterior and posterior to the orbits, forming valuable topographic marks in the pattern. Two short extensions from the apices of the V-shaped grooves turn in toward the center of the shield and terminate abruptly. The lower arms extend posteriorly parallel to the outer edges of the median occipital plate and continue to the posterior edge of the shield. These arms are obscurely marked in some of the specimens, and in one, number 12976, were at first thought to cross over to the median occipital plate and recross over to the external occipitals posteriorly, but this is probably not correct. The passage of the grooves from the central to the external occipital is marked by a sharp posterior deflection of the suture between the two plates, a topographic point noticeable in all the cephalic shields of the Arthrodires.

The sensory grooves form a continuous system near the periphery of the shield on the postorbitals, the marginals and the external occipitals. From this system a branch runs out to the edge on the marginal, slightly anterior to its posterior end, and a second branch runs to the edge near the posterior end of the external occipital and is probably extended across to the anterior dorso-lateral plate.

In Figure 5 and Plate I, Figure 2, there is shown the under surface of the shield as exposed in specimen 12979. This illus-

FIG. 5. Cephalic shield of *Protitanichthys rockportensis*, lower surface. Left side restored from the right. Number 12979 U.M. × about $\frac{2}{3}$

trates how different the pattern of the sutures is upon the two surfaces. There was very extensive overlap, but the limits of the plates are very clearly marked by the edges of the bones and by the course of the fibers. It has been very difficult to harmonize the pattern of the sutures on the two surfaces; it would appear that some of the plates clasp the adjacent plate, overlapping on both surfaces. This demonstrates how uncertain any restoration of one surface from the other must be. The appearance of the lower surface, shown in specimen 12979, is very similar to that of *P. fossatus* as shown in photographic reproduction by Eastman. The pineal plate is almost identical in the two. It is by a study of this specimen and its comparison with others that it can be demonstrated that the preorbital plates meet between the pineal plate and the paired centrals. Specimen 12978 shows the sutures and the sensory grooves of the anterior part of the cephalic shield with exceptional clarity. See Plate III, Figure 1.

No gnathal elements nor any other plates of the cephalic or ventral shields have been found that can be associated with *Protitanichthys.*

The dorsal shield is represented by three median dorsal plates, numbers 12975, 12977 and 13049, all U.M. Specimen number 12977 is preserved in two blocks, having been split through the substance of the bone. It is somewhat distorted by the impression of the rough surface upon which it lay and, unfortunately, the extreme tip of the posterior process was lost. This tip was complete when the specimen was found, but fell between two large blocks of limestone and could not be recovered. The author closely questioned the finder without revealing the significance of the questions and was assured that the whole tip was not over an inch and a half long and came to a sharp point without any terminal thickening. This, with the evidence of the specimen itself, confirms the Coccostean nature of the process as contrasted with the expanded, spoon-shaped termination of the process present in the Dinichthids. The median dorsal plate is slightly convex, with a bifurcate anterior border and a strong median keel on the lower surface. The keel extended from near the anterior border to a point about three centimeters from the posterior border, where it terminated abruptly. See text Figure 6 and Plate II, Figure 1.

The upper surface is shown in number 13049. It is ornamented with tubercles of the same size as those on the cephalic shield, but there are three areas devoid of tubercles, one median and two lateral, which converge toward the posterior edge. No trace of the sensory grooves can be detected on any of the specimens. See Plate II, Figure 2.

Specimen number 13046 U.M. — This is a small, complete spine-like plate with the characteristic tuberculate ornamentation. One end, apparently the attached end, is thinner and spatulate in outline, the main portion is narrow, with nearly parallel edges drawing to a pointed extremity. The plate was found isolated, and no suggestion for its position can be made unless it is a clavicular element. See Plate III, Figure 2.

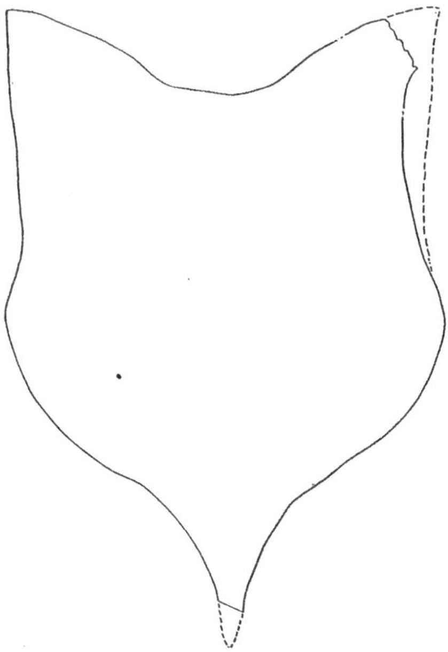

Fig. 6. Outline of the median dorsal plate of *Protitanichthys rockportensis*. Number 12977 U.M. × ½

There are present in the collection several fairly complete plates of the large arthrodire *Holonema*. All are characteristically thin and show the linear sculpture. Plates IV and V show the ornamentation of different regions. So little is known of the plan of the armor of *Holonema* that any assignment of the plates to a definite position must be very tentative. The most interesting point is the occurrence of the genus at this horizon. Most of the recorded specimens have been found in higher beds, none older than the Genesee. Only two specimens have been described, by Hussakof and Bryant,[3] from the Hydraulic Limestone of Milwaukee, which is regarded as the equivalent of the Hamilton and Tully, just above the Marcellus. Since the plates are typically charactetistic of *Holonema*, it must be recognized that this highly

[3] Hussakof, Louis, and Bryant, W. L., *Catalog of the Fossil Fishes in the Museum of the Buffalo Society of Natural Sciences, Bulletin of the Buffalo Society of Natural Sciences*, Vol. 12, pp. 104–105, pls. 32 and 33, 1918.

specialized form was present at the time of the deposition of the Rockport limestone in the Long Lake formation of the Traverse. The Long Lake is regarded by Pohl as equivalent to the Marcellus and as lying below the Milwaukee Limestone. The presence of *Holonema* does not help in fixing the horizon of the Rockport, since this is the lowest bed in which it has been found, unless it be regarded as evidence that the Traverse is equivalent to the Hamilton as represented at Milwaukee.

Specimen number 13040 U.M. — This specimen, which is shown in Plate IV, Figure 1, is one of the most interesting. There are portions of three plates on the block, a central shield-shaped plate, and parts of two lateral plates. The sutures are clearly shown and there is a sensory groove on the left lateral plate. Such a juxtaposition of plates might occur in four places in the pattern of the arthrodire shields, but it seems most probable that these plates are a portion of the anterior part of the cephalic shield. Unfortunately, the median plate has lost the median portion, but there is no indication on the bone preserved, or on the impression, of the presence of a pineal foramen. The outline of the plate is clearly shown either by the bone or by the impression of the lost part. It has the form and relations of a rostral plate in more completely known cephalic shields. The impression shows the mark of a low median keel, deeper posteriorly, where a fragment of bone is still preserved. The course of the sensory groove is that of the well-known lateral groove, which reaches the edge of the shield anterior to the orbit.

Specimen number 13043 U.M. — This is a plate complete on three sides and, apparently, at the lower right-hand corner, as placed on the plate (see Plate IV, Fig. 2), approaches very closely to the edge of the fourth side. The bone is preserved and shows the pattern of the sculpture and the course of a sensory groove very perfectly. The author has attempted to set this plate in the common pattern of the arthrodire cephalic dorsal shield, but without success. It cannot be supposed that the shapes of the plates or the courses of the sensory grooves were closely similar in all the different genera and species, but all descriptions and figures agree within certain limits, and the curvature of the grooves is in a fixed direction. The only position in which this plate could be placed is that of the left external occipital, but there is not the slightest indication, either in shape, thickness of bone, or modification of sculpture, of any articulation with the dorsal shield. Such an articulation was absent in some forms and may well have been absent in *Holonema,* which is notable for the thinness of the plates as well as for the peculiar sculpture.

Specimen number 13051 U.M. — This is a very nearly complete plate, but, unfortunately, only the lower surface is exposed. In a few places the bone is loose and has been removed, thus revealing the beautiful sculpture. Unhappily, the matrix is soft and shaly and any attempt to remove the more adherent bone brings away a thin layer of matrix, which carries the impression. In outline the plate resembles the one figured by Hussakof and Bryant (Plate 30, Fig. 2) and designated by them as the anterior ventro-lateral of *Acanthaspis,* but is about three times the size of their specimen. Especially characteristic is the projection at one corner of the plate. The objection to the assignment to this position is the presence of a sensory groove.

Specimen number 13050 U.M. — This plate is similar to the preceding one, especially in the projection of one corner. It reveals somewhat more of the sculpture, notably an area of reticulate instead of radiating lines near the projection. Only a portion of the very thin bone is preserved, and an effort to remove it was unsuccessful, so that the pattern of the sculpture is not completely shown.

Specimen number 13039 U.M. — This plate is nearly complete and is remarkable for the extreme thinness of the bone. The lower surface is presented; in the few spots where the bone is broken away there is no trace of sculpture revealed by impression, except an extremely fine reticulation near the edge of the plate, such as occurs in similar positions in other plates. The under surface of the plate is marked by a series of broad parallel lines or areas indented with minute pits. It appears as if these were due to a partial solution of the inner surface of the extremely thin bone etching into visibility a sculpture so low that it has left no impression on the matrix; rather a structure of the bone than a surface relief. The elongate, narrow shape of the plate suggests that it may be one of the posterior latero-ventrals.

Specimen number 13048 U.M. — This is the impression of a small, broadly oval plate. Near one end the sculpture is reticulate, shading into an area of small tubercles at the other. The sides of the oval are marked by fine radiating ridges, with the edges of the ridges divided into minute denticulations. These denticulations are reminiscent of *Glyptaspis*, but there is no concentric

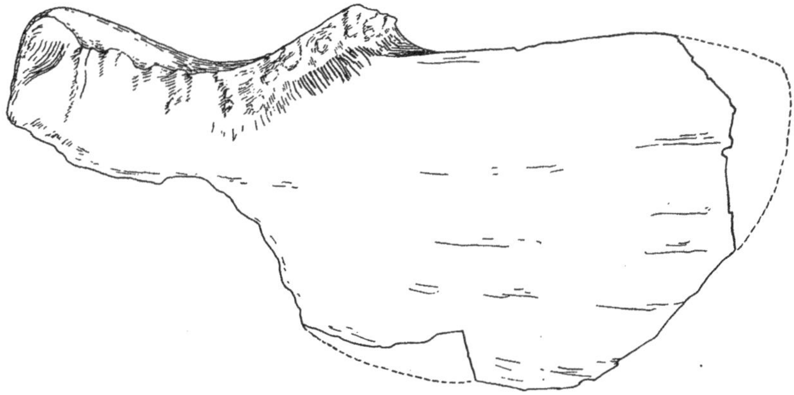

FIG. 7. Outer surface of the left lower gnathal element of *Dinomylostoma* (?), spec. nov. × $\frac{2}{3}$

arrangement of the lines in the center of the plate. It resembles most closely the figure of *Holonema abbreviatum* (Eastman) given by Hussakof and Bryant, or the *Glyptaspis eastmani* of Schwartz,[4] which the former authors place in the genus *Holonema* and perhaps the species *rugosum* of Claypole.

The position of the plate is uncertain, it might be either an anterior or a posterior median ventral. See Plate V, Figure 1.

Specimen number 12974 U.M. — This is a nearly complete lower gnathal element of the left side. It lacks only a small fragment from the posterior end and a small bit of the inner side of the tritoral edge. This specimen has been most kindly examined

[4] Schwartz, C. K., "Middle and Upper Devonian," *Maryland Geological Survey*, p. 700, pl. 73, fig. 1, 1913.

Fig. 8

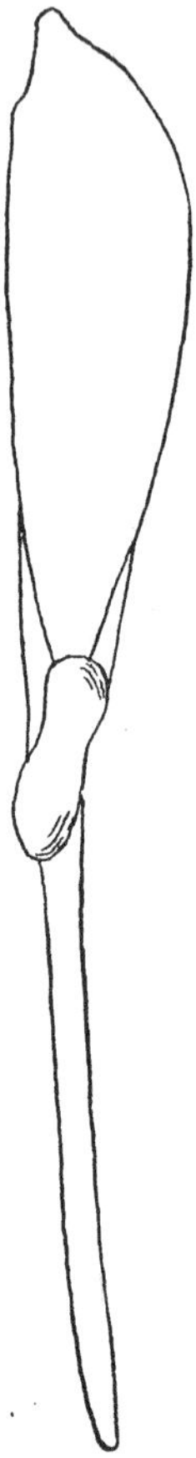

Fig. 9

Fig. 10

Fig. 8. Upper surface of left lower gnathal element of *Dinomylostoma* (?), spec. nov. Number 12974 U.M. $\times \frac{2}{3}$

Fig. 9. Upper surface of lower gnathal element of *Dinomylostoma beecheri*, after Hussakof. $\times$ about $\frac{2}{3}$

Fig. 10. Upper surface of lower gnathal element of *Dinomylostoma eastmani*, after Hussakof. $\times$ about $\frac{2}{3}$

for the author by Dr. Louis Hussakof, who identifies it as belonging to the genus *Dinomylostoma* and as a new species. The form of the element is shown in Figures 7 and 8. The tritoral face is smooth, lacking any tubercles, convex laterally and concave antero-posteriorly. The anterior half of the tritoral face is more nearly horizontal, but is inclined outward and downward; the posterior half rises sharply and is inclined inward and downward. This arrangement gives the whole face a twisted surface, so that any opposed plate moving forward and back upon it would also be shifted from side to side. Both the inner and outer edges of the tritoral face are marked by coarse tubercles. Dr. Hussakof remarks that the presence of tubercles on the inner edge is unknown to him in any other arthrodire. The inner face shows the deep groove below the tritoral face described in other specimens of the genus and there is a deep pit near the anterior edge, evidently for the attachment of a strong ligament connecting the elements of the two sides. As shown in Figures 8, 9 and 10, the outline of the tritoral surface is very different from that of *D. eastmani* and *D. beecheri* and is intermediate between them in breadth.

Specimen number 13056 U.M. — This is a smaller and incomplete lower gnathal element, similar in all preserved parts to number 12974.

Specimens number 13041 U.M. and 13042 U.M. — These specimens are upper anterior gnathal elements of opposite sides, but of different size, evidently from different individuals. Dr. Hussakof is of the opinion that these most probably belong to the same genus as the lower element. These gnathal elements are possibly to be associated with the genus *Holonema* for the following reasons: (1) They are found in the same bed as the rather abundant plates of *Holonema;* (2) Professor Lull of Yale University has kindly examined the type of *Dinomylostoma beecheri* in the Peabody Museum and assures the author that, though the plates associated with the gnathals are exposed on the ventral surface, certain breaks reveal the characteristic sculpture of *Holonema.*

The character of the upper gnathal elements is shown in Figures 11, 12, 13, and in Plate V, Figure 2.

It is assumed that these upper plates belong to the same genus and species as the lower one, because: (1) they are the only gnathals that have been found in the collection; (2) they have the same form of coarse tuberculations around the edges of the tritoral surface; and (3) the faces can be brought into a usable opposition. It is further tentatively suggested that, (1) as these could not belong to *Protitanichthys*, which is of the Coccostid type, (2) as *Dinomylostoma* gnathals are associated with plates having the sculpture of *Holonema*, in *D. beecheri*, (3) as the only other form of body plates found in the collection belong to the form *Holonema*, and (4) as the gnathal elements of *Holonema* have never been identified, these gnathals described as *Dinomylostoma* may belong to the previously described *Holonema*. These elements have been examined for the author by Dr. Hussakof, who writes: "The thickened cusp with the denticles is the biting point. The convex surface is the outer face; the concave side (bluish color), which you thought might be the triturating surface, is the side that was set in muscle, or overlapped the front edge of the head. The broad, flattened

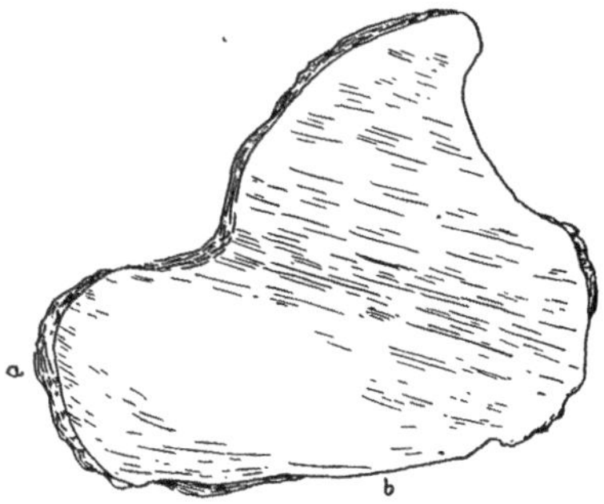

FIG. 11. Anterior upper gnathal element, showing the smooth concave face. Number 13041 U.M. The letters *a* and *b* in Figures 11, 12 and 13 indicate identical points in the three views. × $\frac{2}{3}$

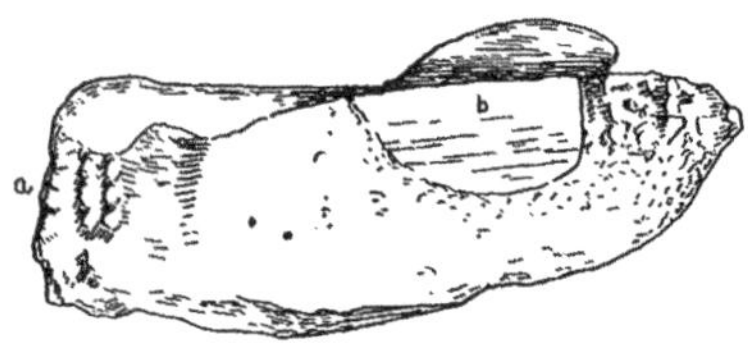

FIG. 12. Lateral view of specimen shown in Figure 11. × $\frac{2}{3}$

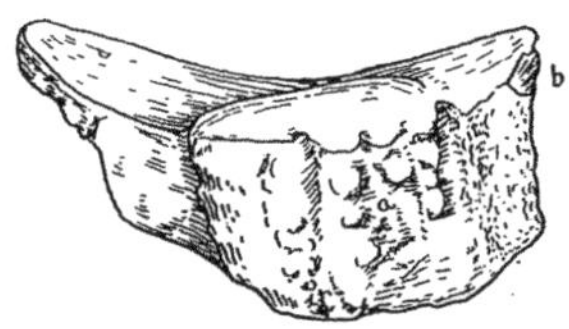

FIG. 13. Lower (anterior?) view of specimen shown in Figures 11 and 12. × $\frac{2}{3}$

expansion is the flange for articulation with the head." Dr. Hussakof sets the elements up as in *Dinichthys herzeri* and *D. lincolni* and notes upon an accompanying sketch that the tubercles on the thickened cusp possibly show wear. The points he mentions can be identified in Figures 11, 12 and 13.

It is with great diffidence that the author ventures to disagree with Dr. Hussakof in a matter dealing with arthrodires, but several points in the structure of the elements compel him to offer an alternative explanation. (1) Dr. Hussakof would set up the elements as in *Dinichthys*, with vertically biting jaws. In *Mylostoma* and *Dinomylostoma* the plates were horizontal and were adapted to crushing. (2) The tubercles are located on all portions of the periphery and so cannot be assumed to be a guide to the tritoral face. On the lower jaw the tubercles are on the side of the tritoral face. (3) The concave face is smooth and very hard and shows evidence of use. It does not have the appearance of having given rise to muscular attachments. (4) The convex face has the appearance of having been attached to cartilage. (5) The smooth convex face cannot be perfectly adjusted to the smooth tritoral face of the lower jaw, but this is equally true of the heavy cusp. The difficulty may be due to lack of correspondence in size of the plates, which are from different individuals. The author suggests that the elements may have been horizontally placed, as in *Mylostoma* and *Dinomylostoma*, with the concave face opposed to the convex tritoral face of the lower jaw. The plates are so placed in Figures 11, 12 and 13, and are so described below.

The tritoral face is smooth and concave both antero-posteriorly and, more sharply, from side to side. The narrow end, assumed anterior, is heavy and the broader, assumed posterior, one quite thin. The straight, assumed inner, side is heavy and mostly with a fine reticulation indicating cartilage attachment, but near the tritoral edge, nearer the posterior end than the anterior, is an area of smooth, hard bone, possibly permitting movement against a similar element on the opposite side. Except for the median portion of the inner edge the entire periphery is marked by coarse tuberculations. The crescent-shaped notch of the posterior edge shows a groove, possibly permitting a movable articulation with a posterior gnathal element. The specimen number 13041 U.M. is perhaps from a larger individual than the lower element, as it projects beyond the tritoral surface of the lower when in closest adjustment.

Because of the peculiar shape of the opposed surfaces, the action of the jaws must have resulted in a sliding motion in the complete act of occlusion. This is borne out by the much broader surface of the upper element and is confirmed by the presence of a multitude of scratches and grooves on the smooth tritoral faces (see Plate V, Fig. 2). These were in many places filled with matrix and were either *ante mortem* or immediately *post mortem* in formation. There is no definite or even dominant direction to the scratches and grooves, which suggests a grinding motion between the upper and lower jaws. This recalls the suggestion made by Dean that there was a rotary motion in the jaws of *Mylostoma,* which was later contested by Eastman. All necessary amount of motion could have been accomplished by the non-rigid attachment of the plates in the cartilage, with no necessary assumption of an anomalous motion of the jaws.

The surface of the plates is very hard, so hard that it cannot be scratched with a steel victrola needle, the cleaning instrument used, without the exercise of great pressure, and the scorings were probably due to the action of the jaws during life. It can hardly be assumed that they were due to the attack of other forms after death, since the plates are as hard as any equivalent structure in possible attacking forms and would offer no occasion for attack.

It is difficult to conceive what sort of food might have been taken that would result in such scorings. The habitat was in all probability the coral reef where the remains are found. There was no form of life that such a durophagous animal would feed upon with hard parts capable of inflicting the injury. The limestone is one of the very pure limestones of northern Michigan used for chemical manufactures, and there is no inclusion of quartz grains. A sample of some size dissolved in hydrochloric acid left only an impalpable slime. It is not possible that the siliceous skeletons of sponges or of plants could have caused the deep scorings. Only one suggestion has been made that would reasonably account for the feature — that the animal was a scavenger and devoured the cadavers of other like forms and of ganoid fishes. The teeth and scales of such forms are sufficiently resistant to do the damage noted. While this possibility is unproved and is unsupported by other observations, it remains as the only explanation that has been offered for the condition of the tritoral surface of the plates.

Specimen number 13045 U.M. — This is a characteristic *Ptyctodus* tooth that cannot be assigned with certainty to any species. The tritoral surface is extremely narrow.

There are present in the collection fragments of ichthyodorulites of uncertain relations, one identified as *Ctenacanthus sp.*

DESCRIPTION OF PLATES

PLATE I

FIG. 1. Photograph of a plastic cast of the impression of the cephalic shield of *Protitanichthys rockportensis*. Specimen number 12980 U.M. × about $\frac{1}{2}$

FIG. 2. Lower surface of cephalic shield of *P. rockportensis*. Specimen number 12979 U.M. Compare with text Figure 5. × about $\frac{1}{2}$

PLATE II

FIG. 1. Median dorsal plate of *P. rockportensis*, showing lower surface and impression of upper surface. Specimen number 12977 U.M. × about $\frac{3}{4}$

FIG. 2. Median dorsal plate of *P. rockportensis*, showing ornamentation of upper surface. Specimen number 13049 U.M. × about $\frac{3}{4}$

PLATE III

FIG. 1. Anterior portion of cephalic shield of *P. rockportensis*, showing sutures and sensory grooves. Specimen number 12978 U.M. × about 1

FIG. 2. Small isolated spine of uncertain position. Specimen number 13046 U.M. × about $\frac{4}{3}$

PLATE IV

FIG. 1. Fragment of armor of *Holonema*, showing three plates, probably the rostral and right and left preorbitals. Specimen number 13040 U.M. × about $\frac{2}{3}$

FIG. 2. One nearly complete plate of *Holonema*, showing the sculpture and a sensory groove. Specimen number 13043 U.M. × about $\frac{3}{4}$

PLATE V

FIG. 1. Cast of the impression of a plate of uncertain position, probably an anterior or posterior median ventral of *Holonema*, showing the fine denticulations on the radiating ridges. Specimen number 13048 U.M. × about $\frac{7}{8}$

FIG. 2. Scored faces of anterior gnathal elements. Scorings filled with chalk. Specimens number 13041 U.M. (below) and 13042 U.M. (above). × 1

PLATE I

Fig. 1

Fig. 2

PLATE II

FIG. 1

FIG. 2

PLATE III

Fig. 1

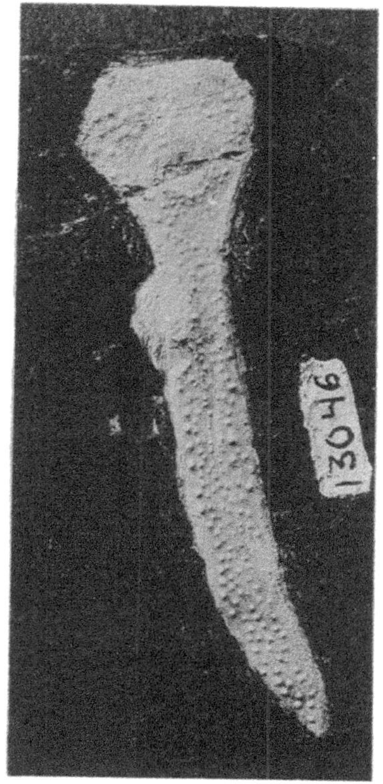

Fig. 2

PLATE IV

Fig. 1

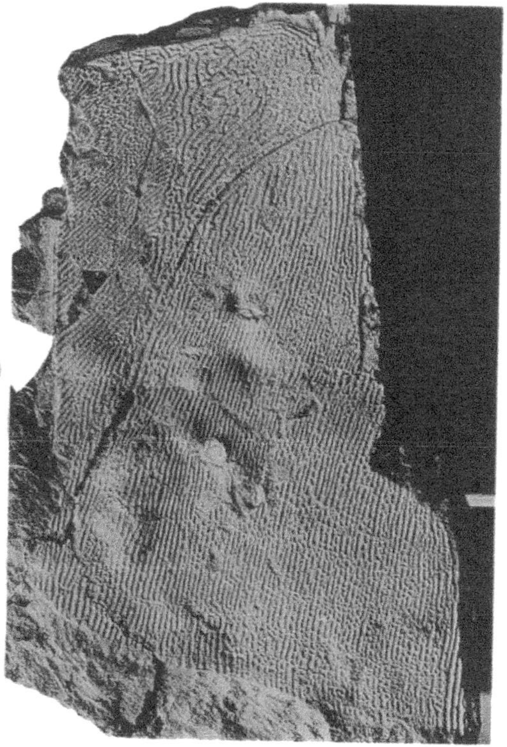

Fig. 2

PLATE V

FIG. 1

FIG. 2

VOL. III, No. 10, pp. 183–185 (1 pl.) DECEMBER 18, 1931

LIFE MODELS OF THE HEADS OF TWO TYPES OF PHYTOSAURS

By E. C. CASE

OUR knowledge of the phytosaurs is still very far from permitting an adequate understanding of the group. Many forms have been described from incomplete material and, without doubt, many of the generic and specific names proposed will be shown eventually to be synonyms. Until fortunate discovery shall place in our hands a complete skeleton with the armor in place we shall be unable to differentiate the various forms and to determine how much of the described differences is to be attributed to systematic difference, how much to sex, age or individual variation.

The abundant material in the possession of the Museum of Paleontology of the University of Michigan, comprising four very perfect skulls, a fifth nearly as good as the others, and a large amount of less complete cranial and axial skeletal material, permits the differentiation of some types with certainty.

It is obvious that, whatever may be the meaning of the elevation of the rostral region, or the character of the nasal opening as to sex, age, or specific differentiation, the feeding habits of the phytosaurs were markedly different in different groups. This is amply attested by the proportions of the skull and the character of the teeth, and has long been recognized. The author has described in detail the skull of *Brachysuchus megalodon* and has pointed out [1] that the relative shortness of the snout, the heaviness of the bones, the large size and weight of the teeth, all indicate that the animal preyed upon heavily armored forms, such as individuals of its own group and the great stegocephalians which lived in the same bodies of water. This evidence is supported by

[1] "On the lower jaw of *Brachysuchus megalodon*," Vol. 3, No. 8, in this publication.

the presence of injuries incurred during life when the powerful jaws were crashed together over some resistant armor.

Opposed to this type was the much lighter type with long, slender jaws and weaker teeth. The length of the jaws permitted a quick snap best adapted to seizing an agile prey.

In both groups the teeth at the anterior end are long and tusk-like, fitted for apprehension and, by their interlocking, the holding of any struggling victim. It is in the posterior teeth that the difference is very apparent. Those of the armor-crushing type are proportionately much heavier than those of the long-nosed fish-eating type.

In an effort to express this difference the author encouraged Mr. Carleton W. Angell, artist in the University Museums, to attempt life models of the heads of two typical forms. The specimen selected for the armor-crushing type was the skull and lower jaw of *Brachysuchus megalodon*, numbers 10336 and 10336B, U.M. The model was criticized during the course of completion by President Alexander G. Ruthven, whose position as an authority upon reptiles is well known, by other workers upon reptiles in the Museum of Zoology, and by the author. The result is shown in Plate I, Figure 1. Mr. Angell preserved most carefully the proportions of the skull and the location of all important points. The dermal covering of the temporal and antorbital openings and the external opening of the lower jaw are shown flattened and somewhat smoother than the skin of the rest of the skull, since, with the mouth open, the muscles would be in a relaxed condition and the skin in this region must have had a greater flexibility than that over the other portions of the skull. The eyes are shown small, in adaptation to the aquatic habitat. The otic region is shown as covered by a flap of integument, as is common in such forms.

The material selected for the model of the smaller fish-eating type was the skull of *Rhytiodon carolinensis* as restored and figured by McGregor.[2] This form shows the greatest extreme of specialization in its peculiar adaptation and is the one for which the

[2] *Memoirs of the American Museum of Natural History*, Vol. 9, Part II. 1906.

PLATE I

FIG. 1. Life model of head of *Brachysuchus megalodon*. × about $\frac{1}{10}$

FIG. 2. Life model of head of *Rhytiodon carolinensis*. × about $\frac{1}{10}$

measurements and proportions were most dependable. The model is shown in Plate I, Figure 2.

Both models have been made one-third natural size. The differences in size and proportion are most strongly brought out in the models. Remains of forms fully as light and slender as *Rhytiodon carolinensis* have been found in the Dockum beds of western Texas and we may be certain that the differences in morphology are evidence of difference of habit and not of geographical separation.

The characters of specimens described as *Mystriosuchus*, *Promystriosuchus*, *Leptosuchus* and *Machaeroprosopus* prove that there were many intermediate types between the extremes shown in the models, and this is perhaps the best evidence we have as yet for the existence of a large number of generically and specifically distinct forms.

Vol. III, No. 11, pp. 187–206 (3 pls., 11 text figs.) December 18, 1931

DESCRIPTION OF A NEW SPECIES OF *BUETTNERIA*, WITH A DISCUSSION OF THE BRAIN CASE

By E. C. CASE

IN THE spring of 1930 Dr. Charles Lawrence Baker of the Rio Bravo Oil Company informed the author of a new locality of Triassic vertebrates in Scurry County, Texas. It had been discovered by Mr. A. N. Huddleston, geologist for the same company, who guided the party from the Museum of Paleontology of the University of Michigan to the locality and rendered all assistance in his power. To both of these gentlemen the director of the Museum of Paleontology wishes to express his appreciation of their disinterested kindness in the matter.

The locality presents a bed of coarse quartz sand of uneven grain, locally very hard and very clean, but with spots of much softer material, including small bits of clay. It is described as "basal" in the Triassic series by the geologists of the oil company.

In the absence of the author, the locality was visited by a party from the Museum of Paleontology to make a reconnaissance. Such material as was available in the brief time at its disposal was collected, and turned out to be entirely the bones of a single species of the genus *Buettneria.* One nearly complete skull, two incomplete skulls showing the basicranial region, isolated clavicles, interclavicles, limb bones and vertebrae were found in close association. The nearly complete skull, number 13055 U.M., is designated as the holotype of a new species, *Buettneria bakeri,* and the two less perfect skulls, numbers 14154 U.M. and 14098 U.M., as the paratypes. The elongation of the pits of the sculpture extending over the frontal and the postorbital regions, the incomplete sensory canal system, the extension of the palatal vacuities anterior to the orbits, the narrower and more elongate skull, and the smaller size distinguish this species from *Buettneria perfecta*, the genotype.

Figures 1 and 2, Plate I, and text Figures 1 and 2 show the

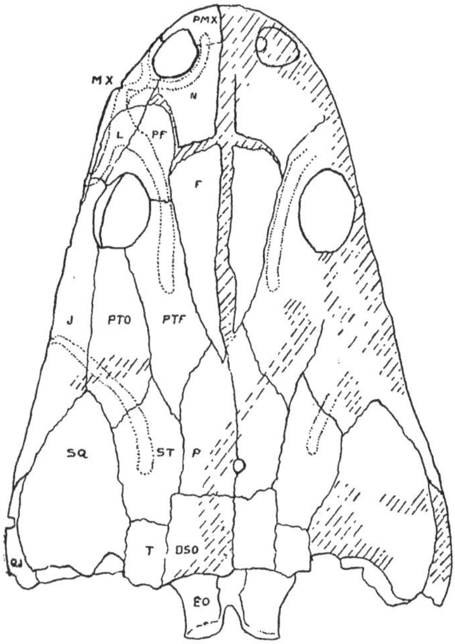

FIG. 1

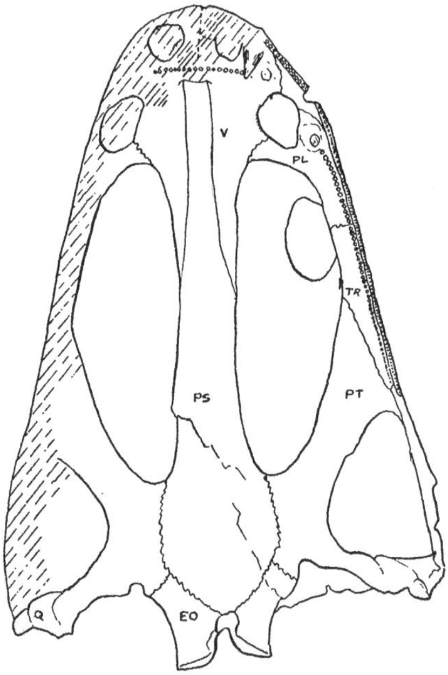

FIG. 2

FIG. 1. Upper surface of skull of *Buettneria bakeri*. Number 13055 U.M. × ¼. The shaded areas show matrix and restored portions. DSO, dermsupraoccipital; EO, exoccipital; F, frontal; J, jugal; L, lachrymal; MX, maxillary; N, nasal; P, parietal; PF, prefrontal; PMX, premaxillary; PTF, postfrontal; PTO, postorbital; QJ, quadratojugal; SQ, squamosal; ST, supratemporal; T, tabulare

FIG. 2. Lower surface of the same skull. × ¼. EO, exoccipital; PL, palatine; PS, parasphenoid; PT, pterygoid; Q, quadrate; TR, transverse; V, vomer

outline of the skull, the character of the sculpture and the course of the sutures and the sensory canals. The restored portions of the skull and the presence of matrix in the cracks are indicated by light cross lines in the text figures and can easily be detected in the photographs. The following table gives exact and comparative measurements.

TABLE SHOWING ACTUAL AND COMPARATIVE MEASUREMENTS OF SKULLS OF *BUETTNERIA*

	B. perfecta	*B. bakeri*
Length / Breadth	$\frac{443.2}{344.2} = 1:0.8$	$\frac{321}{222} = 1:0.69$
Length / Postorbital length	$\frac{443.2}{275} = 1:0.62$	$\frac{321}{178} = 1:0.55$
Length / Length palatal vacuity	$\frac{443.2}{196.8} = 1:0.44$	$\frac{321}{136} = 1:0.42$
Length / Length posterior to palatal vacuity	$\frac{443.2}{132.3} = 1:0.3$	$\frac{321}{76.5} = 1:0.24$
Length / Length anterior to palatal vacuity	$\frac{443.2}{128.7} = 1:0.29$	$\frac{321}{68.6} = 1:0.21$
Length / Breadth at anterior edge orbit	$\frac{443.2}{206.6} = 1:0.47$	$\frac{321}{121} = 1:0.38$
Length / Length orbit	$\frac{443.2}{47.3} = 1:0.1$	$\frac{321}{36.9} = 1:0.12$
Length / Postpineal length	$\frac{443.2}{61.8} = 1:0.14$	$\frac{321}{47.5} = 1:0.15$

Specimen number 13055 is nearly complete, but is preserved in such friable sandstone that it has not been deemed wise to excavate deeply into the brain case. It shows, however, the walls of the brain case formed by the plates from the pterygoid and squamosal bones, the stapes of the left side in slightly disturbed position, and the whole of the posterior face of the skull.

Specimen number 14154 shows the middle of the posterior face and the whole cranial region. The quadrates are missing. Both stapes are in place, the left in undisturbed position. The rising plates of the pterygoids forming the walls of the brain case are complete, that of the left side uncrushed and unbroken, that of the right side slightly crushed. The epipterygoids are preserved in position, that of the right side complete, that of the left side with the posterior edge injured. Portions of the roof and of the parasphenoid have been removed to permit access to the brain case.

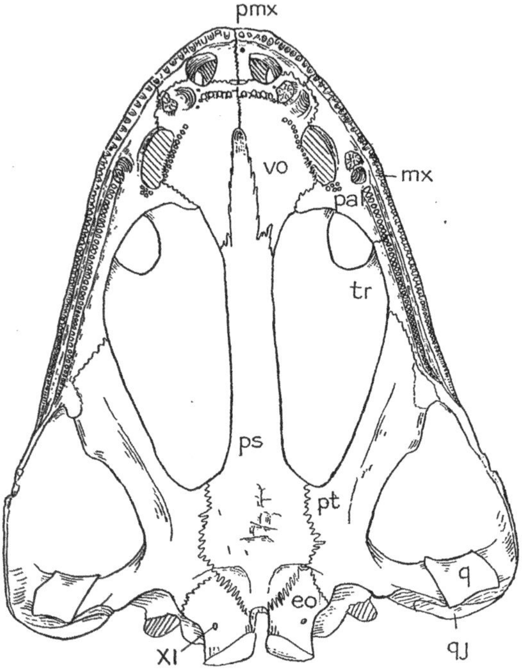

FIG. 3. Lower surface of skull of *B. perfecta*, with the transverse correctly indicated. × ¼. EO, exoccipital; MX, maxillary; PAL, palatine; PMX, premaxillary; PS, parasphenoid; PT, pterygoid; Q, quadrate; QJ, quadratojugal; TR, transverse; VO, vomer

Specimen number 14098 shows the left side of the cranial region complete, with the stapes in undisturbed position. This specimen has been taken apart, so that the details of the brain case can be clearly seen.

External characters of the skull. — The skull, number 13055, is very little crushed, as is shown by the undisturbed plates of the pterygoids and squamosals, but has been somewhat macerated and the bones of the anterior portion have been slightly separated, and there is an oblique crack and slight separation across the postorbital region, so that in the photographs and figures it appears somewhat broader than it was in life. It has the characteristic flat dorsal and palatine surfaces of the genus. The sutures can be traced on one side or the other and are indicated with confidence, except the lachrymal-prefrontal. The pattern is that common to all of the capitosaurian stegocephalians. The sculpture is notably different from that of all described Upper Triassic stegocephalians from North America; the elongation of the sculpture pits, not only on the squamosals, but on the frontal and postorbital regions, approaches in this character the genus *Cyclotosaurus*.

The course of the sensory grooves is indicated in the text figures and can be followed in the photographs. They are exceptionally broad and well marked. The interorbital grooves terminate sharply at about the center of the postfrontals and the temporal grooves terminate equally sharply at the center of the supratemporals. There are no connecting grooves between these two sets.

On the palatine surface the space anterior to the palatine vacuities is proportionately short, permitting the vacuities to extend anterior to the orbits. The vomerine teeth extend transversely in an almost straight line and there are indications of very fine teeth on the inner edges of the choanae. There were strong vomerine tusks; that of the left side is preserved, slanting strongly to the rear. It is probable that the inclination is due to the accidents of preservation. The cultriform process of the parasphenoid is broad and extends as far forward as the transverse row of vomerine teeth, where it rises with the appearance of an opening between it and the vomerine teeth below.

In general arrangement the bones are as in the genotype. The ectopterygoid is clearly outlined and the author takes occasion here to correct the original figure in which he was in doubt concerning this bone. The probable error was pointed out to him some years ago by Dr. D. M. S. Watson and the present specimen confirms the error and permits correction. (See Fig. 3.)

The posterior face of the skull closely resembles that of *B. perfecta*, but the posttemporal openings are present. They are very small, but in specimen number 14098, which has been taken apart, the presence of a perforation is evident.

Cranial region. — The three specimens reveal in undisturbed detail the features of the cranial region. The evidence from the three is mutually confirmatory. In none of the specimens has the author been able to find any evidence of the presence of ossified basioccipital, presphenoid, proötic or epiotic, such as have been reported or suggested by authors in descriptions of similar forms. The advanced degree of secondary chondrification of the walls of the brain case attained in these forms is responsible for the absence of ossified elements which would naturally be looked for.

Descriptions and figures of the cranial region of the Triassic stegocephalians have been given by Fraas (1890), Wiman (1913), Watson (1919), Huene (1922), Case (1922), Wepfer (1923), Hofker (1927), and Branson and Mehl (1929). The very perfect material upon which this paper is based permits a fuller understanding of the less well preserved material described in previous papers. The following detailed description is based largely upon the paratypes of the species. Throughout the study the author has carefully compared the specimens with the specimen of *B. perfecta* and with his original description and finds little to correct; the present description is largely additional matter and explanation of points obscure in the specimen of the genotype.

Exoccipitals. — On the lower surface the exoccipitals are separated by the parasphenoid, which extends to the posterior edge of the floor of the brain case. There is no evidence of an ossified basioccipital, but the two bones have broad inner processes with rough faces which were evidently attached to a cartilage which partially filled the space between them. From the upper surfaces a broad pillar rises to connect with the descending processes of the dermsupraoccipital and the tabulare, and forms the wall of the foramen magnum. This pillar is triangular in section with one, median, face looking into the foramen magnum, one, anterior, looking forward into the brain cavity and one, posterior, looking backward. On the median face, near the base, are two foramina.

(In specimen 14098 there is a third, smaller, one posterior to the other two.) It has been possible to clean out the larger, more posterior, of the foramina in two specimens and to demonstrate that the canal passes directly through the condyle to open on the outer side. The smaller, anterior, foramen leads downward and outward. It has not been possible to clean out the canal, since all the smaller passages are filled with calcite instead of sand grains after a short distance, and the hardness of the calcite is so close to that of the bone substance that it is impossible to distinguish between them with the probe. The canal has been followed to the center of the condylar portion of the bone.

On the anterior face there are two foramina, the lateral much larger than the mesial. The larger canal has been traced downward and a little backward to the center of the bone; from the direction of this canal and that of the anterior foramen on the median face it seems impossible that they could have united. The smaller, mesial canal opens after a short distance into the larger one.

On the lower face there is a large foramen about the center of the condylar portion. The canal rises almost directly into the bone for a short distance and then turns forward in the exoccipital portion of the floor of the brain case just outside of the suture between the exoccipital and the parasphenoid. A fortunate break through the neck of the condyle revealed the calcite filling of this canal, which was cleared out to connect with the foramen, but could not be removed anterior to the break.

Of eight exoccipital bones from six different specimens no two agree exactly in either the number or the position of the main foramina, and there are numerous minute irregularly placed foramina. There was evidently a rather wide range of individual variation. Of all described forms the nearest in agreement with these specimens is that of an isolated exoccipital of *? Metoposaurus stuttgartiensis*, figured and described by Huene (1922), with considerable reservation in his assignment of the nerves and blood vessels to the various canals. So divergent have been the opinions of the various writers regarding the nerves or vessels which occupied the different canals and so inconclusive has been the evidence given in support of the divergent opinions that it seems of little value to recapitulate or discuss them.

There is fairly definite agreement that in the Upper Triassic stegocephalians the hypoglossal nerve was not a cranial nerve, and that the large canal which passes directly outward through the condyle transmitted the IX, X and XI nerves and the jugular vein. This canal has been reported as bifurcate at either the mesial or lateral ends or at both, and as double throughout its extent, but its position is fixed and its function fairly certain. Huene tentatively attributes the anterior foramen of the median face and the mesial foramen of the anterior face (which is much larger in his specimen than in those here described) to the acoustic nerve, and the lateral foramen of the anterior face to a semicircular canal. In view of the course of the canals, as demonstrated in these specimens, and of the practical certainty that the otic capsule was entirely cartilaginous, and not inclosed in bone, these assignments seem less probable.

Various breaks across the exoccipital, the pterygoid and the parasphenoid show that there was a considerable plexus of canals within these bones, a fact recognized by Huene for the exoccipital, and it seems probable that these foramina permitted the passage of the blood vessels which occupied the plexus of canals.

Huene (1922) suggested that the foramen on the under side of the condylar neck may have been for the entrance of the internal carotid artery. The locus of entry of this vessel into the brain case is so variously placed by different writers that the assignment of one or another foramen to its passage is most uncertain.

The fact that the canal from the lower foramen leads forward in the body of the exoccipital would support Huene's suggestion, if any connection with the brain cavity could be shown, but its canal seems to be but a part of the plexus within the parasphenoid and adjacent bones.

The internal carotid arteries have a fixed position for their entry into the skull in most classes of the vertebrates, near the hypophysial region. From the origin of the arteries and their importance the fixity of position is most natural and probable. In the forms of Amphibia with a large parasphenoid bone, there is some departure from the general rule, but, as Watson (1919 and 1926) has demonstrated, the more primitive stegocephalians have a foramen in this region which it is rational to assign to the internal carotids. Even in the highly specialized Anura the entrance is approximately in the normal position. Watson (1919) has figured and described a basisphenoid bone in *Capitosaurus* which he says "is perforated vertically by a foramen, from which a groove on the ventral surface leads to the posterior edge of the bone." Elsewhere in the same paper he refers to the basisphenoid of *Capitosaurus* as "perforated by an obvious carotid."

In the living urodeles the carotids have been described as entering through the parasphenoid at various places, some fairly far back. Sollas (1920), in describing *Lysorophus*, which Williston considered as a Permian urodele, says that "two openings (0.3 mm. in diameter) are visible on the ventral surface of the parasphenoid just in front of the greatest breadth, one on each side and not far from the margin. Their course as they traverse the bone is slightly forwards and a little inwards and they reach the dorsal surface close to the sides of the median trough. These are evidently foramina for the internal carotid arteries.

"In Anura and the Frog they occupy much the same position but lie outside the parasphenoid, just in front of the 'guard' of the dagger-shaped bone."

While it is unwarranted to assume that the condition in such aberrant forms as the urodeles can be taken as a guide to the condition in the Upper Triassic stegocephalians, suggestions have been made of similar conditions. Altogether it seems probable that the internal carotids entered the skull of *Buettneria* in the position previously suggested by the author, anterior to the quadrate ramus of the pterygoid, and passed backward through the canals in the parasphenoid.

The rising process of the exoccipital is bifurcate as in other forms, and is connected by suture with the descending processes of the dermsupraoccipital and the tabulare. There is no perforation in this region in *Buettneria perfecta*, but in *B. bakeri* there is a small opening. The opening is scarcely apparent in the skulls which have been slightly crushed, but in specimen 14098, which has been taken apart, the opening is shown to be larger than would be expected, though still small. The sutures between the exoccipital and the bones above are clearly marked.

Opisthotic. — As in the genotype, there is no distinct opisthotic. In the description of the genus the suggestion was made that a thin plate apparently distinct on the anterior face of the descending process from the tabulare might be the much reduced opisthotic; no indication of such a plate occurs in the present

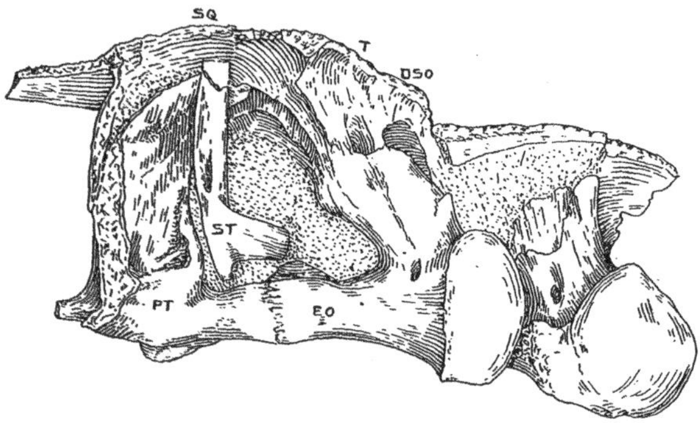

FIG. 4. Oblique view into the left otic opening of number 14154 U.M., showing the position of the stapes and adjacent bones. $\times \frac{2}{3}$. DSO, dermsupraoccipital; EO, exoccipital; PT, pterygoid; SQ, squamosal; ST, stapes; T, tabulare.

specimens, nor can the author detect where such a plate might have been attached or been fused with the tabulare.

Posterior otic opening. — The great posterior opening leading to the otic region is bordered by the exoccipital, pterygoid and tabulare. The squamosal may take a very small part in the upper border, but is apparently excluded from it by the meeting of the tabulare and the rising process of the pterygoid. The relation of the bones in this region is shown in text Figure 4. There is no suggestion of the presence of either a separate proötic or epiotic in the very perfectly preserved material. On the left side of specimen 14154 and on the right side of specimen 14098 the wall of the brain case formed by the meeting of the processes from the squamosal and the pterygoid are preserved complete and uncrushed; both have been examined with minute care and no trace of separate elements can be detected. It is the conviction of the author that the otic capsule was entirely cartilaginous and that the bones usually taking part in the capsule were either absent or reduced to cartilage. The capsule lay in a space partly surrounded by other bones.

The anterior face of the descending portion of the tabulare shows clearly the imprint of a portion of the capsule, and it is this part, if any, which must represent the opisthotic. Seen from the front and below, the face is divided into three areas by two sharp, nearly vertical ridges. The outer area is a concave space on the under side of the process which forms the mesial border of the otic notch. The median area is a depression between two ridges; its upper, anterior, portion is deep and corresponds to the pit in the lower surface of the tabulare described in the genotype. The third, inner, depression is marked off from the median one by the second ridge, which runs vertically on the process and then curves inward on the lower surface of the tabulare. The depression is shaped as if made by the impression of the segment of an ovoid body, i.e. the otic capsule; its inner edge is formed by the border of the opening between the descending processes of the tabulare and the derm-supraoccipital. The surface of the depression is pierced by two foramina which lead into the body of the bone. These are too far posterior and too remote from the median line to be regarded as the openings for the endo- and perilymphatic ducts, but no other openings for these ducts, probably present, have been found. Similar foramina occur in specimen 14098. The anterior wall of the space formed by the processes from the pterygoid and squamosal shows no impression from the capsule.

Stapes. — The stapes is preserved in three specimens, in number 13055 on the right side; in number 14154 on both sides, the left in undisturbed position; in number 14098 on the right side in undisturbed position. The position and form of the stapes are shown in text Figures 4 and 5. The upper end is drawn to a blunt point and lies directly in the otic notch, where it was attached to a tympanic membrane or an area of modified skin. The main shaft of the stapes is broadly elliptical and extends downward and forward. Near the center of the posterior face there is, in specimen 14154, a deep, narrow depression ending sharply below; in the other specimens this depression is less sharply marked. The lower end expands and is separated by a notch on the lower end into two processes. The outer process is smaller and is a direct continuation of the outer edge; it ends within a millimeter of the outer end of the arch covering the canal in the parasphenoid bone described below. This relation was found in both specimens in which the stapes were undisturbed and is very similar to that described by Watson (1919) in *Lydekkerina* and *Laccocephalus* where he says that a lappet from the parasphenoid bone overlies the lower surface of the stapes. The inner process is directed toward the brain cavity at an angle of about 45° to the axis of the shaft and is much heavier than the outer process. The inner end is rough and broadly oval; it was apparently applied to the fenestra ovalis in the cartilaginous otic capsule. There are various small foramina in the bones, but none that could be called the stapedial foramen; perhaps this is represented by the notch in the lower end.

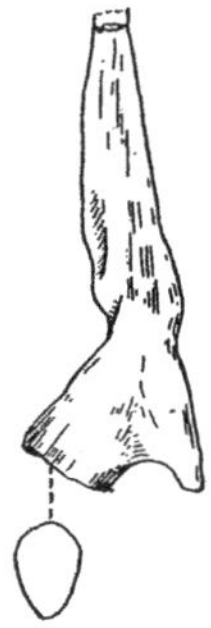

FIG. 5. Stapes of the right side from number 14098 U.M. $\times \frac{2}{3}$

Rising plate of the pterygoid. — This plate is preserved undistorted and complete on the right side of specimen 14098 and the anterior portion is undistorted and complete in specimen 14154. The plate is confined to the quadrate ramus of the pterygoid, rising near the mesial origin of the ramus and running obliquely backward to the outer edge of the otic notch and then turning more sharply outward to the quadrate. The plate is supplemented in the formation of the lateral wall of the brain case by the descending plate from the squamosal described below.

Seen from the posterior view, the plate rises sharply from a point just outside the pterygoid-exoccipital suture forming the outer edge of the great otic opening. The plate is heavy at its origin, but is excavated by the irregular depression described by Wiman as the place of attachment of the pterygoideus muscle.

Beyond this depression the plate is much thinner. It was this portion of the plate which was described in the genotype as bifurcate and as receiving, between an anterior and a posterior portion, the descending plate from the squamosal. Though the condition is still somewhat obscure, there is much in these specimens to confirm the original description. An isolated squamosal bone, number 14099, shows the complete descending plate. The sutural surface on the exposed posterior face shows an overlap downward on the anterior face of the pterygoid at about two thirds of the height of the wall. This suture is hardly to be detected on the other specimens. The descending plate of the isolated squamosal is notched on its outer edge by the large quadrate foramen. Below the notch a long process runs outward toward the quadrate. In specimen 14154 the plate is broken across vertically in its median portion and shows a distinct calcite filling between an anterior and a posterior wall. At the point where the pterygoid plate joins the quadrate there is a thin tongue of bone extending down between the two bones for about half the distance of the contact; this is evidently the extension of the squamosal plate below the quadrate foramen.

The descending plate of the squamosal originates at the posterior end of the squamosal, at the outer portion of the otic notch, and runs outward toward the quadrate; it does not extend forward.

Anterior portion of the pterygoid plate. — This portion of the plate rises to, but does not fuse with, the bones above it; in all the specimens there is a thin line of matrix separating the parts. It has a short contact with the squamosal and the supratemporal; the line of cartilaginous attachment is indicated by the roughened area marked XI in Figure 3 of the description of the genotype. Anterior to the roughened area the upper edge of the plate is parallel to, but free from, the supratemporal for a short distance and then descends sharply to the pterygoid. Just below the upper end of the anterior edge there is a strong process tending directly inward; this is notched on both the upper and lower edges for the passage of nerves or vessels, perhaps for branches of the V or VII nerves. This process, shown in Figure 6, is perhaps the remnant of the proötic, but there is no evidence of even an initial separation from the pterygoid plate.

At the lower end of the anterior edge there is a deep notch; this is in the position described by Watson (1919) in *Capitosaurus* as the place of escape of the vena capitis lateralis and the VII nerve, but in his figure the nerve is marked as passing in front of the epipterygoid. This can be seen on the left side of Figure 6.

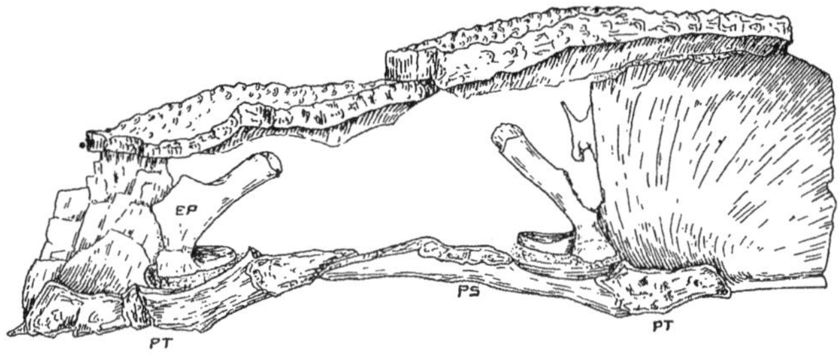

FIG. 6. Anterior view of the brain case of number 14154 U.M., showing epipterygoids and pterygoid plates. × ⅔. EP, epipterygoid; PS, parasphenoid; PT, pterygoid

Epipterygoid. — Just in front of the ascending plate of the pterygoid stands the epipterygoid. This is very peculiar in its form and relation to the adjacent structures. The bones of both sides are present in almost undisturbed position in specimen 14154 as shown in their symmetrical positions. (See Plate II, Fig. 2, and text Fig. 6.) In specimen 14098 the bone of the left side is in undisturbed position. Seen from the side, the bone has a rod-like ascending process which nearly reaches, but does not touch, the parietal. The posterior portion of the lower end is extended backward in a broad process which is closely applied to the edge of the rising plate of the pterygoid just above the notch, but in both specimens the epipterygoid is a separate element. This posterior process is the otic process of the epipterygoid, as described by Watson (1919), but has no such elevation as in his specimen of *Trematosaurus*. Seen from the front, text Figure 6, the lower end of the rising process is shown to originate from a broad, flat, disc-shaped base. In both specimens this base is separated from the pterygoid below by a layer of matrix; evidently it was attached by a fairly thick cartilage. This is the element which was found in a fragmentary state in the genotype skull and tentatively regarded as a proötic. It is the element which puzzled Branson and Mehl (1929) in their description of *Koskinodon.* It is obvious from their description and from their Figure 5 that the plate labeled L and called "epipterygoid" is the basal plate of the epipterygoid; that the part labeled U, "unnamed bone," is the rising process of the epipterygoid; that the part labeled T, "unit imperfectly fused with the 'epipterygoid,'" is a part of the rising plate of the pterygoid. The part labeled S, "stapes," is probably a splinter of the rising plate of the pterygoid or a displaced fragment of the stapes. The recognition of the true form of the epipterygoid and the stapes explains their less well-preserved material and relieves them from the necessity of their rather forced conclusions concerning any relation between the epipterygoid and the auditory apparatus and function.

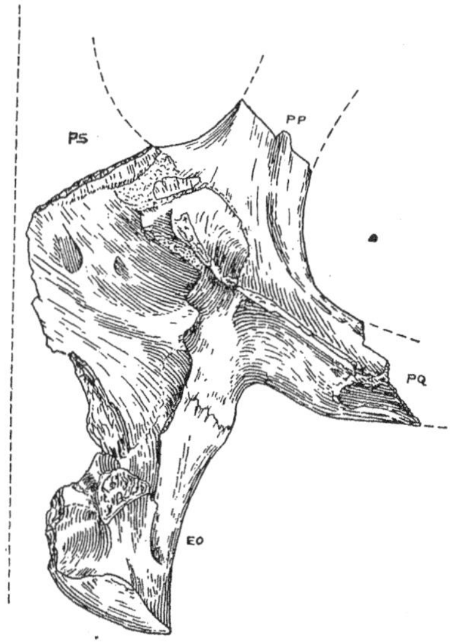

FIG. 7. Right side of the floor of the brain case shown in number 14098 U.M.; EO, exoccipital; PP, palatine ramus of pterygoid; PQ, quadrate ramus of pterygoid; PS, parasphenoid. × $\frac{2}{3}$

Floor of the brain case (text Figure 7). — The floor of the brain case is very similar to that of the genotype, but shows some additional and clarifying details. The description is drawn very largely from specimen 14098. The exoccipital portion is thickened on its outer border and is continued forward as far as the suture with the pterygoid as a raised edge, forming the side of this portion of the brain cavity. This border forms the major portion of the lower edge of the otic opening. There are no foramina in this part of the exoccipital, but the floor is penetrated by the canal which has its outer opening on the lower surface of the condyle. The anterior end of the canal cannot be determined, but it is possible that it communicates with the foramina at the base of the anterior end of the rising plate of the pterygoid, described below.

The parasphenoid forms the main portion of the floor of the brain cavity. It is thin in its median portion, but thickens where it joins the pterygoid and the exoccipital. Just anterior to the exoccipital-pterygoid suture is the oblique canal which was tentatively assigned to the internal carotid in the description of the genotype. The upper border of the outer opening of this canal is elevated and extended into a small process which is directly opposite to, and almost in contact with, the lesser process of the lower end of the stapes, as described above. The canal with its arched covering has the same course as in the genotype.

A similar condition is described by Watson (1919) in *Laccocephalus:*

> Very nearly or quite in the posterior end of the suture between the pterygoid and the parasphenoid a medium-sized foramen enters the latter bone and travels forward and inwards in its substance until it turns upwards towards the brain cavity. Comparison with the skull of *Capitosaurus* . . . shows that this canal must be for the internal carotid artery, for in the latter type the basisphenoid is perforated by obvious carotid foramina, and there is a small foramen at the junction of the exoccipital, pterygoid, and parasphenoid which is the only possible entrance of the carotid, and agrees with the posterior end of the canal in *Laccocephalus.*

Wepfer (1923), in his account of the skull of *Mastodonsaurus cappelensis*, describes (p. 62) and figures (Fig. 10) two grooves on the floor of the brain case in the exact position of these two canals which he suggests may be the impressions marking the course of nerves, the facialis-acusticus.

Just in front of the arch the floor is sharply depressed and forms a groove which leads into a similar groove on the pterygoid and finally into the foramina on the pterygoid just within the epipterygoid. This is the groove which Watson (1919) ascribes to the vena capitis lateralis in *Capitosaurus.*

The pterygoid portion of the floor lies within the rising plate and the epipterygoid. The lower end of the anterior edge of the rising plate is notched, as described above, and from the upper edge of the notch a shelf runs inward toward the opening of the internal carotid artery. This shelf is perforated by a foramen which opens into a much larger one below the shelf. The common

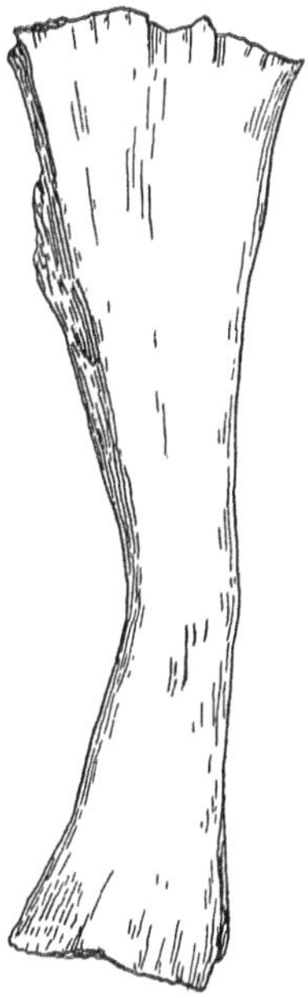

FIG. 8. Lateral view of right femur. Number 12946 U.M. × ½

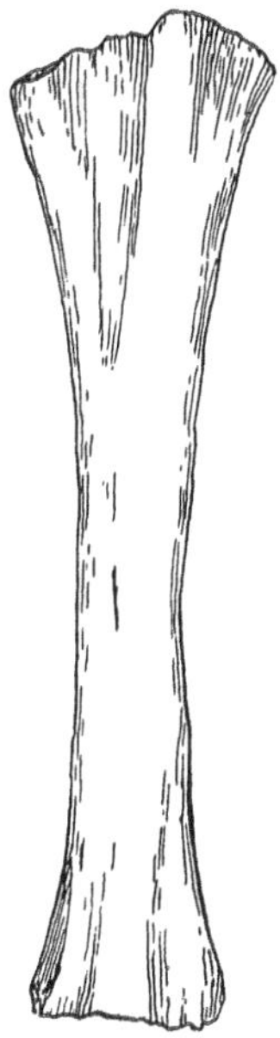

FIG. 9. Anterior view of right femur. Number 12946 U.M. × ½

canal leads into the body of the bone and may communicate with the canal in the exoccipital.

It is apparent that, though the skull of *Buettneria* is similar to Capitosauridae in form, in the position of the orbits and in the general relations of the bones, it is very much more advanced in the secondary chondrification of the brain case, approaching in this respect the *Metoposauridae* and the *Mastodonsauridae*. The same thing may be said for all other forms of stegocephalians described from the Triassic of North America; this is as would be expected from the Upper Triassic age of the beds.

Two nearly complete interclavicles, numbers 13027 U.M. and 13029 U.M., and a complete clavicle, number 13028 U.M., were found associated with the skulls. The interclavicles, Plate III, Figures 1 and 2, are notable for the greater extent and sharper contour of the portion anterior to the faces for the clavicles than

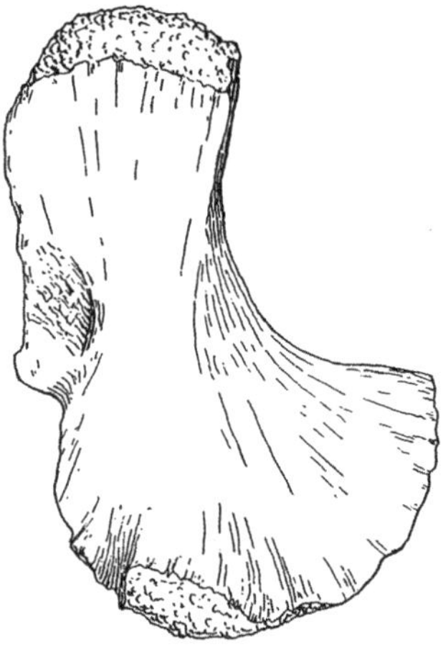

FIG. 10. Anterior face of humerus. Number 12940 U.M. × ⅔

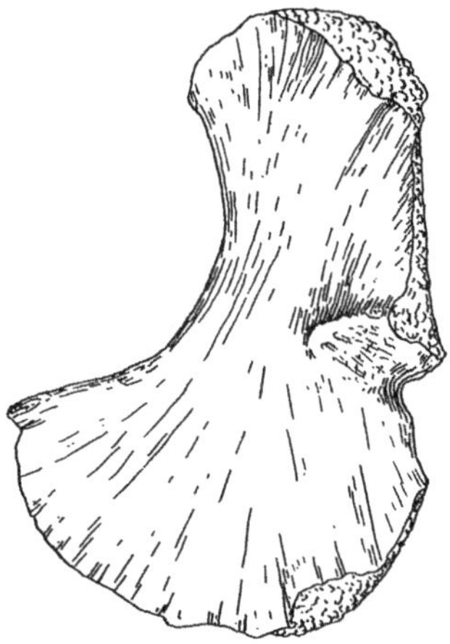

FIG. 11. Posterior face of humerus. Number 12940 U.M. × ⅔

in other interclavicles found, and for the more restricted area of reticulate sculpture. A much larger portion of the surface of the bone is marked by linear sculpture than in *B. perfecta.* In this regard it corresponds to the sculpture of the skulls.

The clavicle shows a similar restriction of the area of reticulate sculpture.

Several femora were found with the skulls. These show few distinctive characters beyond the presence of a strong adductor ridge. The articular faces were evidently, even in the largest specimens, largely cartilaginous. Figures 8 and 9 show the form of specimen number 12946 U.M., which is one of the largest. It is uncrushed and undistorted. The major axes of the articular faces are nearly at right angles to each other.

There are two humeri in the collection, numbers 12940 U.M. and 12941 U.M., both from the left side. Figures 10 and 11 show the form of number 12940, which is undistorted; it is about two thirds of the size of the other specimen. These humeri were collected at a different locality, some miles north and west of Amarillo, Texas, but evidently belong to the genus *Buettneria*, the only large stegocephalian which has been found in Texas, and are, so far as the author knows, the only humeri of stegocephalians recovered from the Triassic beds of North America. The figures give a good idea of the form. The articular ends are rough, but convex instead of concave, as in the femora. The head is well developed, with a rough surface which extends down the outer side to the triangular face of the lower end of the deltoid ridge. The posterior face of the lower end of the ridge is convex and rugose and the anterior face is flat and equally rugose. The distal end of the humerus is expanded into a strong entepicondylar process, but, as is characteristic of all stegocephalians, there is no foramen. The radial tuberosity is prominent, but is not distinct from the region of the articulation of the ulna. On the outer side of the radial tuberosity there is a slight prominence in the position of the ectepicondylar process.

SUPPLEMENTARY NOTE

(September 11, 1931)

In the summer of 1931 the author was permitted to excavate the deposit near Snyder, Texas, through the courtesy of the owner, Mr. P. L. Fuller. Four very perfect skulls and a large amount of other material, almost entirely of *Buettneria bakeri*, were recovered. A description of the deposit and the specimens found will be prepared and published as soon as possible. The author desires in this note to express publicly his appreciation of the kindness of Mr. Fuller in permitting and helping the work of the Museum of Paleontology of the University of Michigan.

LITERATURE CITED

BRANSON, E.B., AND MEHL, M.G. Triassic Amphibians from the Rocky Mountains. University of Missouri Studies, Vol. 4, No. 2 (1929).

CASE, E.C. New Reptiles and Stegocephalians from the Upper Triassic of Western Texas. Publication 321, Carnegie Institution of Washington (1922).

DREVERMANN, F. Über einen Schädel von *Trematosaurus*, Vorläufige Notiz. Centralblatt für Mineralogie, Geologie und Paleontologie, Nr. 20 (1914).

—— Über einen Schädel von *Trematosaurus brauni* Burmeister. Senckenbergiana, Bd. 2, Hefte 3–4 (1920).

FRAAS, EBERHARD. Die Labyrinthodonten der schwäbischen Trias. Paleontographica, Bd. 36 (1890).

HOFKER, J. VON. Archegosaurus decheni Goldfuss. Untersuchung des Schädelbaues. Paleontologia Hungarica, Vol. 2, Fascicule 2 (1927).

HUENE, F. VON. Beiträge zur Kenntniss der Organisation einiger Stegocephalen der schwäbischen Trias. Acta Zoologica, Bd. 3, Seite 392 (1922).

SCHROEDER, H. Ein Stegocephalen-Schädel von Helgoland. Jahrbuch der königlich-preussischen geologischen Landesanstalt für 1912. Bd. 33, Teil II, Heft 2 (1913).

SOLLAS, W. J. On the Structure of Lysorophus as Exposed by Serial Sections. Philosophical Transactions of the Royal Society of London, Series B, Vol. 209 (1920).

WATSON, D. M. S. The Structure, Evolution, and Origin of the Amphibia. — The "Orders" Rachitomi and Stereospondyli. Philosophical Transactions of the Royal Society of London, Series B, Vol. 209 (1919).

WEPFER, E. Der Buntsandstein des badischen Schwarzwalds und seine Labyrinthodonten. Monographien zur Geologie und Paleontologie. Serie II, Heft 1 (1923).

WIMAN, CARL. Über das Hinterhaupt der Labyrinthodonten. Bulletin Geological Institute, Upsala. Bd. 12 (1913–14).

PLATE I

FIG. 1. Upper surface of skull of *Buettneria bakeri*, number 13055 U. M. × about $\frac{1}{3}$

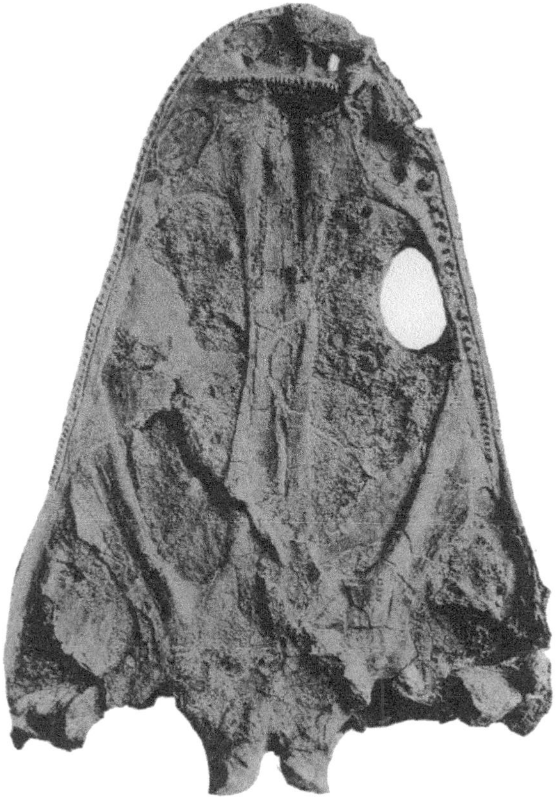

FIG. 2. Lower surface of skull shown in Figure 1. × about $\frac{1}{3}$

PLATE II

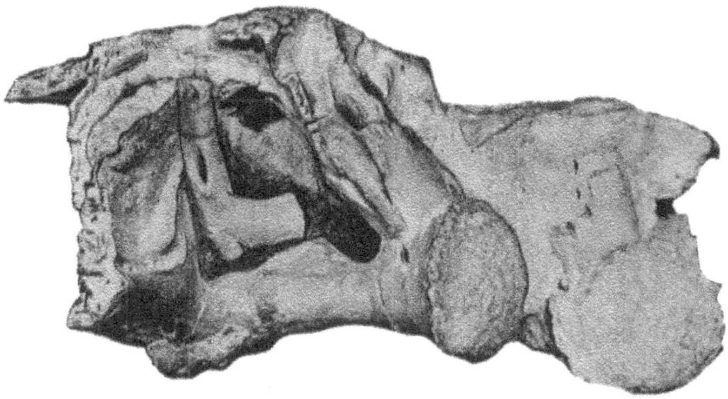

Fig. 1. Photograph of otic region in number 14154. Compare text Figure 5. × about $\frac{1}{3}$

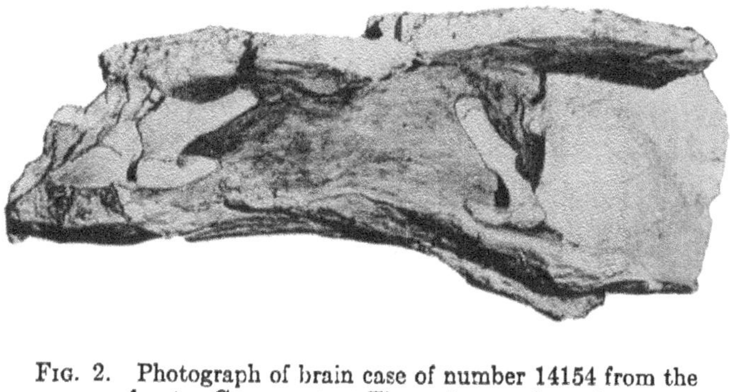

Fig. 2. Photograph of brain case of number 14154 from the front. Compare text Figure 6. × about $\frac{1}{3}$

PLATE III

Fig. 1. Interclavicle, number 13027 U. M. × about $\frac{2}{3}$

Fig. 2. Interclavicle, number 13029 U. M. × about $\frac{2}{3}$

VOL. III, No. 12, pp. 207–232 (7 pls., 9 text figs.) DECEMBER 18, 1931

ON *CALLIXYLON NEWBERRYI* (DAWSON) ELKINS ET WIELAND

By CHESTER A. ARNOLD

IN THE literature on the geology of the states of Indiana, Kentucky, Ohio and Michigan mention is frequently made of the occurrence of petrified wood in the black shales which are commonly assigned to the Upper Devonian period. Although this wood has been known for many years, comparatively few attempts have been made to study it anatomically or to determine whether more than one form is present.

HISTORICAL

Probably the earliest account of wood in the black shales is a paper read before the fourth New Haven meeting of the American Association for the Advancement of Science in 1850. George C. Schaeffer (1850), professor of natural philosophy and chemistry at Centre College, Danville, Kentucky, described fossilized logs that rested upon the black shale between Lebanon and the "Knobs." His account is interesting not only because it is one of the oldest records of fossil wood in this country, but also because the description is sufficiently clear for the material to be recognized as Callixylon.[1] Part of his account is as follows:

> The medullary rays are numerous and distinct, and the whole appearance of the wood not unlike that of a beech or a sugar maple. No rings of annual growth can be found. The diameter of some of the trunks must have been at least two or three feet.
>
> This tree is distinguished from Araucaria by having three, four, or more

[1] Callixylon is characterized by the presence of mesarch primary wood bundles at the margin of the pith and close to or in actual contact with the secondary wood in which the pits on the radial walls of the tracheids are arranged in radially aligned clusters; the spaces on the walls between the groups are unpitted (Zalessky, 1911).

dots side by side on each fiber, and by having the dots in patches, not uniformly distributed.

The dots themselves seem to be peculiar; under a low power, they seem to consist of a covering dehiscent from the center. But by careful management and higher powers, they seem to consist of cells not unlike those forming the stomates of plants, only more in number, — about four to each dot, and different in shape.

He says, further, that it may be a species allied to Voltzia.

A few years later Newberry secured from the septarian concretions in the black shale of Ohio some silicified wood fragments which he submitted to Sir William Dawson. Dawson (1871) described this material as *Dadoxylon Newberryi*. Later Newberry (1889) described the material and stated that he had found it at Delaware, in Huron County, and in southern Ohio, but he was not explicit in stating just where the material came from which was submitted to Dawson. It is probable that Dawson's description was based upon material from more than one locality, although sections designated as types were recently found among the collections of the Peter Redpath Museum at McGill University. Rominger (1873–76) mentions similar material in the black shales of northern Michigan, which he considered as being of Genesee age, and Knowlton (1889) and Penhallow (1900) include this species in systematic accounts of the fossil woods of America. However, their contributions to our knowledge of this wood are comparatively slight.

In 1914 Elkins and Wieland (1914) described a specimen from the black shales as *Callixylon Oweni*. Since this account states that the material was found by Dr. Wieland in the Yale collection, its Indiana source is probable, but it might have come from any one of several localities. They also refer Dawson's *Dadoxylon Newberryi* (*Cordaites Newberryi* of Knowlton, 1889) to Callixylon, although the resemblance was pointed out previously by Zalessky (1911) when he instituted the genus, and later by White (Prosser, 1912, p. 520). Aside from mere references to one or both of these two species in subsequent literature, no further contributions to our knowledge of the petrified wood in the black shale have since been made.

AGE OF THE BLACK SHALE

It is outside the scope of this account to deal at length with the age of the black shales, which has been a matter of considerable controversy. The generally accepted Upper Devonian age for these shales has been challenged by several competent authorities within recent years. Swartz (1929) has shown that the upper part of the Chattanooga shale in northeastern Tennessee and adjacent Virginia is of lower Mississippian age, whereas the thin basal portion is Devonian. Ulrich (1926) places the Antrim shale similarly, with its basal member as Upper Devonian and the upper portion, which includes the fossil plant remains, in the Mississippian. He also includes the New Albany shale in the Mississippian (1911), although no one has ever attempted to show that this formation is divisible into the two parts.

Though the plant remains of the black shales might be indicative of Upper Devonian age, the evidence derived from them is not conclusive, since the vertical range of Callixylon is not known. However, the occurrence of this one species, both the petrified wood and the branch impressions, over so wide a territory affords strong evidence that the shales known as the Antrim, Huron, Ohio, New Albany and Chattanooga are one and the same formation and are identical in age.

LOCALITIES AND OCCURRENCE

During the recent investigation the black shales were examined in southern Indiana; in the vicinity of Lebanon, Kentucky; in southern Ohio; and in the vicinity of Columbus and Delaware of the same state. No additional material was found in Ohio or Kentucky, although about a year previous to the inauguration of the search an excellently preserved piece of wood was secured from the Chattanooga shale at the mouth of Forbush Creek near Mill Springs, Kentucky. The best localities were found in Scott and Clark counties, Indiana, and material is abundant, though widely scattered, throughout the New Albany shale horizon in this part of the state.

Since the New Albany shale is quite uniform throughout its

entire depth in Indiana, it is impossible to associate the petrified wood with any particular level. It is quite probable that it extends to the top of the horizon, because at the locality at Henryville (shortly to be mentioned) the rocks overlying the New Albany shale outcrop only a short distance above the shale where numerous specimens occur. The field work consisted of an examination of the exposures of this shale wherever they could be found. Specimens are most frequently found as loose débris, but occasionally they occur *in situ*. In most quarries the petrified logs are removed as impediments as soon as they are discovered.

The Indiana specimens are quite variable in size. The largest specimen encountered during the course of the present study is nine feet long, three feet broad at the base, and tapers to about eighteen inches at the top. The expanded basal portion shows the attachment of two and possibly three large roots. This specimen, therefore, is the lower portion of a large tree (Plate I). It was found lying prostrate and half buried in the shale in a quarry two and one-half miles east of Vienna in Scott County. Segments of other logs, two and three feet in diameter, were observed in the same quarry.

Numerous specimens were also discovered in a small valley about five miles northeast of Henryville, Clark County. Though none of them were found *in situ*, the shale is extensively exposed in that locality and the tributaries of the small creek coming down the steep sides of the valley had been instrumental in dislodging many specimens, which had rolled down into the creek below.

To judge from the size of the trunks and the rarity of other fossil plant material (except Sporangites) in the New Albany shale, it appears that these logs were driftwood which had floated to its present location. On none of the logs are branches visible, but buried branch bases are revealed if the logs are broken open. Specimens found *in situ* are covered with a thin layer of coal, which apparently was derived from the carbonization of the bark layer. In no specimen is any of the bark petrified.

A series of ridges which extend both lengthwise and crosswise give to the exterior of these logs a striking appearance. These ridges are of varying length; some of them may be but a few inches long, but others may extend for half the circumference. The distance between them is also variable, but most of them are from two to six inches apart. The elevation of the ridges varies, which gives the surface a knotty aspect. The cross ridges give to the edge of a specimen a scalloped appearance, which is somewhat exaggerated if the specimen is slightly flattened. The exterior weathers to a light brown, whereas the interior is a lignitic black.

These ridges appear to be associated with quartz-filled cracks which seem to have their origin at the places of attachment of the branches. Quartz-filled cracks are abundant in the vicinity of the buried branch bases and it seems that the crystallization of the quartz in these areas caused a cracking and subsequent bulging along the lines of the ridges.

IMPRESSIONS OF THE STEM

Throughout the black shales in the region under consideration long, jointed impressions are frequently encountered, to which Dawson (1871) applied the name *Calamites inornatus*. He gives the following diagnosis:

> Ribs continuous, as in C. transitionis, but flat and broad, the breadth of each being one quarter of an inch in a stem four inches in diameter. Nodes distant, prominent in the flattened stems, owing to their greater density compared with the internodes.

He gives the distribution of this form as Cayuga Lake, New York; Kettle Point, Ontario; and Gaspé, Quebec. In 1911 Kindle found a specimen in the Huron shale near Milan, Ohio (Prosser, 1912, p. 522). White (quoted by Prosser, *ibid.*) compares this specimen to Nathorst's genus *Pseudobornia*.

During the course of the present investigation specimens of this form were found in practically every locality in which the shale was examined and they are very abundant in the Antrim shale at Alpena, Michigan. These specimens have a width varying from three fourths of an inch to six inches and are straight or very slightly curved. The distance between the so-called nodes varies slightly, but it is usually about two inches. The nodal lines, instead of extending entirely around the stems, as in Calamites, consist, especially in the larger stems, of segments which extend only part way. A segment adjacent to another, then, will not lie on the same circumference, but slightly above it, and with its end projecting back beyond the end of the segment just below (Plate II, Fig. 3). The ribs mentioned by Dawson do not show on all specimens, but where they do (Plate II, Fig. 1) they are clearly not ribs of the Calamites type as Dawson interpreted them, but are probably due to a longitudinal buckling of the surface when the stem was compressed.

Some of the specimens from the Antrim shale bear short branches which depart from the main stem at an angle of about 45 degrees (Plate II, Fig. 2). All of them are covered with a thin layer of coal, which in one specimen yielded some small fragments of silicified wood, thus showing that silicification had commenced before coalification was complete.[2] These small fragments were sufficient, however, to reveal the Callixylon type of pitting and rays of the *C. Newberryi* type. This, then, shows that these calamite-like structures are the impressions of the smaller branches of Callixylon, and without doubt the same as the large trunks from Indiana and Kentucky. The carbonized stem remains, which are common in the lower Portage of New York (Arnold, 1929, Plate II, Fig. 3) and which were interpreted as possible Callixylon remains, appear to be different.

PRESERVATION

All the Callixylon remains so far observed in the black shales are silicified and differ in this respect from the material from the Upper Devonian of New York, which, contrary to statements frequently made, is calcified. The silica is precipitated in the cell cavities and there is very little alteration of the cell walls. Small fragments of the wood can be desilicified without loss of cell structure. They can then be infiltrated with collodion, sectioned with the sliding microtome and stained with safranin in the same way as living wood. The large amount of organic matter that remains in this material renders it quite soft as compared with most silicified wood. It splits readily into thin, flat radial plates,

[2] These fragments were discovered by Professor H. H. Bartlett, to whom the author is indebted for permission to publish this information.

the surfaces of which present a woody appearance to the unaided eye. With the aid of a single lens the radial alignment of the pit clusters is distinguishable. The softness of this material and the facility with which it splits render the preparation of thin transverse sections exceedingly difficult. The best sections were prepared by Walton's cellulose peel method (Walton, 1928).[3]

ANATOMY

Pith

The pith of the main stem is large, varying from one to three centimeters in diameter. At places it is septate, but for the most

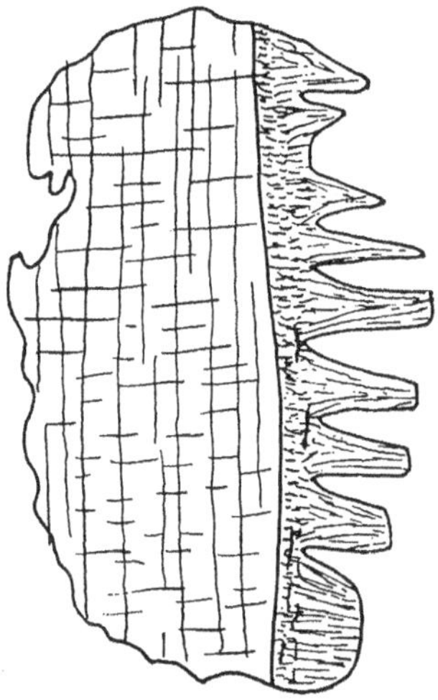

FIG. 1. Diagram of radial section, showing septation of pith. × 1

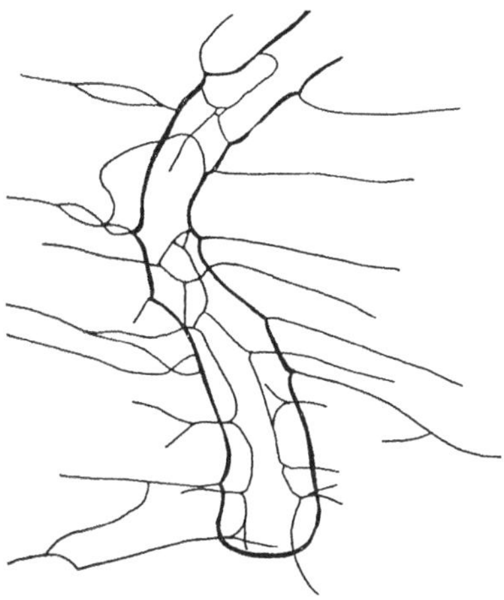

FIG. 2. Tracheid-like cell in pith, but it has failed to develop a secondary wall. × 100

part is continuous. In the septate portions the septa are regularly spaced (text Fig. 1), and the lacunae between them extend to

[3] The procedure with this material was as follows: surface ground smooth; treated for 30 seconds with full-strength commercial hydrofluoric acid; washed in water; immersed 1 minute in waste alcohol, then 1 minute in 95 per cent alcohol; rinsed with absolute alcohol; then, without being allowed to dry, etched surface covered with 5 per cent solution of pyroxylin dissolved in ether-alcohol mixture.

within three or four millimeters of the inner edge of the wood. The pith of the lateral branches is small, seldom having a diameter of more than five millimeters; at a distance of ten centimeters from the pith of the main trunk it has approximately the same diameter as at its point of origin.

The central portion of the pith appears to be loosely arranged and in most specimens it has disappeared. As the periphery of the pith is approached, it becomes slightly firmer and the cells become elongated in a radial direction. This feature is quite pronounced immediately around the primary bundles (Plate VII, Fig. 1).

Scattered throughout the pith, except in the central region, are numerous tracheids (Plate IV, Fig. 3). Except when near the outside, where there may be two or three in a group, these tracheids are isolated cells with closely wound spiral or scalariform markings. Occasional cells of the shape of tracheids have failed to develop a secondary wall (text Fig. 2).

Primary bundles

The exact number of bundles in any one section has not been determined because it has been impossible to secure a complete transverse section. To judge from the number contained in a sector, it is estimated that there are at least twenty in the main part of the stem, but they might range as high as thirty, since in many specimens they are irregularly spaced (Plate VII, Fig. 1).

Many of these bundles are separated from the secondary wood and are completely surrounded by pith tissue (Plate IV, Fig. 2). In only rare specimens is a strand separated from the secondary wood by a distance greater than its own diameter and usually the intervening pith zone is only one or two cells in extent. Other bundles may be partly separated, but still retain their connection with the secondary wood by a narrow isthmus of xylem. When these imbedded strands commence to bend outward previously to forming a trace, they do not come in contact with the secondary wood immediately, but remain surrounded by a zone of parenchymatous tissue even after they have reached the inner zone of the secondary wood.

In no other American species of Callixylon has this separation of the bundles been observed, although it is known in *C. Trifilievi* (Scott, 1912). In this species, however, only a few of the bundles are isolated and the majority of them are in contact with the secondary wood.

The continuation of the reparatory strand at the margin of the pith without the formation of a leaf gap is, so far as can be determined, identical with that of *C. Zalesskyi* and *C. erianum*. Also no connection can be observed between the isolated tracheids in the pith and the organized bundles at or near the margin.

In the lateral branches with smaller pith the primary bundles are correspondingly fewer, and most of them are in contact with the secondary wood (Plate IV, Fig. 1).

Secondary tissues

The secondary tissues show no great differences from those previously described for this or any other species of the genus. However, a few points worthy of consideration are noted.

Penhallow (1900), in diagnosing this species, says that there is no evidence of growth rings [4] (obscure or obsolete) in a radial extent of seventeen millimeters. This character is used by Elkins and Wieland (1914) as a distinguishing feature of *C. Oweni*, in which there are six growth rings in fifteen millimeters of radial extent. Examination of the type sections of these respective forms shows that the figures given above are correct for these particular specimens, but in larger specimens the spacing of the rings is so irregular that measurements taken from a small area are of no value. Growth rings were observed in practically all specimens examined, but they are usually absent from the first fifteen or twenty millimeters of secondary wood immediately surrounding the pith, and at other places they are absent for a greater distance. Generally they are spaced at intervals of three to seven millimeters, although the spacing is entirely too irregular to constitute a specific character. The radial extent of the "summer wood" is always small (Plate III), never more than two or three

[4] The term "growth ring" is used in this account in preference to the less accurate term "annual ring," which is more commonly used.

cells. Concentrically placed zones of crushed tissue may, upon casual examination, present a deceptive appearance of growth rings.

Because of the very fine preservation of the material, details of pit structure show in a striking manner. The pits are circular when not crowded and the crossed apertures are usually visible (Plate IV, Fig. 4). The more crowded pits may assume a hexagonal outline from mutual pressure. Most of the groups contain from six to fifteen pits, but variations may range from one to fifty.

The presence of "bars of Sanio" between the pit clusters was discussed in a former account (Arnold, 1929). These structures had not been previously recognized in this genus, although they are readily distinguishable in the type sections of *C. Oweni*. The authors of that species (Elkins and Wieland, 1914) mention a thickening of the tracheid walls and a "spindle shaped interspace" between the pit groups as visible in the tangential section, but farther on in the account the statement is made that the bars of Sanio are not present even in a rudimentary condition. This, of course, is a misinterpretation of these thickenings which, however, are quite different from the bars as they are commonly seen in most woods.

Considerable tangential pitting occurs in places. These tangential pits are small and range from 7.7 to 9.3 microns in diameter as compared with the radial pits, which range from 10 to 15 microns. They are round, rather distantly spaced, and are not grouped. They are common in the portions of the secondary wood adjacent to the pith and also in what appears to be the "summer wood," although this latter point is not established with finality. The radial pits in the same tracheids with the tangential pits are grouped in the usual fashion.

Rays

Penhallow (1900) describes the rays of this species as being of medium height, one- or two- or rarely three-seriate, 24 to 55 microns broad, and the oval or round cells all thin-walled. Examination of the type sections shows that his diagnosis is essentially correct, but lacking in detail because of the small amount of material upon which the species was based.

Examination of the new material shows much variation in height and seriation of the rays and in the shape and contents of the ray cells (Plate V, Figs. 1–2). Within a single specimen a considerable variation of structure exists. The majority of the rays are biseriate, at least in part. Many of them are entirely so and rays which are three- and four-seriate are not uncommon in some specimens (Plate V, Fig. 3). Rare instances were noted of rays five and six cells wide at the widest place. At first it was suspected that specimens with the very wide rays might represent distinct species, but prolonged examination showed that such features are merely variations which are likely to be found in all the material.

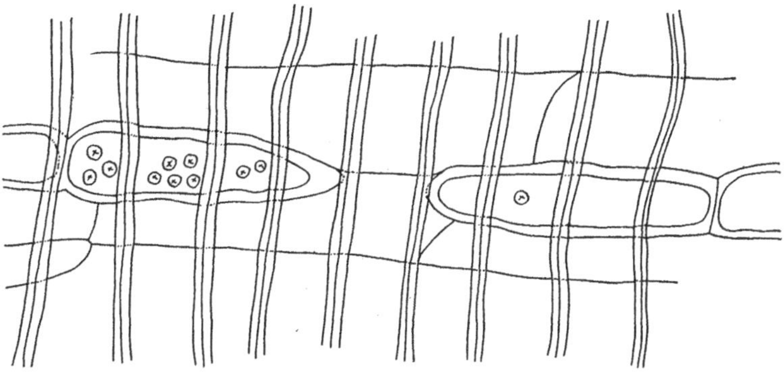

FIG. 3. Ray tracheids, showing rounding of ends of cells. × 300

Many of the ray cells contain black material. In some of them the material is collected at the ends, but others are full. This mass of blackish material has usually shriveled somewhat, thus leaving a space between it and the cell wall. This gives an exaggerated aspect to the slight narrowing of the cells at the ends, and this feature was probably observed by Penhallow when he said that the ends of ray cells are "conspicuously narrower." In some of the ray cells are black globular bodies (Plate VII, Fig. 3), though others contain a foamy substance. Penhallow refers to the material in the cells as starch and resin, but its chemical nature has never been determined.

Occasional ray tracheids are encountered. They occur as thick-walled cells bearing full-bordered pits of the same type, only small, that occur in the wood tracheids. These ray tracheids are frequently isolated or several may occur in a group (Plate VI, Fig. 2; text Figs. 3–4). They are widely scattered and in many of them none will appear in a radial section of considerable size. They are more common in some specimens than in others, but because of their sporadic occurrence it is never safe to assume their

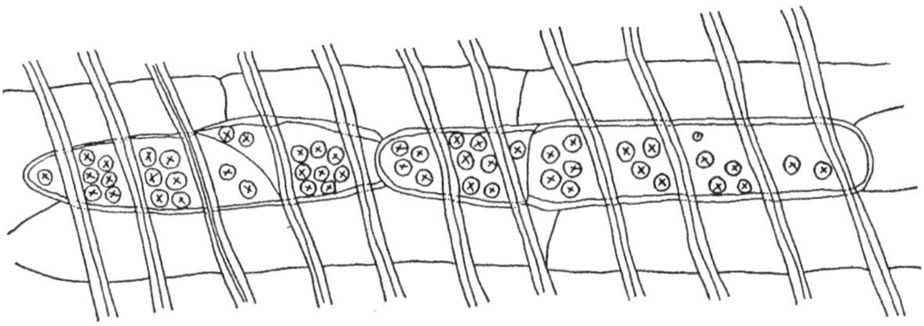

FIG. 4. Ray tracheids arranged in series and showing full-bordered pits on the side walls. × 300

absence except after a prolonged examination of considerable material. In one specimen in which they were not observed in the main trunk, they were found in abundance in a lateral branch. Many of these structures are rounded or much narrowed at the ends, so that adjacent ray tracheids may touch by only a small portion of their ends. Some of them are very short, not longer than high.

The scattered ray tracheids and the fact that they were overlooked by previous authors is responsible for the statements in connection with *Callixylon Newberryi* and *C. Oweni* that the ray cells of these species are thin-walled. No ray tracheids can be found in the type sections of the former and they cannot be recognized with certainty in the latter.

Fungous mycelium is abundant in the tracheids and ray cells of some of the material (Plate V, Fig. 4).

THE IDENTITY OF *CALLIXYLON NEWBERRYI* AND *C. OWENI*

It was only after a considerable number of specimens of wood from the black shale had been studied that it became apparent that such descriptive characters as the variations of the diameter of the tracheids and the number of pits per group, the contents of the ray cells, presence or absence of growth rings in a small section, etc., do not indicate different species. If such characters are to be used it would become necessary to establish a separate species for almost every specimen examined.

Examination of the type sections of both *Callixylon Newberryi* and *C. Oweni*[5] made it obvious that the sections of *C. Newberryi* were too small and that the material from which they had been prepared was too badly compressed for them to be satisfactory in identifying unknown material. Of the four sections representing the type specimen one is an oblique cut which is practically useless. Of the others the transverse section measures 3 × 12 millimeters, the radial 7 × 14 millimeters, and the tangential $5\frac{1}{2}$ × 13 millimeters (text Fig. 5). The longest dimension of the transverse section is in the radial direction. Small areas of them can be studied with some satisfaction, but they give no idea of the amount of variation which occurs in these woods and consequently descriptions based upon them tend to stress features which are not so applicable to the material as a whole. Penhallow's description, however, is about as accurate and as complete as the material would permit.

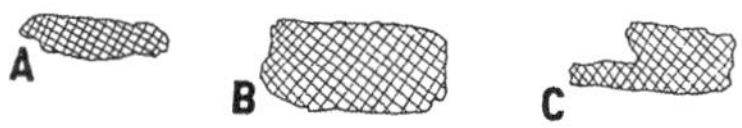

FIG. 5. Diagram showing exact size of the type sections of *C. Newberryi*. *A*, transverse; *B*, radial; *C*, tangential

The sections of *C. Oweni* are much better prepared and are from better material, so that the description is more satisfactory. The authors of that species, however, did not have access to the types of *C. Newberryi*, and hence Penhallow's meager description was their only basis for comparison. As a consequence,

[5] The type sections of *C. Newberryi* were very kindly lent to the author by Dr. T. H. Clark of the Peter Redpath Museum. Dr. Wieland has very generously donated the type sections of *C. Oweni* to the Museum of Paleontology of the University of Michigan.

the differences cited by Dr. Wieland (Elkins and Wieland, 1914, p. 75) are not so obvious when the specimens are compared, and they do not show characters justifying separation into two species.

Dr. Wieland states that the growth rings might serve as a basis for separation, but the examination of more material shows that the spacing of the growth rings is not a constant feature and it is easy to locate areas which show no rings in blocks much larger than that from which the sections of *C. Oweni* were made. Dr. Wieland suggests this possibility when he says: "It is also possible that the specimen studied by Penhallow may have been taken from an area between two such successive rings of unusual width. This is merely a conjecture. The far greater chance is that both specimens are average trunk wood and do differ by presence or absence of growth rings." The present author is inclined to the opinion that the first alternative is the correct one, and that Penhallow's section was made from one of the larger areas which contained no growth zones.

It is unfortunate that the poorer specimen must remain the type of the species when the specimen described as *C. Oweni* is more nearly representative of the material as a whole, but at the same time the association of Newberry's name with it is appropriate, since he first made it an object of study.

After it had become apparent that the material represented by the types of *C. Newberryi* and *C. Oweni* belongs to one species, the question immediately arose concerning the specific relationships of the newly discovered material. The most variable features in these woods are the width and the seriation of the rays, and at first it seemed possible that the specimens with the very broad rays might represent distinct species. Although in many of the specimens, in fact the majority of them, the rays appear narrower (Plate V, Fig. 2), there are no accompanying differences in primary wood structure or in surface features. Also, the narrower rays are invariably associated with a slight amount of flattening of the wood, which caused a slight buckling of the horizontal walls of the ray cells, but which did not affect the structures with stronger walls. This flattening gives the rays a deceptive appearance upon casual observation (Plate VII, Fig. 2).

Furthermore, since such features as presence or absence of growth rings, contents of the ray cells, and variations in size of the pit clusters are not diagnostic, the present author has concluded that the material so far studied from the black shales represents but a single species to which, according to the rule of priority, the name CALLIXYLON NEWBERRYI (Dawson) Elkins et Wieland must be applied, and that whatever differences do occur are within the limits of this one species.

When one is dealing with species, it is customary to emphasize differences rather than similarities. It is indeed true that with but a limited amount of material at hand for study certain differences will be noted. Also, when previous descriptions must be relied upon entirely, as was the case when *C. Oweni* was described, these differences are likely to be overemphasized. If a quantity of material is available for first-hand comparison these apparent differences will seem less conspicuous and what at first seems to be a possible specific character finally shows only as a variation which might distinguish it from another individual, but which is still well within the limits of the species.

It is only to be expected that wood with structure as complicated as that of Callixylon should vary considerably, but if the differences shown are all within the limits of one species, a momentary reflection on the pronounced similarities in all the material is sufficient to force the conclusion that a unified group is represented. Of course there still remains the possibility that we are dealing here with a group of several species which differ mostly with respect to foliar or floral characters, but in that event specific considerations must remain until such structures are discovered.

Before discussing the affinities between *C. Newberryi* and other closely related forms the revised diagnosis will be given.

CALLIXYLON NEWBERRYI (Dawson) Elkins et Wieland

Fossil coniferous wood, Schaeffer, Proc. Am. Assn. Adv. Sci. 1850.
Calamites inornatus, Dawson, Journ. Geol. Soc. London. 1859.
Dadoxylon Newberryi, Dawson, Canada Geol. Surv. 1871.
Cordaites Newberryi, Knowl. Proc. U. S. Nat. Mus. 1889.
Pseudobornia inornatus, White, Geol. Surv. Ohio. 1912.
Callixylon Oweni, E. & W. Am. Journ. Sci. 1914.
Callixylon Newberryi, E. & W. Am. Journ. Sci. 1914.

Arborescent; trunks straight and slender, probably reaching a height of forty feet or more; branches small and straight, caducous; outer decorticated surface ornamented with cross and longitudinal ridges of varying lengths and varying distances apart. Secondary wood gymnospermous, composed entirely of tracheids and rays; tracheids variable in size and shape and from 25 to 75 microns in diameter; wall between the lumina of two adjacent tracheids 10 to 15 microns thick; bordered pits on radial walls in groups of variable sizes, but generally numbering from 6 to 15; opposite terminals of the pit aperture at right angles; tangential pits smaller than radial pits, not grouped, not present in all tracheids; rays variable in height and width, characteristically biseriate wholly or in part, but frequently three- to many-seriate or wholly uniseriate; ray cells clear or wholly or partly filled with black or foamy material, sometimes slightly contracted at the ends; ray tracheids scattered, frequently isolated, sporadic; growth rings present, but not strongly developed; "summer wood" 2 to 4 cells in extent, rings irregularly spaced and sometimes absent over considerable radial extent. Primary wood bundles mesarch, in contact with secondary wood or separated from it by a narrow zone of pith; variable in size; 20 to 30 in the main stem. Pith of main stem 1 to 3 centimeters in diameter, but smaller in lateral branches; imperfectly septate; cells thin-walled and radially elongated when adjacent to primary wood; pith tracheids isolated and scattered.

COMPARISONS WITH OTHER FORMS

The closest relative of *C. Newberryi*, so far as our knowledge of fossil plants permits us to judge, is *Pitys antiqua* Witham. This species from the lower Carboniferous rocks of Scotland (Scott, 1902) has a large imperfectly discoid pith, which is sometimes as much as two inches in diameter. Around the periphery, and for the most part imbedded in it, are numerous mesarch primary xylem strands (text Fig. 6), sometimes forty or fifty (text Fig. 7). The secondary wood has the Dadoxylon structure, with two or more rows of hexagonal bordered pits on the radial walls of the tracheids. The rays are numerous and large, frequently being as much as 6 cells wide, but uniseriate rays are also common (text Fig. 8).

The two points of similarity between Callixylon and *P. antiqua* which are here emphasized are the broad rays and the imbedded primary xylem strands in the pith. *C. Trifilievi* differs from *P. antiqua* by having narrow rays, but the biseriate rays are char-

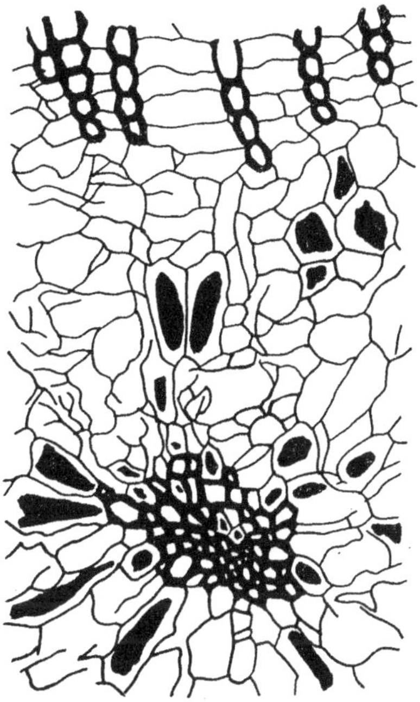

FIG. 6. Primary wood bundle of *Pitys antiqua* Witham, which is very similar to bundle shown in Plate IV, Figure 2. After Scott, 1902. × about 70

acteristic of *C. Newberryi*. Then the homology is further strengthened by the fact that many of the rays of *C. Newberryi* are 3 and 4 cells wide and, rarely, as much as 5 or 6.

In *C. Newberryi* more of the bundles (probably half of them) are in contact with the secondary wood than in *P. antiqua*, but those in the pith of both species so closely resemble each other (text Fig. 6) as to be practically indistinguishable. In this respect *C. Newberryi* resembles *P. antiqua* even more than it does the other known species of Callixylon, although this separation has been observed to exist to a limited extent in *C. Trifilievi*.

Also, in discussions of *C. Trifilievi* considerable stress has usually been placed upon the narrowing of the secondary wood segments

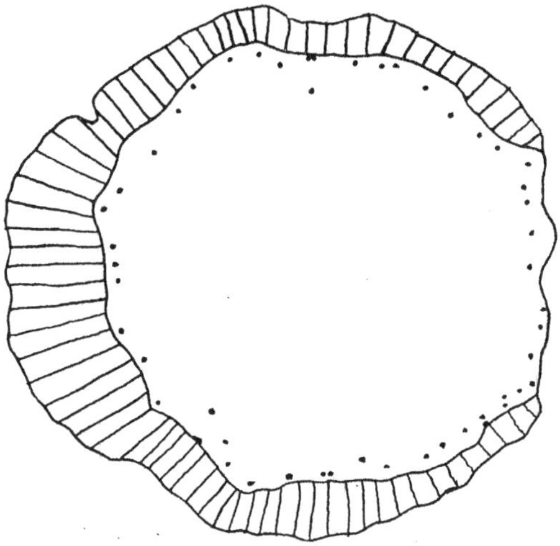

FIG. 7. Diagram of cross-section of pith of *Pitys antiqua* Witham, showing size of pith and arrangement of primary bundles. After Scott, 1902. × 1

into wedges as they approach the pith. This narrowing is not so pronounced in *C. Newberryi* and the inner edge of the secondary wood between the primary bundles presents a fairly straight line, which is intercepted by numerous rays as they join the pith.

Recent investigations of the various species of Pitys show that there are bundles scattered throughout the extent of the pith (Scott, 1923). Such structures are not present in *C. Newberryi*, although their counterparts might be recognized in the isolated pith tracheids previously mentioned.

Considering *C. Newberryi* and *P. antiqua* in all their aspects, they are certainly close relatives, the outstanding difference being the pitting on the radial walls. Were it not for this one feature it seems doubtful whether they could be regarded as belonging to separate genera, and the only other difference which might warrant serious consideration is the fact that in *C. Newberryi* more of the primary bundles are in contact with the secondary wood.

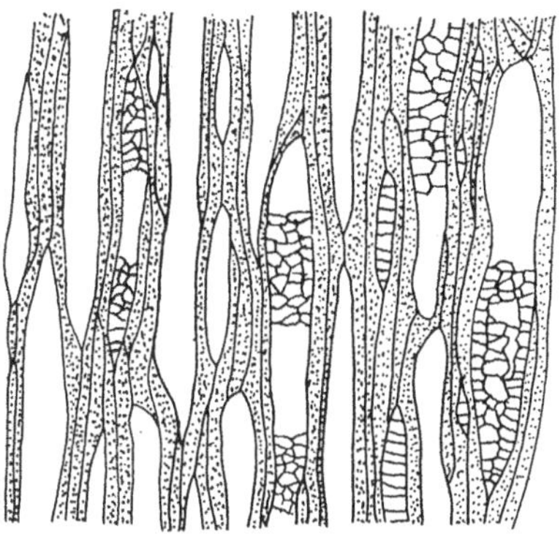

FIG. 8. Tangential section of *Pitys antiqua* Witham, showing broad rays very similar to that shown in Plate V, Figure 3. × about 20

But this feature is so inconstant and variable that it would indicate nothing more than a specific difference.

Of all the known species of Callixylon *C. Newberryi* seems to be the closest to Pitys. All the other species have narrow rays. A possible exception is *C. mentethense* (Arnold, 1929), but this form is imperfectly known and might be the same as *C. Newberryi*. Biseriation is frequent in *C. Zalesskyi* and *C. erianum* (Arnold, 1929), but it is seldom more than partial and then it rarely results in any appreciable widening of the rays in the biseriated portions.

Archeopitys Eastmanii Scott & Jeffrey is quite similar to Pitys. It was originally distinguished from Pitys by the occurrence of bundles throughout the pith, but these have been found in Pitys also and it is possible that these two forms are congeneric. A point of interest in connection with *A. Eastmanii*, however, is that it came from the base of the Waverley shale, which immediately overlies the New Albany black shale in central Kentucky. The two horizons are separated by an unconformity which, should the New Albany shale prove to be of Mississippian age, is not so great as was formerly supposed.

Though the results of the present investigation serve to link Callixylon more closely with the Pityeae, it is likewise to be distinguished more than has been previously recognized from forms representing the true Cordaiteae. The tall, slender, sparsely branched trunks of *C. Newberryi* are quite different from those of Cordaites as we have come to recognize it from the restorations by Grand'Eury (1877). True cordaitean trees, though tall and slender, show more diffuse branching. Also the wood structure, though still not closely resembling that of the Pteridospermae, is farther from Cordaites than was formerly supposed. It would seem, then, in view of the close resemblance between Pitys and Callixylon, along with the enormous development of these forms during pre-coal measure times, that the Pityeae might deserve ordinal rank.

PROBABLE HABIT

The recent identification of impressions such as have commonly been referred to *Calamites inornatus* Dawson (*Pseudobornia inornatus* White) as the branches and smaller trunks of *C. Newberryi* may be instrumental in throwing still further light on the habit of the plant. On one specimen from the Antrim shale at Alpena there are two blunt projections somewhat less than an inch in length (Plate VI, Fig. 1). It is possible that these are the leaves of *C. Newberryi* because of their similarity to the leaves of *Pitys Dayi* Gordon (Scott, 1923), which are described as short, fleshy structures without a true lamina. More material must be examined, however, before their identity can be established. They are situated at the nodes, where leaves would be expected to occur, and have left a thin carbonaceous residue.

As previously mentioned, the largest trunk discovered during the course of the present study was nine feet long. From a width of three feet at the base it narrows to one and one-half feet at the top. In the exhibition hall of the United States National Museum there is a similar trunk from the New Albany shale of Lebanon, Kentucky, which was identified as *C. Newberryi* by its surface features. It is twenty feet high and tapers from about eighteen inches at the base to about four inches at the top. It is perfectly straight, and with the exception of a slight irregularity on one side near the top, there is nothing that could be interpreted as exposed branch bases. Superimposing, now, this specimen from Kentucky upon our specimen from the same horizon in Indiana, we would have a trunk twenty-nine feet high (text Fig. 9). With this as a basis for estimation and assuming that the branch-bearing tip extended the trunk a few feet farther, it is plausible to picture the Indiana trunk as the relic of a tree which once rose to a height of forty feet. Specimens having an even greater diameter are known. A trunk in the Woodford chert of Oklahoma is said to be five feet in diameter at some distance above the base, so that calculations similar to those above would give to this Oklahoma tree a minimum height of sixty feet. The absence of preserved branches on the surface of these trunks, even on the smaller ones, indicates their caducous character. It is possible that these branches were borne near the apex of the straight slender trunks, but failing to remain attached for any length of time they never attained a diameter of more than a few centimeters. Since the branches died and

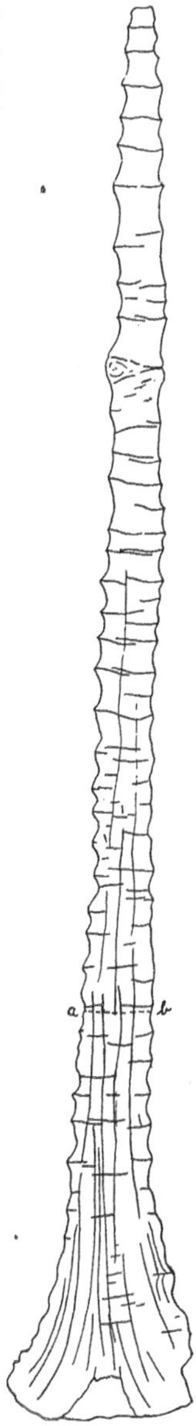

Fig. 9. Partial restoration of trunk of *C. Newberryi*. Lower portion, below dotted line *ab*, from specimen shown in Plate I; upper portion from specimen in the United States National Museum. Total height 29 feet

fell away before the trunk reached any considerable size, their bases are not exposed on the surface of the lower portion of the tree.

CONCLUSION

As stated in the opening pages of this account, the exact age of the black shales that contain the silicified remains of *C. Newberryi* still remains a question. The genus occurs in the Genesee shale, the Genundewa limestone and the lower Portage rocks of central and western New York (Arnold, 1929), in the Upper Devonian of the Donetz Basin of south Russia, and questionable remains of it are known in the Upper Devonian of Germany (Kräusel and Weyland, 1929). Hence Callixylon has so far been treated as a genus of Devonian gymnosperms which existed solely within that period. Since Callixylon has never been reported from any undisputed lower Carboniferous horizon, it can be said that *C. Newberryi*, considering it only in the light of what is already known concerning the distribution of the genus, is indicative of the Devonian age for the black shales. Should *C. mentethense* from the Genundewa limestone be conspecific with *C. Newberryi*, this interpretation would be strengthened.

Though Callixylon may be recognized for the time being as a Devonian genus, it is necessary to admit the possibility of its having extended into early Carboniferous times before it became extinct. At the same time there is considerable negative evidence that it failed to survive very far into Carboniferous times.

Interest in Devonian vegetation has been considerably increased within the past decade through several important discoveries, of which those of the Rhynie chert are probably the most outstanding (Kidston and Lang, 1917–21). Although fossils having the dimensions of trees were previously known from Devonian rocks (White, 1907), the discovery of Eospermatopteris at Gilboa, New York (Goldring, 1924, 1927), shows the size attained by some of these early pre-coal measure plants. Also of significance was the discovery of the Elberfeld flora in Germany (Kräusel and Weyland, 1929), which has been instrumental in connecting the early

Devonian flora with that of later Devonian times in such a way as to show that the two floras meet and overlap instead of being separate, as was formerly supposed (Scott, 1926).

There is a recent account of fossil trees in the Elkins and the Valleyhead sandstones (of Chemung age) in the Tygart Valley of West Virginia (Reger, 1928), to which the present author feels strongly inclined to take exception. A recent examination (which of necessity was brief) of these specimens, together with the original description of them, leaves one strongly of the opinion that there is insufficient evidence for the interpretation of these structures as trees and that they must be accounted for in some other way. Though these early times are famous for their bizarre forms of vegetation, it is at the same time unwise to postulate such structures as are present in the Elkins sandstone as trees, when so many essential diagnostic features are obscure or entirely lacking.

Should the Mississippian age of the black shales become finally established, the contained plant fossils would lose none of their evolutionary significance. The fact that these plants are congeneric with plants from Upper Devonian horizons in other localities, and their close affinity with Pitys and Archeopitys from the lower Carboniferous of both this and other countries, shows a continuity of the family which commences low down in the Upper Devonian and extends well up into the Carboniferous period.

These recent studies of Callixylon have served to emphasize two points which, it is believed, are significant. The first is the wide distribution of the genus. As it is at present known in the United States, it extends from 35 degrees north latitude in Oklahoma to about 46 degrees at Alpena, Michigan. In Europe, it extends to approximately 50 degrees in Russia. Whether these are its extreme north and south limits is doubtful. The second point is the great abundance of Callixylon remains and the relative scarcity of other forms of vegetation in the Callixylon-bearing horizons. The almost complete dominance of this genus in the New Albany, the Ohio and the Antrim shales is a remarkable fact and the same thing is true of the Genesee and Portage formations of New York. These facts, then, indicate the past occurrence of great and widespread forests of Callixylon during pre-coal measure times which, though probably lacking in such diversity of forms as constituted the coal measure forests, were fully as large and as widespread.

LITERATURE CITED

1. ARNOLD, C. A. On the Radial Pitting in Callixylon. Am. Journ. Bot., Vol. 16, pp. 391–393 (1929).
2. —— The Genus Callixylon from the Upper Devonian of Central and Western New York. Pap. Mich. Acad. Sci., Arts and Letters, Vol. 11, pp. 1–50 (1929). Published 1930.
3. DAWSON, J. W. The Fossil Plants of the Devonian and Upper Silurian Formations of Canada. Canada Geol. Surv. 1871 (1872).
4. ELKINS, M. G., AND WIELAND, G. R. Cordaitean Wood from the Indiana Black Shale. Am. Journ. Sci., Vol. 188, pp. 68–78 (1914).
5. GOLDRING, W. The Upper Devonian Forest of Seed Ferns in Eastern New York. New York State Mus. Bull., No. 251, pp. 50–71 (1924).
6. —— The Oldest Known Petrified Forest. Sci. Monthly, Vol. 24, pp. 514–529 (1927).
7. GRAND' EURY, M. Mémoire sur la flore carbonifère du département de la Loire et du centre de la France. Mém. a l'Acad. des Sci. de l'Inst. de France, Vol. 24 (Paris, 1877).
8. KIDSTON, R., AND LANG, W. H. On Old Red Sandstone Plants Showing Structure, from the Rhynie Chert Bed, Aberdeenshire, I–IV. Trans. Roy. Soc. Edinburgh, Vols. 51–52 (1917–21).
9. KNOWLTON, F. H. A Revision of the Genus Araucarioxylon of Kraus. Proc. U. S. Nat. Mus., Vol. 12, pp. 601–617 (1889).
10. KRÄUSEL, R., UND WEYLAND, H. Beiträge zur Kenntnis der Devonflora, III. Abhandl. der senckenbergischen naturf. Gesell., Vol. 41, pp. 315–360 (1929).
11. NEWBERRY, J. S. Devonian Plants from Ohio. Journ. Cincinnati Soc. Nat. Hist., Vol. 12, pp. 53–55 (1889).
12. PENHALLOW, D. P. Notes on American Species of Dadoxylon. Trans. Roy. Soc. Canada, Vol. 6, pp. 51–97 (1900).
13. PROSSER, C. S. The Devonian and Mississippian Formations of Northeastern Ohio. Geol. Surv. Ohio, Ser. 4, Vol. 15 (1912).
14. REGER, D. B. The Tygart Valley Devonian Trees of West Virginia. Am. Journ. Sci., Vol. 215, pp. 49–57 (1928).
15. ROMINGER, C. Black Shales of Michigan. Geol. Surv. Mich. Vol. 3, pp. 63–67 (1873–76).
16. SCHAEFFER, G. C. Fossil Coniferous Wood from the Lower Devonian Strata, Lebanon, Marion County, Kentucky. Proc. Am. Assn. Adv. Sci. (4th New Haven Meeting), 1850, pp. 193–194 (1850).

17. SCOTT, D. H. On the Primary Structure of Certain Paleozoic Stems with the Dadoxylon Type of Wood. Trans. Roy. Soc. Edinburgh, Vol. 40, pp. 331–373 (1902).

18. —— The Structure of Mesoxylon Lomaxii and M. poroxyloides. Ann. Bot., Vol. 26, pp. 1011–1030 (1912).

19. —— Studies in Fossil Botany, II (London, 1923).

20. —— New Discoveries in the Middle Devonian Flora of Germany. New Phytol., Vol. 25, pp. 373–379 (1926).

21. —— AND JEFFREY, E. C. On Fossil Plants Showing Structure from the Base of the Waverley Shale of Kentucky. Phil. Trans. Roy. Soc. London, B, Vol. 205, pp. 315–373 (1914).

22. SWARTZ, J. H. The Age and Stratigraphy of the Chattanooga Shale in Northeastern Tennessee. Am. Journ. Sci., Vol. 217, pp. 431–448 (1929).

23. —— The Devonio-Mississippian Boundary in the Southeastern United States. Science, Vol. 70, p. 609 (1929).

24. ULRICH, E. O. Revision of the Paleozoic Systems of North America. Bull. U. S. Geol. Surv., Vol. 22, pp. 281–680 (1911).

25. —— Black Shale Problem in Michigan. Paper read before the eighteenth annual meeting of Paleont. Soc., Madison, 1926.

26. WALTON, J. A New Method of Preparing Sections of Fossil Plants Contained in Coal Balls or in Other Types of Petrifactions. Nature, Vol. 122, p. 571 (1928).

27. WHITE, D. A Remarkable Fossil Tree Trunk from the Middle Devonic of New York. New York State Mus. Bull., No. 107, pp. 327–340 (1907).

28. ZALESSKY, M. D. Étude sur l'anatomie du Dadoxylon Tchihatcheffi Goeppert sp. Mém. du Comité Géol. de Russe, N. S., Vol. 68, p. 29 (1911).

PLATE I

Callixylon Newberryi (Daws.) E. & W. Trunk on exhibition in the University Museums, University of Michigan. From New Albany shale, Scott County, Indiana. Height 9 feet

PLATE II

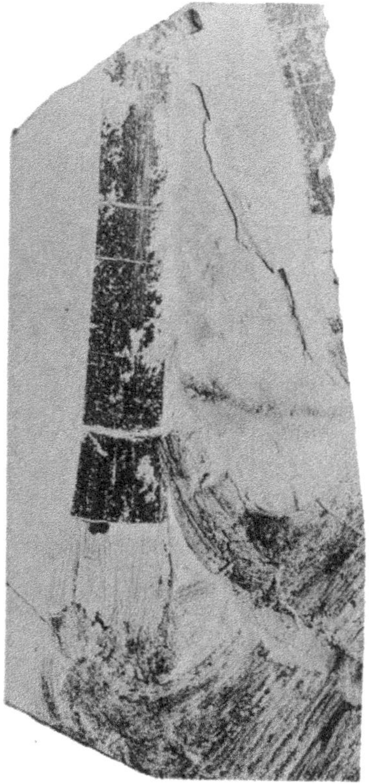

Fig. 1. Impression of branch. $\times \frac{1}{3}$

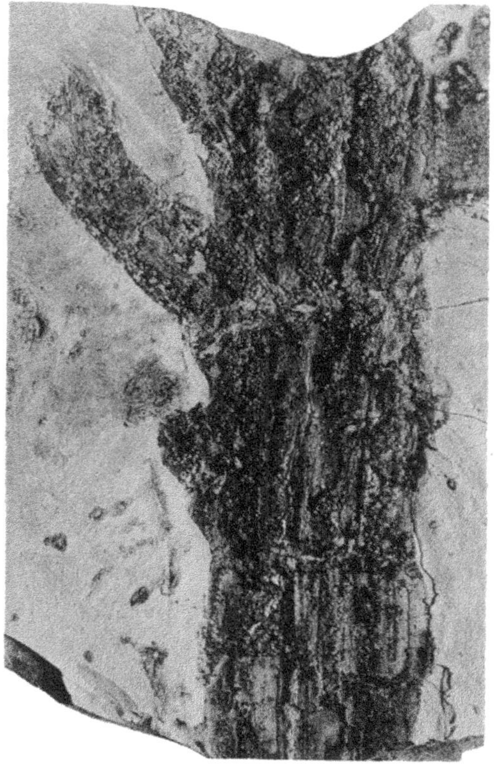

Fig. 2. Impression of branch or small trunk bearing smaller lateral branches. $\times \frac{1}{2}$

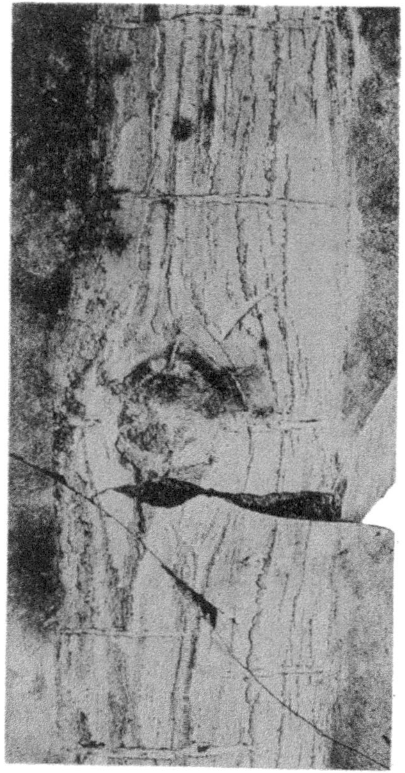

Fig. 3. Impression of branch or small trunk showing nodal lines and imprint of branch. $\times \frac{1}{2}$

Callixylon Newberryi (Daws.) E. & W. Antrim shale, Alpena, Michigan

PLATE III

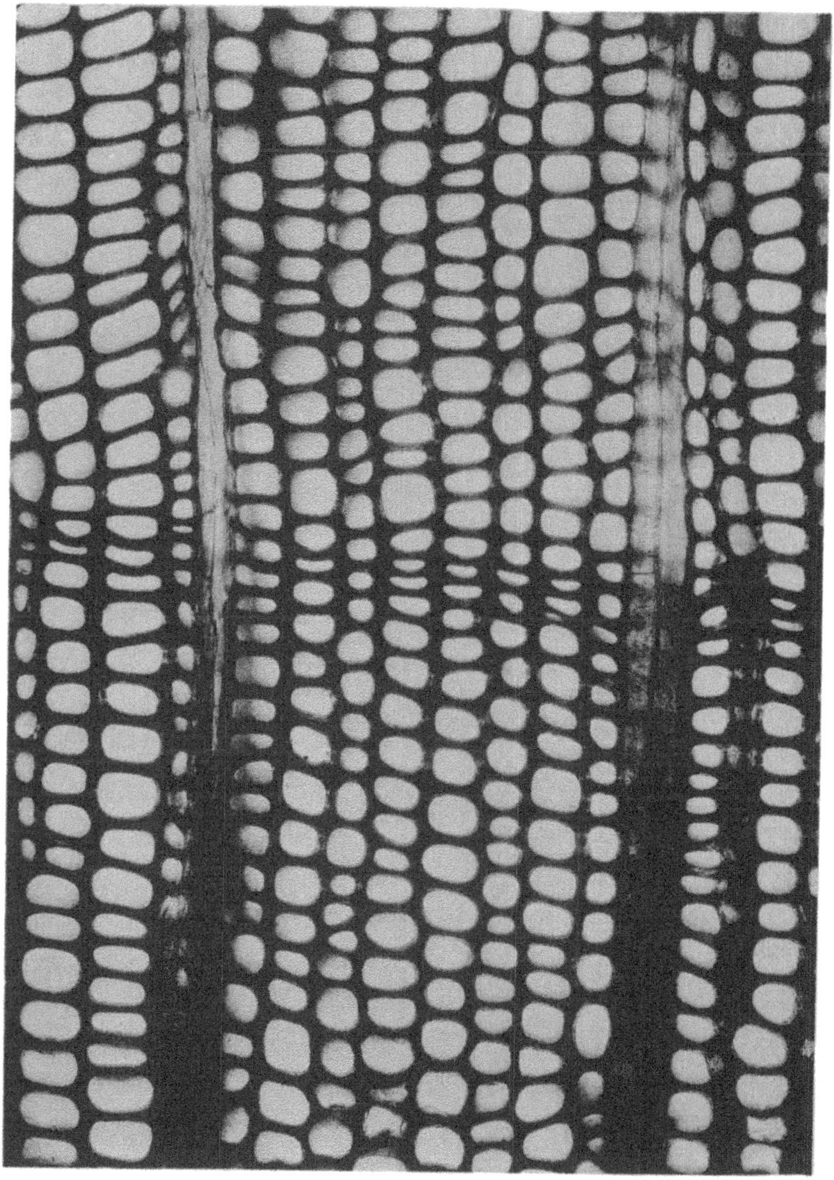

Callixylon Newberryi (Daws.) E. & W. Transverse section through portion of trunk shown in Plate I. It shows the narrow "summer wood" zone and the very fine state of preservation. × 165

PLATE IV

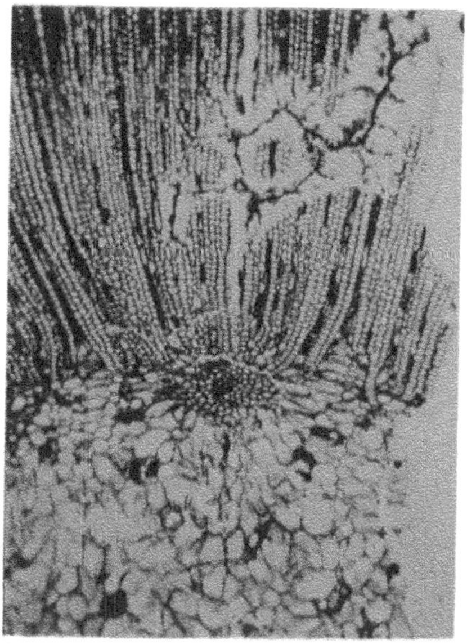

FIG. 1. *Callixylon Newberryi* (Daws.) E. & W. Primary wood bundle of small branch in contact with the secondary wood. New Albany shale, Clark County, Indiana. × 26

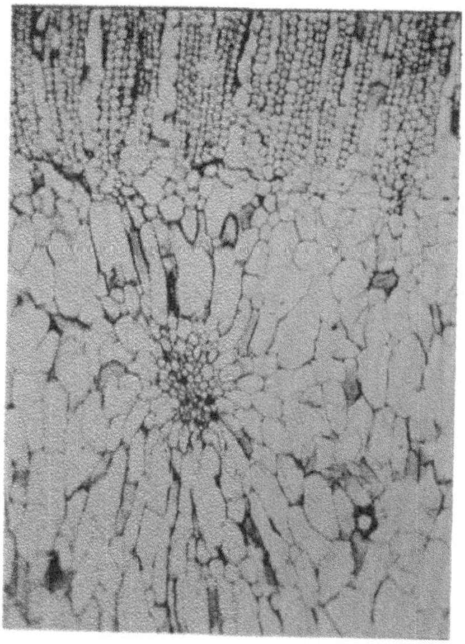

FIG. 2 *C. Newberryi* (Daws.) E. & W. Primary wood bundle of main trunk completely surrounded by pith tissue. From same specimen as Figure 1. × 26

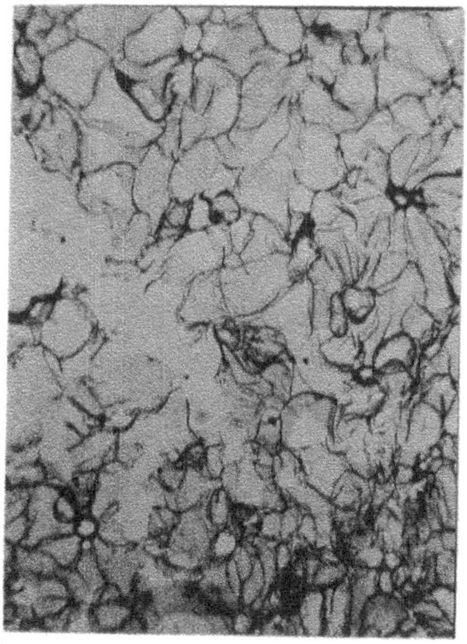

FIG. 3. *C. Newberryi* (Daws.) E. & W. Portion of lateral branch showing pith tracheids. From same specimen as Figure 1. × 37

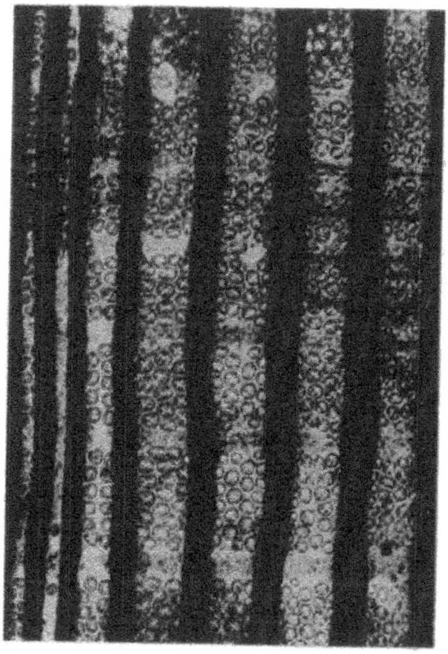

FIG. 4. *C. Newberryi* (Daws.) E. & W. Radial section, showing alignment of pit groups and crossed pit apertures. Chattanooga shale, Mill Springs, Kentucky. × 185

PLATE V

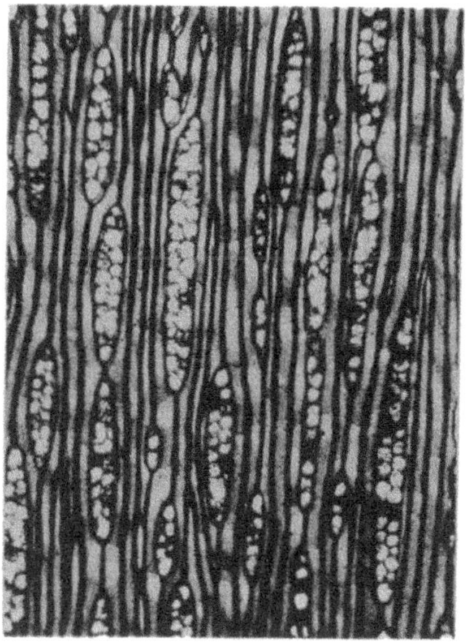

FIG. 1. *Callixylon Newberryi* (Daws.) E. & W. Tangential section showing biseriation of rays. New Albany shale, Scott County, Indiana. × 48

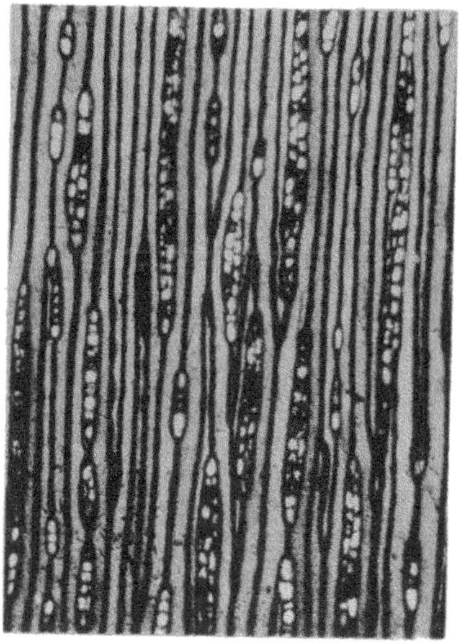

FIG. 2. *C. Newberryi* (Daws.) E. & W. Tangential section showing slight narrowing of rays due to flattening of trunk during fossilization. New Albany shale, Indiana. × 48

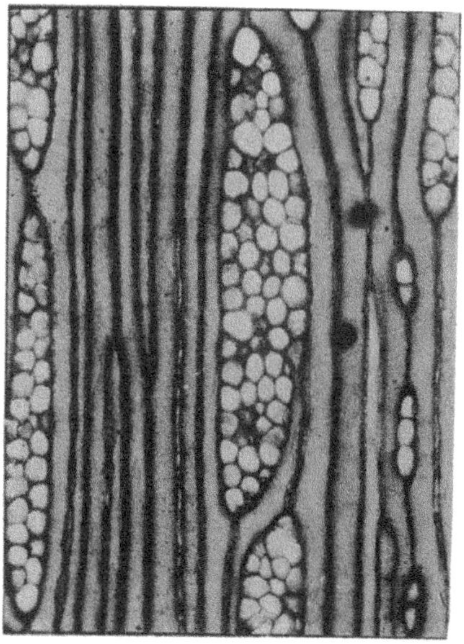

FIG. 3. *C. Newberryi* (Daws.) E. & W. Tangential section showing broad ray similar to those of *Pitys antiqua*. From same specimen as Figure 1. × 67

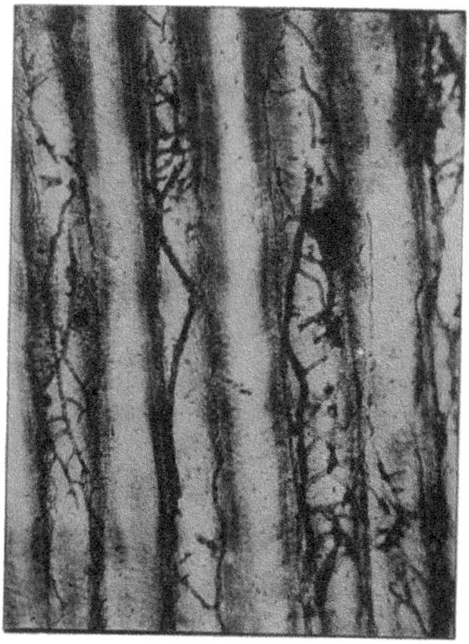

FIG. 4. *C. Newberryi* (Daws.) E. & W. Fungous hyphae in tracheids. New Albany shale, Clark County, Indiana. × 170

PLATE VI

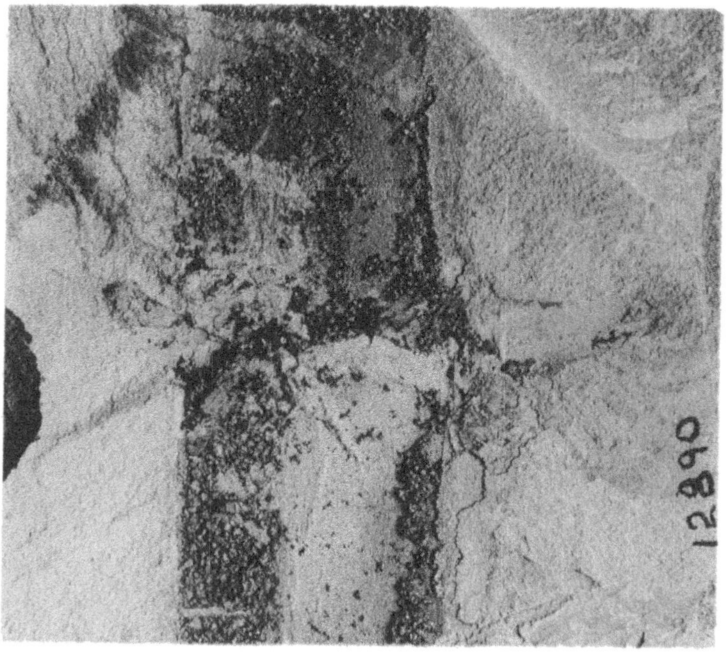

FIG. 1. *Callixylon Newberryi* (Daws.) E. & W. Impression of branch bearing two structures, probably leaves. Antrim shale, Alpena, Michigan. × 1

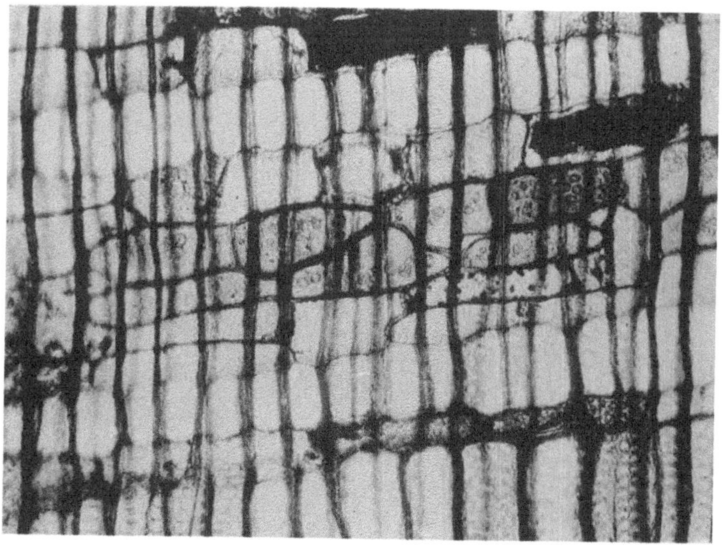

FIG. 2. *C. Newberryi* (Daws.) E. & W. Radial section showing ray tracheids with bordered pits. New Albany shale, Clark County, Indiana. × 150

PLATE VII

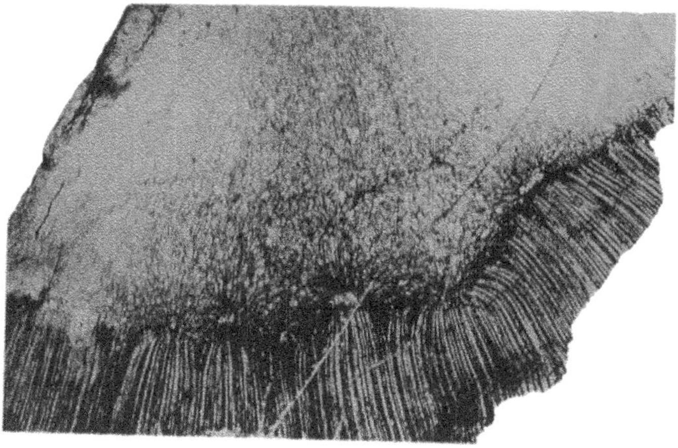

FIG. 1. *Callixylon Newberryi* (Daws.) E. & W. Sector of central portion of main trunk showing large pith, disposition of primary bundles, and inner portion of secondary wood. New Albany shale, Clark County, Indiana. × 4

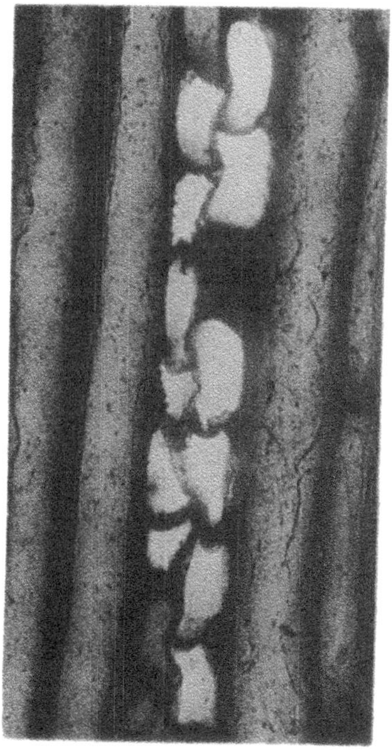

FIG. 2. *C. Newberryi* (Daws.) E. & W. Tangential section showing buckling of ray-cell walls due to compression of wood during fossilization. New Albany shale, Scott County, Indiana. × 360

FIG. 3. *C. Newberryi* (Daws.) E. & W. Radial section showing globular objects in ray cells. New Albany shale, Clark County, Indiana. × 105

VOL. III, No. 13, pp. 233–275 (14 pls.) APRIL 9, 1932

A DESCRIPTION AND STRATIGRAPHIC CORRELATION OF THE FENESTELLIDAE FROM THE DEVONIAN OF MICHIGAN *

By CHARLES F. DEISS

CONTENTS

ACKNOWLEDGMENTS

THE writer wishes to express his thanks to the men who have so kindly aided him during the course of this study: to Professor E. C. Case for untiring help, direction, and encouragement during the composition and revision of this paper; to Dr. R. S. Bassler for invaluable suggestions for the preparation of the material and for the study of the specimens, as well as for making available for comparison the collections of Bryozoa in the United States National Museum; to Dr. E. O. Ulrich for aid concerning the anatomy of the Fenestellidae; to Professor G. M. Ehlers for opening up the literature of the subject, especially that dealing with previous work in Michigan; and to Dr. W. P. Clark for criticism of the nomenclature used in designating new species.

* Published by permission of the Director of Geological Survey of Michigan.

Introduction

The family Fenestellidae occurs with especial luxuriance and in large numbers in the Devonian rocks of Michigan. Although it is represented in such abundance, but two species have been described from the state, *Fenestella eximia* Winchell and *F. filitexta* Winchell. Dr. R. S. Bassler[1] was kind enough to point out this fact to the author, and since practically no work has ever been done upon specimens of this family from Michigan, the decision was made to undertake the task of identifying and describing the Fenestellidae in the hope that the results would clarify to some extent the Devonian Stratigraphy in the Southern Peninsula of Michigan.

The material examined was collected from the Devonian strata of the northern part of the Southern Peninsula of Michigan during the summer of 1926 by a party representing the Geological Survey of Michigan. The personnel of this party consisted of Dr. E. O. Ulrich, of the United States Geological Survey, Professor E. C. Case of the University of Michigan, Assistant Professor G. M. Ehlers, of the same university, Mr. S. W. Warthin, Jr., and Mr. C. F. Deiss.

It is the author's hope eventually to complete a description of the Bryozoa of the Devonian rocks of Michigan. This paper is a preliminary study in the work, and deals only with the different forms of the Fenestellidae.

The investigation of the forms described in the following pages was the basis for a dissertation submitted in partial fulfillment of the requirements for the degree of Doctor of Philosophy at the University of Michigan, 1928.

The problem was suggested to the author by Dr. R. S. Bassler in the winter of 1926. He said: "The situation in the Michigan Devonian Fenestellidae is as follows: many names have been applied to eastern Hamilton fenestellids without adequate descriptions of figures. I have considerable material from the eastern localities, but have never found time to work out the species to my satisfaction. Therefore for the Hamilton fenestellids we

[1] Personal communication.

have little material, although for the underlying formations we have much named material."

In a recent paper Pohl [2] has suggested that the Traverse group fills a hiatus between the Onondaga beneath and the basal Hamilton (Marcellus) above. The Geological Survey of Michigan has accepted Pohl's correlation. All the specimens described in this paper are from the Traverse group.

Descriptions of Family and Genera

Family Fenestellidae King

King, A Monograph of the Permian Fossils of England, pp. 28–49 (1850).

Original description. — "Zoaria forming reticulate expansions, celluliferous on one side only, composed of rigid branches united to each other by regular non-poriferous bars called dissepiments; or they may be sinuous and anastomose; or even remain free. Primitive portion of zoœcia of oblong, quadrate or hexagonal outline. Superior hemiseptum usually present, the inferior one less frequently. Primitive aperture semi-elliptical, being truncated at the posterior side. Superficial apertures rounded, with peristome, and, in the perfect state, closed by centrally perforated opercula."

Genus Fenestella Lonsdale

Lonsdale, in Murchinson's Silurian System, Pt. II, p. 677 (1839).

Original description. — "Zoaria flabellate to infundibuliform, poriferous on the inside; branches nearly straight, and connected with each other at rhythmical intervals by non-poriferous dissepiments. Zoœcia in two rows, separated by a more or less developed median keel. Type *Fenestella antiqua*, Miller. Range, Cincinnatian to Permian."

Genus Hemitrypa Phillips

Phillips, Figures and Descriptions of the Paleozoic Fossils of Cornwall, Devon, and West Somerset, p. 27 (1841).

Original description by Simpson. — "Zoaria funnel-shaped or undulating foliar expansions; branches rigid. Zoœcia in two

[2] Pohl, Erwin, "The Middle Devonian Traverse Group of Rocks In Michigan." *Proc. U. S. Nat. Mus.*, Vol. 76. Art. 14, p. 33, 1929.

ranges, their apertures separated by a moderately developed keel. The latter is elevated at regular intervals into small pillars, which, when the superstructure they support is worn away, appear as spinelike prominences. The superstructure consists of straight or zigzag longitudinal bars, of which one is placed over each branch upon the row of pillars, and another, usually somewhat thinner, suspended midway between the branches. These bars are then connected by transverse processes, so as to leave regular, small, generally hexagonal openings, corresponding in number and position with the zoœcial apertures beneath them."

Type, *Hemitrypa oculata* Phillips. Range, Clinton to St. Louis.

Genus Isotrypa Hall; (emended) Ulrich

Hall, New York State Geologist, Annual Report 4, p. 37 (1885).
Ulrich, Geol. Surv. Ill., Vol. 8, Text, p. 395 (1890).

Emended description. — "Zoaria infundibuliform; branches connected by dissepiments. Keel at first very thin, then abruptly expanded; at intervals corresponding to the true dissepiments the thickened summits are connected by strong processes, giving to the superstructure an appearance very much like that of the reverse face of the frond. Zoœcia in two ranges."

Type, *Isotrypa conjunctiva* Hall. Range, Niagara to Hamilton.

Genus Phyllopora King

King, Annals and Magazine of Natural History, Sec. Ser., Vol. 3, p. 389 (1849).

Original description. — "A Fenestellidia consisting of infundibuliform, folded, perforated fronds or foliaceous expansions; cellules on the whole of the outer or under surface of the fronds, and planted more or less approximately to a position at right angles to the plane of the capillary tubular basal plate; cellule apertures with plain margins and parallel to the surface of the fronds."

Type, *Phyllopora ehrenbergi* King. Range, Devonian to Permian.

Genus Polypora McCoy

McCoy, A Synopsis of the Carboniferous Limestone Fossils of Ireland, p. 206 (1845).

Original description. — "Zoaria in most respects like *Fenestella*, but differing in having from two to six or even eight rows of cells, and in wanting the characteristic median keel. The latter is sometimes represented by a row of strong tubercles."

"Such species approach *Fenestralia* Prout" (Ulrich).

Type, *Polypora dendroides* McCoy. Range, Niagara to Coal Measures.

Genus Ptiloporina Hall

Hall and Simpson, Geological Survey of the State of New York. Paleontology, Vol. 6, p. 172, pl. 43, figs. 7–9 (1887).

Original description. — "Bryozoum having the same manner of growth and general appearance as *Ptiloporella*, but differing from that genus in having three or more ranges of cell apertures, not separated by a carina. Branches of two sizes, the smaller or secondary branches proceeding laterally from the larger or primary ones, either from one or both sides."

Type, *Ptiloporina sinistralis* Hall. Range, Upper Helderberg.

Genus Semicosinium Prout

Trans. St. Louis Acad. Sci., Vol. I, p. 443 (1859).

Original description. — "Zoaria funnel-shaped, poriferous on the outer side. Dissepiments wide, very short, the branches appearing on the non-poriferous side to anastomose. Here also, the fenestrules are subrhomboidal or rounded. Zoœcia in two ranges. Median keel very high and more or less expanded at the summit."

Type, *Semicosinium rhomboideum* Prout. Range, Niagara to Hamilton.

Descriptions of Species

Genus Fenestella Lonsdale

For generic description see page 235.

Fenestella depressa Hall

Hall and Simpson, Geological Survey of the State of New York. Paleontology, Vol. 6, p. 111, pl. 45, figs. 16–17 (1887).

Original description. — "Bryozoum infundibuliform. Branches slender, width above a bifurcation, .30 mm., increasing to .45 mm., rounded, finely striated. Interstices narrower than the branches. Dissepiments very slender, width .18 mm., rounded, much depressed, nine in the space of 5 mm. Fenestrules oval or subquadrangular, length .40 mm., width about .20 mm.

"On the celluliferous face the dissepiments are rounded, much depressed. Fenestrules having the same appearance as on the opposite face. Cell apertures in two ranges, opening directly outward, twenty-six in the space of 5 mm., separated by less than the diameter of an aperture: margins elevated, strongly indenting the borders of the fenestrules. Surface between the ranges of apertures angular, nodose."

Onomatype No. 13251 U.M.

Horizon and locality. — Upper Helderberg group, Falls of the Ohio River.

Fenestella eximia Winchell **problematica,** var. nov.

(Plate I, Fig. 1)

Winchell, The Grand Traverse Region, Appendix, p. 92 (1886).

Original description. — "Rays angulated along the middle, bearing two rows of pores with salient margins, opening obliquely; twenty-three pores in the distance of ten fenestrules. Length of fenestrule .51 mm., breadth .37."

This description is inadequate and the type is unknown. For these reasons the following description is given of a form which approximates the description of *Fenestella eximia* more closely than any other form observed.

It is suggested that the species described by Winchell may possibly be the same as the one described here, but at this time it seems safer to give this form the tentative name *F. eximia problematica*, var. nov.

Description. — Obverse: branches slender; subangular,

smooth; straight to slightly sinuous. Apertures open laterally, separated by more than the diameter of an aperture; twenty-two in 5 mm. Dissepiments slender, subangular; slightly expanded at junction of, and depressed below plane of, branches; nine in 5 mm. Fenestrules oval to subquadrangular; average length 0.50 mm., width from 0.18 to 0.28 mm.

Reverse. — Not seen.

Holotype No. 10505 U.M.

Horizon and locality. — Zone 2 of Location 10.[3]

Fenestella idalia Hall

(Plate I, Figs. 4–6)

Hall and Simpson, Geological Survey of the State of New York. Paleontology, Vol. 6, p. 52, pl. 21, figs. 6–9 (1887).

Original description. — "Bryozoum infundibuliform, broadly spreading, nearly flat for a short distance above the base. Branches moderately slender, rounded, compactly arranged, granulose striations on each branch; sometimes so strongly granulose as to obscure the striae. Bifurcations frequent. Interstices usually about the width of the branches. Dissepiments very slender, sub-angular; from six to nine in the space of 5 mm.; very slightly expanding at their junction, and on the same plane with the branches. Fenestrules sub-quadrangular, occasionally oval.

"On the celluliferous face the branches are angular, or sub-carinated. Dissepiments angular and much depressed. The fenestrules appear narrower than on the opposite face. Cell apertures usually in two ranges, circular; opening somewhat laterally and separated by less than the diameter of an aperture: margins elevated, indenting the borders of the fenestrules. Width of branch above a bifurcation .20 mm., increasing to .50 mm.; diameter of dissepiments .16 mm.; length of fenestrules on different specimens varying from .40 mm. to .50 mm., width from .25 to .40 mm."

[3] All locations are those of the party of the Geological Survey of Michigan of 1926. See page 273 for list and description of locations.

Horizon and locality. — Lower Helderberg group, near Clarksville, New York.

Plesiotype No. 10482 U.M.

Horizon and locality. — Zone 1 of Location 17.

Fenestella nodosa Prout

Trans. St. Louis Acad. Sci., Vol. 2, p. 410 (1866).

Original description. — "Bryozoum fan-shaped expansion, with small, round, longitudinal rays, [branches] regularly distributed, and without the appearance of dissepiments or nodes to the naked eye. Branches uniform, straight, round, bifurcating from one to one and a half lines apart, presenting under the microscope regular lines of nodes or tubercles, slightly longer than broad, about their own length apart and having no correspondence with the dissepiments, being to the number of 14–15 in each of the two lines in the length of the rays. Dissepiments depressed, short, round, thickened at junction with longitudinal rays. Fenestrules mostly quadrangular, but occasionally sub-oval from thickening of the dissepiments at their junction. Cells or chalices small, tapering upwards, placed in a line above the insertion of dissepiments and about midway between the obverse and reverse. In worn specimens the cells become distinct, showing three to each fenestrule, while the tubercles disappear entirely."

"Reverse striated where worn."

Onomatype No. 13252 U.M.

Horizon and locality. — Hamilton group of Illinois.

Fenestella vera Ulrich var. **acuta**, var. nov.

(Plate I, Figs. 2–3)

Geol. Surv. Ill., Vol. 8, p. 535 (1890), pl. 44, figs. 1–1a (1849).

Original description. — "Zoarium a somewhat undulating flabellate expansion; largest fragment seen over 6 cms. in length. Obverse with branches rather straight, slender, ridge-shaped, bifurcating at distant intervals, about .35 mm. in width and twenty-four in 1 cm. Dissepiments short, sub-carinate, varying in width from about two-thirds to almost the width of the branches.

Fenestrules elliptical or sub-quadrangular, about .50 mm. by .20 mm.; fourteen in 1 cm. Carina prominent, not sharp, bearing small nodes about twice their diameter apart, and six or seven in 2 mm. Zoœcia in two ranges. Apertures small, circular, opening obliquely into the fenestrules, generally four to each fenestrule, and twenty-three in 5 mm. On the reverse the branches are usually zigzag, on the same level and scarcely stronger than the dissepiments. The fenestrules have a somewhat hexagonal shape, less marked in the older portion of the zoarium, and form quite regular diagonal series. Surface of both branches and dissepiments very minutely granulose."

Horizon and locality. — Hamilton group, Buffalo, Iowa.

This form differs from the foregoing in three respects: (1) The carina on the obverse face is thick, sharp, and prominent with large nodes; (2) There are twenty-one apertures in 5 mm.; (3) The fenestrules are from 0.50 to 0.62 mm. in length, and average 0.30 mm. in width.

Holotype No. 10504 U.M.

Horizon and locality. — Shale zone of Location 14.

Fenestella compacta, sp. nov.

(Plate I, Figs. 7–8)

Zoarium unknown, fragments flattened.

Obverse. — Branches slender, angular, subparallel, bifurcations distant. Average width of branch 0.20 mm.; fourteen in 5 mm. Dissepiments short, angular, subequal to branches in thickness, depressed; eleven or twelve in 5 mm. Fenestrules very narrow, subquadrangular. Apertures open directly, separated by more than the diameter of an aperture; twenty-three to twenty-five in 5 mm. Peristome moderately developed and slightly indenting borders of fenestrules. Carina low, angular.

Reverse. — Not seen upon specimen. In thin section branches appear to be rounded, striated, and of about same width as those on opposite face. Dissepiments slender, broadly expanded at junction with branches. Fenestrules oval; average length 0.30 mm., width from 0.16 to 0.26 mm.

Holotype No. 10481 U.M.
Horizon and locality. — Shaly limestone of Location 35.

Fenestella paridistans, sp. nov.

(Plate II, Figs. 1–3)

Zoarium unknown, appears flabellate in large fragments.

Obverse. — Branches very slender, angular, practically parallel, bifurcations few, distant. Average width of branch 0.22 mm.; twelve to thirteen in 5 mm. Dissepiments extremely slender, subangular, depressed; ten in 5 mm. Fenestrules oval to subquadrangular; length and width about one to two, at places as one to three. Apertures open laterally; separated by more than diameter of an aperture; twenty in 5 mm. Peristomes distinct but weak, only slightly indenting borders of fenestrules. Carina low, subangular, in places sharp and thin; summit with a row of elongate, blunted nodes, unequally distant from each other.

Reverse. — Branches rounded, about equal in width to those of obverse face, coarsely granulose with low granular nodes irregularly distributed upon them. Dissepiments subequal, or stronger than branches; rounded, granulose, occasionally nodose and in places raised above plane of branches. Fenestrules oval to circular to subquadrangular; length 0.32 to 0.48 mm., width 0.18 to 0.26 mm.

Holotype No. 10480 U.M.
Horizon and locality. — Zone 1 of Location 9.

Fenestella variifenestrula, sp. nov.

(Plate II, Figs. 4–6)

Zoarium not seen, fragments flattened.

Obverse. — Branches slender, angular, subparallel, average width 0.23 mm.; thirteen in 5 mm.; bifurcations infrequent. Dissepiments thin; depressed below plane of branches; 0.20 mm. in diameter; eleven in 5 mm.; rounded to subangular. Fenestrules vary widely in shape, from thin oval to nearly square. Apertures open laterally; twenty-five in 5 mm., separated by more than diameter of an aperture. Peristomes weakly developed. Carina angular, low, with a row of minute nodes.

Reverse. — Branches slender, rounded, coarsely granulose, with strong granular, conical nodes, irregularly disposed. Dissepiments rounded to subangular; in some places slightly depressed below, in others slightly raised above plane of branches; strong; granular, but rarely nodose. Fenestrules oval to circular to subquadrate; length from 0.26 to 0.40 mm., width from 0.10 to 0.26 mm.

Holotype No. 10479 U.M.

Horizon and locality. — Shaly limestone of Location 35.

Fenestella nodicula, sp. nov.

(Plate II, Figs. 7–8)

Zoarium unknown, fragments flattened.

Obverse. — Branches slender, angular, subparallel, straight; average width 0.28 mm.; eleven in 5 mm.; bifurcations distant. Dissepiments slender, angular; slightly depressed below plane of branches; average width 0.18 mm.; ten in 5 mm. Fenestrules flat oval to subquadrangular; sizes equal to those on reverse face. Apertures open laterally, separated by more than diameter of an aperture; three to a fenestrule, twenty in 5 mm. Peristome weakly developed. Carina low, strong, angular, with minute granular nodes in single row on summit.

Reverse. — Branches rounded, straight; from 0.18 to 0.28 mm. in thickness, striated and with large conical nodes irregularly disposed. Dissepiments slender, rounded, smooth; depressed below plane of branches; average diameter 0.18 mm. Fenestrules oval to subquadrangular; length from 0.40 to 0.42 mm., average width 0.20 mm.

Holotype No. 10478 U.M.

Horizon and locality. — Shaly limestone of Location 35.

Fenestella minutiserrata, sp. nov.

Zoarium infundibuliform.

Obverse. — Branches slender, straight, subparallel. Dissepiments extremely thin and but slightly expanded at junction with branches; four to five in 5 mm.; slightly depressed below plane of branches. Fenestrules quadrangular, of same size as those on reverse face. Apertures open laterally and alternately, separated by more than diameter of an aperture; nineteen to twenty in 5 mm. Peristomes of alternate apertures give serrated appearance to branches. Carina very faint; minutely nodose.

Reverse. — Branches strong when compared with those on obverse face; rounded, coarsely striated, faintly carinated in places; bifurcation every second fenestrule. Dissepiments more slender than branches; not depressed below plane of branches and but slightly expanded at junction with them. Fenestrules quadrangular to elongate oval; length from 0.95 to 1.55 mm., width from 0.25 to 0.52 mm.

Holotype No. 10536 U.M.

Horizon and locality. — Gray limestone of Location 41.

Fenestella incerta, sp. nov.

(Plate II, Fig. 9; Plate III, Figs. 1–2)

Zoarium unknown, fragments flattened.

Obverse. — Branches strong, angular, straight; width from 0.25 to 0.30 mm., bifurcations distant. Dissepiments slender, rounded; much depressed below plane of branches; eight to nine in 5 mm. Fenestrules subquadrangular to oval. Apertures open laterally, separated by more than the diameter of an aperture; twenty in 5 mm., average three to a fenestrule. When apertures are disposed upon dissepiments they appear to open directly. Peristome strongly developed, indenting edges of fenestrules. Carina strong, low, rounded; bearing strong conical nodes upon summit.

Reverse. — Branches rounded to subangular; straight, subparallel; minutely granulose, with extremely large nodes irregularly disposed and with smaller, but prominent, nodes disposed at junction of branches and dissepiments. Dissepiments of same width as branches; rounded, granular, in places subangular; not depressed below plane of branches. Fenestrules quadrangular, rarely oval; length from 0.45 to 0.60 mm., width from 0.20 to 0.28 mm.

Holotype No. 10477 U.M. Paratype No. 10534 U.M.
Horizon and locality. — Zone 1 of Location 9.

Fenestella foraminosa, sp. nov.

(Plate III, Figs. 3–4)

Zoarium unknown, fragments flattened to flexuous.

Obverse. — Branches moderately slender, average about 0.25 mm. in width; angular, subparallel; bifurcations distant. Dissepiments slender, subangular, short, width 0.18 mm.; depressed below plane of branches; twelve in 5 mm. Fenestrules narrow oval to subquadrangular. Apertures open directly; twenty-five in 5 mm., occasionally twenty-seven in 5 mm., separated by more than the diameter of an aperture. Peristomes very strongly developed, indenting borders of fenestrules. Carina low, angular; minutely nodose on summit.

Reverse. — Branches more slender than on obverse face; smooth, rounded; with a large node near junction of dissepiments. Dissepiments rounded, occasionally faintly carinated, more slender than branches and expanded at junction with them; slightly depressed below plane of branches. Fenestrules oval to subquadrangular; length from 0.26 to 0.40 mm., width from 0.18 to 0.23 mm.

Holotype No. 10498 U.M. Paratypes Nos. 10519, 10520, 10521, 10522, U.M.

Horizon and locality. — Gray limestone of Location 36.

Fenestella megalopora, sp. nov.

(Plate III, Figs. 7–9)

Zoarium unknown, fragments flattened.

Obverse. — Branches slender, angular, subparallel; average width 0.30 mm.; nine in 5 mm. Dissepiments very slender, rounded, to subangular to flattened to concave; depressed below plane of branches; six to seven in 5 mm. Fenestrules oval to subquadrangular and of same size as on opposite face. Apertures large, opening laterally; eighteen to twenty in 5 mm., separated by less than half the diameter of an aperture; additional apertures at bifurcations of branches. Peristomes weakly developed in proportion to size of apertures; thin, not prominent and but slightly indenting margins of fenestrules. Carina prominent, sharply angular; summit bears a row of high conical spines in some places and in others spines are flattened so that they appear as continuations of carina.

Reverse. — Branches sharply angular to rounded; more slender than those on opposite face; strongly carinated in mid-line and with an occasional prominent, granular node at junction with branches and dissepiments. Dissepiments rounded to subangular and usually carinated; carina of dissepiments uniting with that of branches; dissepiments not depressed below and usually raised above plane of branches. Fenestrules oval to subquadrangular; length from 0.60 to 0.75 mm., width from 0.26 to 0.40 mm.

Holotype No. 10497 U.M.

Horizon and locality. — Top zone of Location 28.

Fenestella incisa, sp. nov.

(Plate III, Figs. 5–6)

Zoarium unknown, fragments flattened, undulating.

Obverse. — Branches slender, angular, average width 0.34 mm.; eleven to twelve in 5 mm.; bifurcations distant. Dissepiments more slender than branches, average thickness 0.22 mm.; ten in 5 mm.; strongly expanded at junction with branches. Fenestrules oval; average length 0.30 mm., average width 0.18 mm. Apertures small, opening directly; twenty-two to twenty-four in 5 mm., separated by more than diameter of an aperture. Peristomes strongly developed, indenting borders of fenestrules. Carina prominent, angular, thin at summit, which bears a row of sharp spinules; four in 1 mm.

Reverse. — Branches rounded, slender, coarsely granulose, with low, granular nodes opposite dissepiments. Dissepiments about two thirds as wide as branches; rounded, granular, occasionally nodose; usually not depressed below plane of branches. Fenestrules oval; length from 0.26 to 0.38 mm., width average about 0.18 mm.

Holotype No. 10509 U.M.

Horizon and locality. — Top zone of Location 28.

The obverse face of this specimen was not seen macroscopically; the characters described above were taken from a thin section.

Fenestella longispinosa, sp. nov.

(Plate IV, Figs. 1 and 3)

Zoarium unknown, fragments flexuous.

Obverse. — Branches strong, angular, slightly sinuous to straight; average width 0.30 mm.; twelve in 5 mm. Dissepiments slender, from 0.10 to 0.20 mm. in width; subangular, depressed below plane of branches; ten in 5 mm. Fenestrules of same size and shape as on reverse face. Apertures open directly, separated by more than diameter of an aperture; two to a fenestrule; twenty-three to twenty-five in 5 mm. Peristomes strongly developed, indenting margins of fenestrules. Carina strong, angular, low, with sharp conical nodes disposed upon summit; seven nodes in 2 mm.

Reverse. — Branches rounded to subangular, coarsely granulose, straight; bifurcating every fourth fenestrule; low granular nodes disposed at junction of branches and dissepiments. Dissepiments comparatively slender, rounded to subangular; faintly depressed below plane of branches. Fenestrules oval; length from 0.26 to 0.36 mm., average width 0.22 mm.

Holotype No. 10476 U.M.

Horizon and locality. — Shaly limestone of Location 35.

Genus Hemitrypa Phillips

For generic description see page 235.

Hemitrypa variosa, sp. nov.

(Plate IV, Figs. 2 and 5)

Zoarium unknown, fragments flexuous.

Obverse. — Branches not seen on specimen. In thin section they appear angular, slender; bifurcations average 2.1 mm.

distant; twelve to thirteen branches in 5 mm. Dissepiments more slender than branches; angular, carinated; nine to ten in 5 mm. Fenestrules elongate-oval. Apertures open laterally; twenty-five in 5 mm., separated by more than diameter of an aperture. Carina strong, thin, expanded at summit. Thin transverse processes extend laterally from summit of carina, uniting in mid-line and forming a pseudo-carina. Interstices between carina, pseudo-carina, and transverse processes hexagonal to circular to quadrate; average 0.12 mm. in diameter; twenty-four to twenty-six in 5 mm.

Reverse. — Branches slender, rounded, smooth; average width 0.20 mm. Dissepiments stronger than branches; average width 0.24 mm.; rounded, smooth; not depressed below plane of branches. Fenestrules circular to oval to subhexagonal; length from 0.22 to 0.34 mm., width from 0.16 to 0.22 mm.

Holotype No. 10475 U.M.

Horizon and locality. — Blue shale zone of Location 46.

Genus Isotrypa Hall

For generic description see page 236.

Isotrypa tropozomena, sp. nov.

(Plate IV, Figs. 4 and 6)

Zoarium unknown, fragments flattened.

Obverse. — Branches strong, zigzag, angular; average width 0.40 mm.; eight in 5 mm. Dissepiments stronger than branches; sharply carinated; average width 0.50 mm. Fenestrules broadly oval to circular. Apertures open directly, separated by the diameter of an aperture; twenty to twenty-three in 5 mm.; disposed in oval rows surrounding fenestrules. Carina strong, thin with expanded summit. Transverse processes extend laterally, connecting carina of adjacent branches; usually occur above a dissepiment and equal in number to dissepiments; six in 5 mm. Carina and transverse bars of equal size; smooth, rounded, and of same size and appearance as branches and dissepiments of reverse face.

Reverse. — Branches smooth, subangular, with a low broad node at junction of dissepiments. Dissepiments and branches subequal in size. Dissepiments rounded, smooth; not depressed below plane of branches. Fenestrules circular; length from 0.55 to 0.70 mm.

Holotype No. 10474 U.M.

Horizon and locality. — Middle shale zone of Location 14.

Isotrypa megista, sp. nov.

(Plate V, Figs. 1–3)

Zoarium unknown, fragments flattened and flexuous.

Obverse. — Branches slender, zigzag, bifurcations distant; average 0.26 mm. in thickness; nine in 5 mm. Dissepiments angular, more slender than branches; six to seven in 5 mm. Fenestrules broadly oval; length from 0.56 to 0.70 mm., width from 0.33 to 0.37 mm. Apertures open directly, separated by more than the diameter of an aperture; twenty in 5 mm. Carina thin, expanded at summit, which is angular. Transverse bars connect the carina of adjacent branches. Transverse bars comparatively slender; slightly depressed below plane of carinae; eight to ten in 5 mm. Low node present at junction of carina with transverse bar. Interstices between transverse bars and carinae circular to subquadrangular; average length 0.38 mm.

Reverse. — Branches angular, slender, with mid-row of granular nodes; usually zigzag, occasionally straight; from one fifth to one third wider than carina on obverse branches; average 0.30 mm. in width. Dissepiments more slender to faintly broader than branches; angular to rounded; depressed below plane of branches; usually minutely nodose. Fenestrules broadly oval, occasionally subrhomboidal or hexagonal; average length 0.50 mm., width from 0.30 to 0.40 mm.

Holotype No. 10473 U.M. Paratype No. 10528 U.M.

Horizon and locality. — Middle shale zone of Location 14.

Isotrypa angulata, sp. nov.

(Plate V, Figs. 4–6)

Zoarium unknown, fragments flattened.

This species differs from *I. megista* in (1) the carina and transverse bars, which are more sharply angular; (2) the absence of a node at the intersection of the transverse bars and carinae; and (3) the absence of nodes on the reverse branches, which are minutely granulose.

Holotype No. 10472 U.M.

Horizon and locality. — Blue shale zone of Location 47.

Isotrypa anomala, sp. nov.

(Plate V, Figs. 7–8)

Zoarium unknown, fragments flattened, flexuous.

Obverse. — Branches slender, average width 0.20 mm.; eight to nine in 5 mm. Dissepiments vary from 0.20 to 0.48 mm. in width; five to six, usually six, in 5 mm.; seen only in thin sections, where they appear broadly rounded, depressed below plane of branches. Fenestrules elliptical; length from 0.36 to 0.42 mm., width from 0.18 to 0.28 mm. Apertures open directly, separated by more than diameter of an aperture; twenty-three to twenty-five in 5 mm. Carina slender, thin with expanded summit, which is slightly angular. Carinae of adjacent branches connected by transverse bars of same size as carina and not depressed below planes of carinae. Six to seven bars in 5 mm. Interstices between transverse bars and carinae subquadrangular to oval; length from 0.40 to 0.55 mm., width from one third to four fifths of length.

Reverse. — Branches rounded, smooth, straight to sinuous; bifurcations distant. Dissepiments vary in width from slightly less to slightly more than that of branches; rounded to faintly carinated, usually depressed below plane of branches. Fenestrules elongate-oval; length from 0.40 to 0.62 mm., width from 0.26 to 0.30 mm.

Holotype No. 10471 U.M. Paratypes Nos. 10526, 10547 U.M.

Horizon and locality. — Zone 1 of Location 9.

Isotrypa anomala var. **sinuosa,** var. nov.

(Plate V, Fig. 9; Plate VI, Figs. 1–2)

Zoarium unknown, fragments flattened, flexuous.

Obverse. — Branches slender, from 0.30 to 0.34 mm. in width; slightly sinuous. Dissepiments slender, rounded; six to seven in 5 mm. Fenestrules narrow, elliptical. Apertures open directly, separated by more than diameter of an aperture; twenty-three in 5 mm. Branches, dissepiments, fenestrules, and apertures observed only in thin sections. Carina thin, summit expanded and sharply angular. Transverse bars of same thickness as the branches, sharply angular, angular part connecting with sharp summit of carina; eight bars in 5 mm. Interstices between carinae and transverse bars suboval to quadrate; length from 0.42 to 0.56 mm., width from 0.26 to 0.42 mm.

Reverse. — Branches slender, rounded, with minute granular nodes; straight to zigzag. Dissepiments stronger than branches; broadly expanded at junction with them, rounded; with minute granular nodes of same appearance as those on branches. Fenestrules elliptical; length from 0.44 to 0.54 mm., width from 0.24 to 0.32 mm.

Holotype No. 10470 U.M. Paratypes Nos. 10545, 10546 U.M.

Horizon and locality. — Zone 1 of Location 49.

Isotrypa rara, sp. nov.

(Plate VI, Figs. 3–4)

This species differs from *Isotrypa anomala* in that (1) there are twenty instead of twenty-three to twenty-five apertures in 5 mm.; and (2) the obverse carina and transverse bars and the reverse branches are more irregularly disposed and more granular.

Holotype No. 10469 U.M.

Horizon and locality. — Zone 8 of Location 40.

Isotrypa ovata, sp. nov.

(Plate VI, Figs. 5–7)

Zoarium unknown, fragments flattened, flexuous.

Obverse. — Branches slender, width average 0.24 mm., sinuous to straight. Dissepiments slender, much depressed below plane of branches; seven in 5 mm. Fenestrules elliptical; length 0.52 mm., width 0.26 mm. Apertures open directly, separated by more than the diameter of an aperture; twenty in 5 mm. Carina thin, broadly expanded at summit, which is broadly flattened, smooth, faintly striated where worn. Transverse bars more slender than carinae, usually rounded, faintly carinated and raised above the plane of the carinae in places; eight bars in 5 mm. Interstices between carinae and transverse bars subquadrangular to circular; length from 0.40 to 0.76 mm., width from 0.28 to 0.40 mm.

Reverse. — Branches zigzag to sinuous, rarely straight, rounded, slender; width from 0.24 to 0.33 mm., with very low granular nodes. Dissepiments stronger than branches; width from 0.28 to 0.46 mm., rounded, not depressed below plane of branches, with granular, low nodes. Nodes on dissepiments and branches are disposed in oval rows surrounding fenestrules. Fenestrules from 0.32 to 0.46 mm. in length, average width 0.32 mm.

Holotype No. 10468 U.M.

Horizon and locality. — Zone 1 of Location 17.

Isotrypa hexagona, sp. nov.

(Plate VII, Figs. 1–2)

Zoarium unknown, fragments flattened.

Obverse. — Branches average 0.32 mm. in width, slightly sinuous. Dissepiments stronger than branches; average 0.42 mm. in width, rounded; five in 5 mm. Fenestrules elliptical; average length 0.50 mm., average width 0.22 mm. Apertures open directly, separated by more than diameter of an aperture; twenty in 5 mm. Carina thin, expanded at summit, which is thin and extremely sharp; zigzag, never straight. Transverse bars as sharply angular as carina. The angular summits of both bars and carinae unite to form hexagonal interstices between them, with length from 0.40 to 0.60 mm. Eight bars in 5 mm.

Reverse. — Branches strong, flattened to subangular; zigzag to sinuous; with low sharp carina in mid-line. Dissepiments stronger than branches, smooth, rounded; depressed below plane of branches, to slightly carinated and not depressed below plane of branches. Fenestrules sharp-ended ovals; average length 0.50 mm., average width 0.25 mm.

Holotype No. 10467 U.M.

Horizon and locality. — Zone 1 of Location 17.

Isotrypa vibrata, sp. nov.

(Plate VII, Figs. 3–5)

Zoarium unknown, fragments flattened.

Obverse. — Branches average 0.30 mm. in width; zigzag. Dissepiments much stronger than branches; average width 0.40 mm.; five in 5 mm. Fenestrules elliptical to broadly oval; average length 0.60 mm., width 0.30 mm. Apertures open directly, separated by more than the diameter of an aperture; twenty-five in 5 mm. Carina thin, expanded at the summit, rounded, faintly granular. Transverse bars more slender than carina, in places slightly thicker than the carina and faintly angular; eight bars in 5 mm. Interstices between carinae and transverse bars elongate-oval to subquadrangular; length from 0.40 to 0.61 mm., average width 0.28 mm.

Reverse. — Branches very slender, minutely granulose; extremely zigzag; anastomosed. Dissepiments rarely present. Anastomosed portions of branches strong; width from 0.52 to 0.64 mm.; four, rarely five in 5 mm.; slightly raised above plane of other parts of branches. Fenestrules rhomboidal to oval; average length 0.65 mm., average width 0.45 mm.

Holotype No. 10466 U.M.

Horizon and locality. — Zone 1 of Location 17.

Isotrypa oxytropis, sp. nov.

(Plate VII, Figs. 6–7)

Zoarium unknown, fragments flattened, flexuous.

Obverse. — Branches 0.18 to 0.30 mm. in width; sinuous to zigzag. Dissepiments wider than branches; average 0.32 mm., six in 5 mm., very short, rounded, depressed. Fenestrules oval; average length 0.52 mm., average width 0.26 mm. Apertures open directly, separated by more than the diameter of an aperture; twenty to twenty-two in 5 mm. Carina thin, expanded at summit, which is sharply angular, with a conical node at junction of transverse bar and carina. Transverse bars of same width as carina, sharply angular, connected with angular summit of carina inclosing quadrangular to hexagonal interstices. Seven transverse bars in 5 mm. Interstices vary from 0.40 to 0.70 mm. in length and from 0.30 to 0.40 mm. in width.

Reverse. — Branches angular; usually zigzag; anastomosed. Dissepiments strong, angular, in some places depressed, in others raised above the plane of the branches. Large pores 0.20 mm. in diameter occupy dissepiments. Pore margins strong, elevated, giving appearance of node on dissepiment. Fenestrules oval; average length 0.50 mm., average width 0.28 mm.

Holotype No. 10465 U.M.

Horizon and locality. — Bell Shale of Location 31.

Isotrypa gigantica, sp. nov.

(Plate VIII, Figs. 1–2)

Zoarium unknown, fragments flattened.

Obverse. — Branches slender, zigzag. Apertures open directly; separated by more than diameter of an aperture. Carina thin, expanded and rounded to gently angular at summit, sinuous to zigzag; in places carina of adjacent branches appear to anastomose, giving rise to very short, wide transverse bars; both transverse bars and branches granulose. Transverse bars and carina subequal in width when carinae are not anastomosed. Interstices between transverse bars and carinae hexagonal to subrhomboidal to oval; length from 0.70 to 0.80 mm., width from 0.32 to 0.50 mm.

Reverse. — Branches strong, rounded to angular, slightly granulose; zigzag; rarely sinuous. Dissepiments formed by anastomosed portions of branches; strong, short, rounded, slightly granulose; faintly raised above plane of branches. Fenestrules rhomboidal to suboval. Length from 0.50 to 0.55 mm., width and length subequal.

Holotype No. 10500 U.M.

Horizon and locality. — Zone 8 of Location 40.

Isotrypa isopeda, sp. nov.

(Plate VIII, Figs. 3–5)

Zoarium unknown, fragments flattened.

Obverse. — Branches average 0.22 mm. in thickness; sinuous. Dissepiments stronger than branches, short, rounded; average 0.28 mm. in width; six in 5 mm. Apertures open directly, separated by more than diameter of an aperture; twenty-five to twenty-seven in 5 mm. Fenestrules 0.48 to 0.62 mm. in length, average width 0.26 mm.; elongate-oval in shape. Carina thin, expanded and sharply angular at summit; straight. Transverse bars of same width as carina, sharply angular, not depressed below plane of carinae; ten bars in 5 mm. Interstices subquadrangular to oval; average length 0.40 mm., average width 0.32 mm.

Reverse. — Branches irregularly zigzag to sinuous, slender; of equal width to branches of opposite face, sharply angular, anastomosed in places. Dissepiments normally slender, rounded, faintly depressed, occasionally slightly angular and not depressed below plane of branches. Anastomosed portions of branches rounded, broad; on same plane as branches. Fenestrules truncated oval to smooth oval; average length 0.60 mm., average width 0.30 mm.

Holotype No. 10464 U.M.

Horizon and locality. — Blue shale zone of Location 47.

Genus PHYLLOPORA King

For generic description see page 236.

Phyllopora aequirotundata, sp. nov.

(Plate VIII, Figs. 6–7)

Zoarium unknown, fragments flattened, flexuous.

Obverse. — Branches gently rounded, sinuous to zigzag; average 0.80 mm. in width. Dissepiments from 0.40 to 0.55 mm. in width; four in 5 mm., gently rounded, depressed below plane of branches. Fenestrules narrow oval; length from 0.60 to 0.75 mm., width from 0.22 to 0.40 mm. Zoœcia disposed in three to five ranges, on both branches and dissepiments; mid-line of dissepiments not celluliferous. Apertures small; twenty in 5 mm.; distant longitudinally by diameter of an aperture or more, transversely by less than the diameter of an aperture.

Reverse. — Branches more slender than on opposite face, rounded to faintly angular, smooth to minutely granulose; zigzag to sinuous. Dissepiments and branches subequal in width, rounded, not depressed below plane of branches. Fenestrules broadly oval; average length 0.94 mm., average width 0.72 mm.

This species differs from *Phyllopora superba* Ulrich [4] in having smaller branches, dissepiments and fenestrules, and in the greater number of apertures in 5 mm.

Holotype No. 10499 U.M. Paratype No. 10543 U.M.

Horizon and locality. — Zone 1 of Location 17.

Genus Polypora McCoy

For generic description see page 237.

Polypora ambiplana, sp. nov.

(Plate IX, Figs. 1–2)

Zoarium unknown, fragments flattened.

Obverse. — Branches from 0.40 to 0.54 mm. in width, six in 5 mm.; straight to slightly sinuous; rounded. Dissepiments rounded, depressed below plane of branches; four to five in 5 mm. Fenestrules oval to subquadrangular; average 0.70 mm. by 0.50

[4] Ulrich, E. O., "Paleozoic Bryozoa," Geol. Surv. Ill., Vol. 8, Text, Pt. 2, p. 613; pl. 44, figs. 6–6c, and pl. 55, figs. 9–9a (1890).

mm. Apertures in three to five ranges; twenty in 5 mm., separated by less than diameter of an aperture.

Reverse. — Branches rounded to faintly angular, with low nodes usually at junction of dissepiment and branch. Dissepiments vary in width from equal to twice as wide as branches. Fenestrules subquadrate to subhexagonal to oval; average length 0.78 mm., average width 0.50 mm.

Holotype No. 10463 U.M.

Horizon and locality. — Zone 1 of Location 49.

Polypora finitima, sp. nov.

(Plate IX, Fig. 3)

Zoarium unknown, fragments flattened.

Obverse. — Branches strong, sinuous. Dissepiments of same width as branches; six in 5 mm. Fenestrules broadly oval to circular. Apertures disposed in three to four ranges, separated longitudinally by less than diameter of an aperture; twenty-two to twenty-four in 5 mm.

Reverse. — Branches angular; sinuous to zigzag, rarely straight; granulose, with low granular nodes usually at junction of branch and dissepiment. Dissepiments subangular to rounded; usually depressed below plane of branches; width from less than to twice that of branches. Fenestrules oval to circular to subhexagonal; length from 0.35 to 0.62 mm., width from 0.30 to 0.42 mm.

Holotype No. 10506 U.M. Paratype No. 10502 U.M.

Horizon and locality. — Bell Shale of Location 31.

Polypora amorpha, sp. nov.

(Plate IX, Figs. 4–5)

Zoarium unknown, fragments flattened, flexuous, flabellate.

Obverse. — Branches angular; 0.30 to 0.50 mm. in width, low angular ridges between ranges of apertures, elongated nodes on summit of ridges, irregularly disposed. Dissepiments slender; average width 0.20 mm.; rounded to subangular, slightly depressed below plane of branches; six in 5 mm. Fenestrules elongate, irregularly oval, narrower than those on reverse face. Apertures in two to four ranges; twenty in 5 mm., separated longitudinally by less than diameter of an aperture.

Reverse. — Branches rounded, in places subangular; straight to sinuous; width slightly less than on obverse face; bifurcations distant. Dissepiments comparatively slender; rounded to gently angular, smooth; not depressed below plane of branches. Fenestrules oval to subquadrangular to subhexagonal; length from 0.65 to 0.80 mm., width from 0.26 to 0.40 mm.

Holotype No. 10460 U.M.

Horizon and locality. — Zone 6 of Location 29.

Polypora magnifica, sp. nov.

(Plate IX, Fig. 6; Plate X, Fig. 1)

Zoarium unknown, fragments flattened.

Obverse. — Branches angular; width from 0.32 to 0.74 mm.; conical nodes irregularly distributed between ranges of apertures; branches straight to faintly sinuous. Dissepiments rounded, depressed below plane of branches; average 0.40 mm. in width; four, rarely five, in 5 mm. Fenestrules broadly oval to subquadrangular, slightly smaller than on opposite face. Apertures in three to six ranges; twenty in 5 mm., separated by less than diameter of an aperture. Margins of lateral ranges indenting borders of fenestrules.

Reverse. — Branches slender, width average 0.26 mm.; rounded to subangular, smooth, with low node near junction of dissepiment and branch. Nodes irregularly disposed. Dissepiments of same width as branches; rounded to subangular; usually not depressed below plane of branches, and occasionally slightly raised above it. Fenestrules quadrate to very broadly oval; length 0.80 to 1.10 mm., width 0.65 to 0.80 mm.

Holotype No. 10495 U.M. Paratype No. 10542 U.M.

Horizon and locality. — Zone 1 of Location 49.

Polypora brevissima, sp. nov.

(Plate X, Figs. 2–3)

Zoarium unknown, fragments flattened.

Obverse. — Branches angular, slender, width from 0.30 to 0.50 mm.; with low nodes irregularly distributed between ranges of apertures. Dissepiments vary from 0.40 to 0.68 mm. in width; rounded to flattened, very short; slightly depressed below plane of branches; four to five in 5 mm. Fenestrules elongate-oval; length from 0.60 to 0.80 mm., width from 0.27 to 0.34 mm. Apertures in two to four ranges; nineteen to twenty-one in 5 mm., separated longitudinally by less than diameter of an aperture.

Reverse. — Branches slender; angular to broad, flattened, rounded at a bifurcation; faintly striated where worn. Dissepiments of same size as on obverse face; rounded, smooth; slightly depressed below plane of branches. Fenestrules elongate-oval, occasionally subquadrangular; length from 0.60 to 0.75 mm., width from 0.32 to 0.40 mm.

Holotype No. 10494 U.M.

Horizon and locality. — Zone 1 of Location 49.

Polypora pyramidata, sp. nov.

(Plate X, Figs. 4–5)

Zoarium unknown, fragments flattened.

Obverse. — Branches slender, average 0.36 mm. in width; bifurcations frequent; straight. Dissepiments average 0.45 mm. in thickness; rounded, slightly depressed; four, occasionally five in 5 mm. Fenestrules oval; length 0.60 to 0.78 mm., width 0.30 to 0.48 mm. Apertures in two to five ranges; twenty in 5 mm., separated by less than diameter of an aperture. Very sharp, prominent, angular spinules irregularly distributed between ranges of apertures.

Reverse. — Branches 0.28 to 0.40 mm. in width; angular, straight to slightly sinuous; faintly striated where worn, smooth, minutely granulose. Dissepiments of same width as branches; smooth, subangular; not depressed below plane of branches.

Fenestrules oval, similar in appearance to those of obverse face; length from 0.64 to 0.92 mm., average width 0.42 mm.

Holotype No. 10493 U.M.

Horizon and locality. — Zone 1 of Location 49.

Polypora indentata, sp. nov.

(Plate X, Figs. 6–7)

Zoarium unknown, fragments flattened.

Obverse. — Branches slender, average width above bifurcation 0.22 mm.; ten in 5 mm.; angular ridge between ranges of apertures, straight to slightly flexuous. Dissepiments average 0.20 mm. in width, six in 5 mm.; slightly angular and depressed below plane of branches. Fenestrules irregularly oval; length from 0.55 to 0.65 mm., average width 0.26 mm. Apertures in two to four ranges; twenty in 5 mm., separated longitudinally by less than the diameter of an aperture. Margins of lateral ranges of apertures indenting borders of fenestrules.

Reverse. — Branches slender, rounded, sharply striated for short distances, with low, longitudinally elongated nodes at junction of dissepiments and branches; straight to sinuous. Dissepiments slightly more slender than branches; smooth, rounded, not depressed below plane of branches. Fenestrules subquadrangular to oval; length from 0.55 to 0.62 mm., width from 0.28 to 0.42 mm.

Holotype No. 10492 U.M.

Horizon and locality. — Zone 6 of Location 29.

Polypora uniplana, sp. nov.

(Plate X, Figs. 8–9)

Zoarium unknown, fragments flattened.

Obverse. — Branches slender, 0.22 mm. above a bifurcation; subangular. Dissepiments angular, slightly depressed; six in 5 mm.; more slender than branches. Fenestrules flat-oval to subquadrangular; average length 0.72 mm., average width 0.28 mm. Apertures in two to four ranges; sixteen to eighteen in 5 mm., separated by less than diameter of an aperture. Low blunted nodes irregularly disposed between ranges of apertures.

Reverse. — Branches slender, subangular, striated where worn. Straight to faintly sinuous, subparallel. Dissepiments slender, average 0.20 mm. in width; rounded, smooth; not depressed below plane of branches, very slightly expanded at the junction with branches. Fenestrules subquadrangular to occasionally oval; length from 0.65 to 0.75 mm., width from 0.30 to 0.42 mm.

Holotype No. 10491 U.M.

Horizon and locality. — Middle shale zone of Location 14.

Polypora minuta, sp. nov.

(Plate XI, Figs. 1–3)

Zoarium unknown, fragments flattened, flexuous.

Obverse. — Branches rounded, slender, flattened, subparallel. Dissepiments more slender than branches, from 0.30 to 0.42 mm. in width; rounded; depressed below plane of branches, so that they appear as grooves between branches; five, rarely six, in 5 mm. Fenestrules elongate-oval to elliptical; length average 0.52 mm., width 0.24 mm. Apertures in two to four ranges, usually three; nineteen to twenty in 5 mm., separated by more than diameter of an aperture. Peristomes of lateral ranges of apertures indent borders of fenestrules.

Reverse. — Branches slender, angular; zigzag to sinuous, rarely straight. Dissepiments equal to branches in width, angular; not depressed below plane of branches. Both dissepiments and branches faintly granulose. Fenestrules oval to subquadrangular to subhexagonal; length from 0.50 to 0.62 mm., width from 0.28 to 0.50 mm.

Holotype No. 10490 U.M. Paratype No. 10538 U.M.

Horizon and locality. — Bell Shale of Location 31.

Polypora allelomorpha, sp. nov.

(Plate XI, Figs. 4–5)

Zoarium unknown, fragments flattened, flexuous.

Obverse. — Branches rounded, slender, width 0.32 to 0.50 mm.; straight, subparallel. Dissepiments rounded, depressed below plane of branches, appear as grooves between branches; very short; six in 5 mm. Fenestrules narrow-oval; width one fourth to one third of length. Apertures in two to four ranges, usually three; twenty-two to twenty-four in 5 mm., separated longitudinally by more than diameter of an aperture.

Reverse. — Branches slender, angular, smooth; zigzag to sinuous; minutely granulose. Dissepiments usually more slender than branches and expanded at junction with them; subangular to sharply rounded; depressed below plane of branches. Where branches anastomose, dissepiments thus formed appear raised above plane of branches. Fenestrules broadly oval to circular; length from 0.50 to 0.66 mm., width from 0.34 to 0.48 mm.

Holotype No. 10507 U.M.

Horizon and locality. — Bell Shale of Location 31.

Polypora labellosa, sp. nov.

(Plate XI, Figs. 6–7)

Zoarium unknown, fragments flattened, flexuous.

Obverse. — Branches rounded, strong, straight, subparallel; bifurcations distant. Dissepiments rounded, equal in width to branches; depressed below plane of branches; six in 5 mm. Fenestrules very narrow-oval; width from one third to one half of length. Apertures in three to four ranges; eighteen to twenty in 5 mm., separated by more than diameter of an aperture; peristomes prominently developed.

Reverse. — Branches slender, sharply angular, with coarse, granular nodes at the summit, usually a larger node at the junction of a dissepiment; never straight, zigzag to sinuous; anastomosed. Dissepiments more slender than branches; angular, granular; depressed below plane of branches. Anastomosed portions of branches raised above plane of branches. Fenestrules predominantly hexagonal, in places quadrangular to suboval, length from 0.50 to 0.70 mm., width and length subequal.

Holotype No. 10462 U.M.

Horizon and locality. — Bell Shale of Location 31.

Polypora tenuosa, sp. nov.

(Plate XI, Figs. 8–9)

Zoarium infundibuliform, fronds rapidly expanding from very short base.

Obverse. — Branches rounded, slender, from 0.30 to 0.44 mm. in width; straight, subparallel. Dissepiments more slender than branches, average 0.18 mm. in width; rounded to subangular; slightly depressed below plane of branches; eleven in 5 mm. Fenestrules flattened-oval; average length 0.32 mm., average width 0.12 mm. Apertures in two to three ranges; twenty-two to twenty-four in 5 mm., separated by more than diameter of an aperture. Peristomes weakly developed, faintly indenting borders of fenestrules.

Reverse. — Branches straight to sinuous, slender, rounded, smooth; thirteen to fourteen in 5 mm. Dissepiments equal in width to branches; rounded, smooth; not depressed below plane of branches. Fenestrules quadrangular to broadly oval to subhexagonal; length from 0.22 to 0.36 mm., width from 0.18 to 0.30 mm.

Holotype No. 10461 U.M.

Horizon and locality. — Zone 1 of Location 49.

Polypora exemplaria, sp. nov.

(Plate XII, Figs. 1–3)

Zoarium infundibuliform, rapidly expanding.

Obverse. — Branches slender, width from 0.22 to 0.40 mm.; subangular, with low, rounded ridges between ranges of apertures; straight, subparallel. Bifurcation every fifth dissepiment in basal portion of frond, more distant in peripheral region of zoarium. Dissepiments rounded, short; average 0.30 mm. in width; seven to eight in 5 mm. Fenestrules oval; average length 0.40 mm., average width 0.20 mm. Apertures in two to three ranges; twenty to twenty-one in 5 mm., separated by more than diameter of an aperture. Peristomes absent.

Reverse. — Branches smooth, rounded, slender, straight to zigzag; eight in 5 mm. Dissepiments rounded, more slender than branches to slightly thicker; in basal portion of frond depressed below plane of branches, in peripheral region on plane of branches; from 0.28 to 0.36 mm. in thickness. Fenestrules broadly oval to circular; length from 0.38 to 0.52 mm., width from 0.22 to 0.32 mm.

Holotype No. 10489 U.M.

Horizon and locality. — Zone 1 of Location 17.

Polypora muricula, sp. nov.

(Plate XII, Figs. 4–6)

Zoarium infundibuliform, rapidly expanding.

Obverse. — Branches angular, strong; width from 0.50 to 0.65 mm.; straight, subparallel; bifurcations frequent in basal portion of frond. Dissepiments short, very strong, rounded, slightly depressed below plane of branches. Fenestrules narrowly oval; average length 0.70 mm., width from 0.20 to 0.30 mm. Apertures in two to three ranges; twenty-two to twenty-four in 5 mm., separated by more than diameter of an aperture. Ranges of apertures divided by low, rounded to slightly angular ridges. Apertures disposed on ends of dissepiments, but never completely covering them. Summits of ridges bear minute, low, granular nodes irregularly distributed.

Reverse. — Branches subangular to rounded; minutely granulose; sinuous to zigzag, rarely straight. Dissepiments of same thickness as branches; rounded, smooth to minutely granulose; slightly depressed below plane of branches; five in 5 mm. Fenestrules broadly oval to subquadrangular; length from 0.70 to 0.85 mm., width from 0.50 to 0.60 mm.

Holotype No. 10488 U.M.

Horizon and locality. — Zone 1 of Location 17.

Polypora modesta, sp. nov.

(Plate XII, Figs. 7–8)

Zoarium unknown, fragments flattened, flexuous.

Obverse. — Branches slender; width from 0.30 to 0.44 mm.; angular, straight, subparallel, with low ridges separating ranges of apertures. Dissepiments angular; slightly depressed below plane of branches; short; average width 0.32 mm.; ten in 5 mm. Fenestrules oval; width from one fourth to two thirds of length. Apertures in two to three ranges; twenty-one to twenty-three in 5 mm., separated by more than diameter of an aperture. Peristomes slightly developed only on anterior edge of apertures.

Reverse. — Branches in thin sections appear angular; sinuous to zigzag; average width 0.24 mm. Dissepiments from slightly less thick to slightly thicker than branches; angular; appear not depressed below plane of branches. Fenestrules circular to oval; length from 0.26 to 0.44 mm., width from 0.18 to 0.32 mm. All characters of reverse face taken from thin section. Reverse face not seen macroscopically.

Holotype No. 10496 U.M.

Horizon and locality. — Shaly limestone of Location 35.

Genus Ptiloporina Hall

For generic description see page 237.

Ptiloporina jugosa, sp. nov.

(Plate XII, Fig. 9; Plate XIII, Fig. 2)

Zoarium unknown, fragments flattened.

Obverse. — Branches of two sizes; in thin section they appear angular, with sharp ridges between ranges of apertures; summits of ridges bear sharp, prominent spines. Dissepiments of about half the thickness of smaller branches; set oblique to branches in places where smaller branches unite with larger ones; six dissepiments in 5 mm. Fenestrules subrhomboidal to distorted oval; length from 0.46 to 0.72 mm., width from 0.26 to 0.40 mm. Apertures in two to three ranges; seventeen to nineteen in 5 mm., separated by less than diameter of an aperture.

Reverse. — Branches of two sizes; larger ones 0.64 mm. in width; smaller ones average 0.30 mm. Larger branches gently rounded to flattened; raised above plane of smaller ones; minutely granulose; slightly sinuous in places. Smaller branches rounded to subangular, faintly carinated in places; minutely granulose;

straight to sinuous; bifurcate from larger branches. Dissepiments of equal width to smaller branches; rounded, minutely granulose; usually oblique to branches; not depressed below plane of smaller branches. Fenestrules oval to subquadrangular; length from 0.50 to 0.62 mm., average width 0.32 mm.

Holotype No. 10487 U.M.

Horizon and locality. — Middle shale zone of Location 14.

Genus SEMICOSINIUM Prout

For generic description see page 237.

Semicosinium rhombicum Ulrich var. **obliquum,** var. nov.

(Plate XIII, Figs. 1, 3, 5)

Zoarium unknown, fragments flattened.

Obverse. — Branches slender, width from 0.20 to 0.30 mm.; sinuous to zigzag; sharply angular. Dissepiments much stronger than branches; width from 0.32 to 0.60 mm.; short, rounded to flattened; much depressed below plane of branches; five to six in 5 mm. Large accessory pore, 0.30 mm. in diameter, disposed on about fifty per cent of dissepiments. Fenestrules oval; length from 0.38 to 0.62 mm., width from 0.18 to 0.28 mm. Apertures open laterally; twenty-five in 5 mm., separated by less than diameter of an aperture. Carina thin, flexuous, sinuous, prominent; not expanded at summit.

Reverse. — Branches broadly angular; straight to faintly sinuous; minutely granulose. Dissepiments from slightly less than width of branches to equal to it; faintly angular; not depressed below plane of branches; minutely granulose. Fenestrules quadrangular to circular; length from 0.32 to 0.40 mm.

Holotype No. 10486 U.M.

Horizon and locality. — Zone 1 of Location 49.

Semicosinium thyene Hall var. **michiganense,** var. nov.

(Plate XIII, Figs. 4, 6)

Zoarium unknown, fragments flattened.

Obverse. — Branches slender, average 0.28 mm. in width; straight to slightly sinuous; sharply rounded to angular; ten in

5 mm. Dissepiments short, rounded to angular, much depressed; width from 0.30 to 0.48 mm.; seven in 5 mm. Fenestrules oval; width from 0.18 to 0.20 mm.; length equal to that of those on reverse face. Apertures open directly, separated by more than diameter of an aperture; from two to three to a fenestrule; twenty-three in 5 mm. Carina thin, flexuous, sinuous, very high, summit but slightly expanded.

Reverse. — Branches zigzag, slender, subangular, minutely granulose; anastomosed in places. Dissepiments subangular, with minute granular nodes; twice width of branches and elevated above plane of branches, in places continuing across them when dissepiments occur opposite each other. Fenestrules circular, oval to subrhomboidal; length from 0.36 to 0.50 mm., width from 0.26 to 0.38 mm.

Holotype No. 10485 U.M.

Horizon and locality. — Shale bed of Location 46.

Semicosinium approximatum, sp. nov.

(Plate XIV, Figs. 1–2)

Zoarium unknown, fragments flattened.

Obverse. — Branches slender, width from 0.22 to 0.36 mm.; flattened to angular, sinuous. Dissepiments very broad, average 0.56 mm.; rounded to flattened, depressed below plane of branches, appear as grooves between branches; five in 5 mm. Fenestrules oval, length 0.28 to 0.42 mm., average width 0.26 mm. Carina thin, sinuous; not expanded at summit; prominent. Apertures open directly; twenty-five in 5 mm., separated by more than diameter of an aperture. Large accessory pores on dissepiments absent in this form.

Reverse. — Branches in thin section appear angular; zigzag to sinuous, slender, anastomosed. Dissepiments, or anastomosed portions of branches, much stronger than normal branches; rounded to angular; not depressed below plane of normal branches. Fenestrules oval to circular to quadrangular; length from 0.30 to 0.38 mm., width and length subequal.

Holotype No. 10484 U.M. Paratypes Nos. 10550, 10551 U.M.

Horizon and locality. — Shaly limestone of Location 35.

Semicosinium flexuosum, sp. nov.

(Plate XIV, Figs. 3–4)

Zoarium unknown, fragments flattened.

Obverse. — Branches moderately strong, average width 0.40 mm.; angular, sinuous. Dissepiments from 0.40 to 0.60 mm. in width; rounded; depressed below plane of branches; five in 5 mm. Fenestrules oval; length from 0.44 to 0.70 mm., width from 0.24 to 0.32 mm. Apertures open directly; twenty in 5 mm., separated by more than diameter of an aperture. Carina thin, expanded and flattened at summit, prominent, flexuous, sinuous; in places summit appears rounded.

Reverse. — Branches slender, rounded, smooth, straight to sinuous. Dissepiments more slender than branches; broadly expanded at junction with, and not depressed below plane of, branches. Fenestrules oval to subquadrangular; length from 0.70 to 0.92 mm., width from 0.50 to 0.62 mm.

Holotype No. 10483 U.M. Paratype No. 10548 U.M.

Horizon and locality. — Zone 1 of Location 17.

Conclusions

The forms of fenestellid Bryozoa in the Devonian rocks of Michigan represent but a small number of the total fauna. Any group as small as the family Fenestellidae does not give sufficient evidence for stratigraphic correlation, but is useful as contributing evidence to that furnished by the complete fauna.

Such forms as *Fenestella vera* Ulrich var. *acuta*, *F. variifenestrula*, *F. incerta*, *F. foraminosa*, *Isotrypa tropozomena*, *I. megista*, *I. anomala*, *I. anomala* var. *sinuosa*, and *Polypora amorpha* occur in certain similar lithologic horizons on both sides of the state, but these horizons cannot be correlated as the same stratigraphic zone upon such meager evidence.

The species *Fenestella longispinosa*, *Polypora minuta*, *P. allelomorpha*, and *Semicosinium approximatum* occur in both the Bell Shale at Location 31, which lies at the base of the Lower Traverse group, and in the shaly limestone at Partridge Point, Location 35, which represents the highest beds of the Upper

FAUNAL CHART

SPECIES	LOCALITIES															
	9.1	10.2	14 Shale	17.1	28 Top	29.6	31 Bell Shale	32	35	36	38 Bell Shale	40.8	41	46	47	49.1
Fenestella compacta, sp. nov.	..	..	..	..	..	..	..	..	+	..	..	..	..	..	+	+
depressa Hall	..	+	..	+	..	..	..	..	+	..	..	..	..	+	..	+
eximia Winchell problematica, var. nov.	..	+	..	..	..	..	..	..	..	..	..	..	..	..	..	+
foraminosa, sp. nov.	..	..	+	..	..	..	..	..	..	..	+	..	..	..	..	+
idalia Hall	..	..	..	+	..	..	..	..	..	..	..	..	..	..	..	+
incerta, sp. nov.	..	+	+	+	..	..	..	..	..	..	..	+	..	..	..	..
incisa, sp. nov.	..	..	..	..	+	..	..	..	+	..	..	..	..	..	..	..
longispinosa, sp. nov.	..	..	..	..	..	..	+	..	+	..	..	+	..	..	+	+
megalopora, sp. nov.	..	..	+	..	..	..	..	..	..	..	..	..	..	..	..	..
minutiserrata, sp. nov.	..	..	..	..	..	..	..	..	..	..	..	..	+	..	..	..
nodicula, sp. nov.	..	..	..	..	..	..	..	+	+	..	..	+	..	..	..	..
nodosa Prout	+	+	..	..	..	..	..	..	+	..	..	..	..	..	+	..
paridistans, sp. nov.	+	..	..	..	..	..	..	..	..	..	..	..	..	..	..	..
variifenestrula, sp. nov.	+	..	..	+	..	..	..	..	+	..	..	+	..	..	+	+
vera Ulrich var. acuta, var. nov.	+	..	+	+	+	+	+	..	..	..	..	..	..	..	..	+
Hemitrypa variosa, sp. nov.	..	..	..	..	..	..	..	..	..	..	..	..	..	+	..	..
Isotrypa angulata, sp. nov.	..	..	..	+	..	..	..	..	..	..	..	+	..	..	+	..
anomala, sp. nov.	+	..	+	+	..	..	+	..	..	..	..	+	..	..	..	..
anomala var. sinuosa, var. nov.	+	..	..	+	..	..	+	..	..	..	..	..	..	..	+	+
gigantica, sp. nov.	..	..	..	..	..	..	..	..	..	..	..	+	..	+	..	..
hexagona, sp. nov.	..	..	..	+	..	..	..	..	..	..	..	..	..	..	..	..
isopeda, sp. nov.	..	..	..	..	..	..	..	..	..	..	..	..	..	..	+	..
megista, sp. nov.	..	..	+	+	..	+	+	..	..	..	..	+	..	..	..	..
ovata, sp. nov.	..	..	..	+	..	..	..	..	..	..	..	..	..	..	..	..
oxytropis, sp. nov.	..	..	..	..	..	..	+	..	..	..	..	+	..	..	+	+
rara, sp. nov.	..	..	..	..	..	..	..	..	..	..	..	+	..	..	..	..
tropozomena, sp. nov.	..	..	+	..	..	..	..	..	..	..	..	..	..	..	..	+
vibrata, sp. nov.	..	..	..	+	..	..	..	..	..	..	..	..	..	..	..	..
Phyllopora aequirotundata, sp. nov.	..	..	..	+	..	..	..	..	..	..	..	..	..	..	..	+
Polypora allelomorpha, sp. nov.	..	..	..	..	..	..	+	..	+	..	..	..	..	..	..	..
ambiplana, sp. nov.	..	..	..	..	..	..	..	..	..	..	..	..	..	..	..	+
amorpha, sp. nov.	..	..	+	+	+	+	..	..	..	..	..	+	..	..	..	+
brevissima, sp. nov.	..	..	..	..	..	..	..	..	..	..	..	..	..	..	..	+
exemplaria, sp. nov.	..	..	..	+	..	..	..	..	..	..	..	..	..	+	..	..
finitima, sp. nov.	..	..	..	..	..	..	+	..	..	..	..	..	..	..	..	..
indentata, sp. nov.	..	..	..	..	..	+	..	..	..	..	..	..	..	..	..	..
labellosa, sp. nov.	..	..	..	..	..	..	+	..	..	..	..	..	..	..	..	..
magnifica, sp. nov.	..	..	..	..	..	..	..	..	..	..	..	..	..	..	..	+
minuta, sp. nov.	+	..	..	..	..	..	+	..	+	..	+	+	..	..	..	..
modesta, sp. nov.	..	..	..	..	..	..	..	..	+	..	..	..	..	..	..	+
muriculata, sp. nov.	..	..	..	+	..	..	..	..	..	..	..	..	..	..	..	..
pyramidata, sp. nov.	..	..	..	..	..	..	..	..	..	..	..	..	..	..	..	+
tenuosa, sp. nov.	..	..	..	..	..	..	..	..	..	..	..	..	..	..	..	+
uniplana, sp. nov.	..	..	+	..	..	..	..	..	..	..	..	..	..	..	..	..
Ptiloporina jugosa, sp. nov.	..	..	+	..	..	..	..	..	..	..	..	..	..	..	..	..
Semicosinium approximatum, sp. nov.	..	..	..	..	..	..	+	..	+	..	+	..	..	..	..	..
flexuosum, sp. nov.	..	..	..	+	..	..	+	..	..	+	+	..	..	..	..	..
rhombicum Ulrich var. obliquum, var. nov.	..	..	..	..	..	..	..	..	+	..	..	..	..	..	..	+
thyene Hall var. michiganense, var. nov.	..	..	..	..	..	..	..	..	..	..	..	..	..	+	..	..

Traverse group. This long vertical range may represent similar depositional conditions which were favorable to these species.

These forms may have lived in other places during the whole of the time in which the intermediate strata were being deposited. Then with a change in conditions which created favorable environment the species returned and thrived until conditions were again altered. If this hypothesis is correct, similar conditions must have prevailed at four intervals during Devonian time in Michigan, because the same species occur at two different intermediate horizons between the Bell Shale and the beds at Partridge Point.

It is thought that further study of the other groups of the fauna of these horizons may suggest a similar condition.

With the exception of three species of *Fenestella*, a new variety of *Fenestella vera* Ulrich, a new variety of *Semicosinium rhombicum* Ulrich, and a new variety of *Semicosinium thyene* Hall, all the forms described are new. This great difference in the fauna from that of any other known region suggests that during Devonian time Michigan was more or less an isolated basin. It appears that this difference between the faunas of Michigan and those of adjacent regions is due to geographic isolation and does not represent a difference in age.

Dr. Erwin Pohl [5] has been studying some of the other groups of the Devonian fauna from Michigan and his conclusions correlate definitely with those of the author.

Stratigraphic Summary

The following correlations are purely tentative and are suggested only as possibilities of what the situation may be in the Traverse Group. The evidence from the fenestellid Bryozoa is insufficient to do more than suggest possible correlations.

Western side

On the western side of the state the material examined represents three zones. The blue shale occurring in the middle portion of the Petoskey Portland Cement Co. quarry, Location 14, is the lowest. This bed is from forty to fifty-five feet below the second

[5] Pohl, Erwin, *Proc. U. S. Nat. Mus.*, Vol. 76, Art. 14, p. 1, 1929.

of the three zones, which is the blue shale zone exposed as the lowest bed in the quarry of the Charlevoix Rock Products Co., Location 9, and in the quarry of the Antrim Lime Co., Location 17. The third and highest zone is a more or less crystalline limestone, yellow to gray in color, with shaly partings, occurring in the Charlevoix City quarry, Location 10, Zone 2. This zone is from thirty-seven to forty feet above the blue shale of Locations 9 and 17.

The blue shale of Zone 1, Locations 9 and 17, occurs about two feet below a two-foot bed of brown to yellow limestone with wavy, black carbonaceous streaks. This wavy layer occurs only as boulders in the drift high above the quarry at Location 14, but is limited to a narrow band, indicating that the rock is in place near the surface at this point. There is no exposure of the blue shale just below the wavy layer at this place. There is, however, a small flat developed at a slightly lower elevation than the band of boulders composed of the wavy limestone, and adjacent to it, which suggests that the rock exposed on the slope was soft and easily eroded. Since this flat occupies the theoretical position of the blue shale, it is probable that it was formed by the rapid erosion of an exposure of the shale.

The blue shale zone in the middle portion of the quarry walls at Location 14 is interbedded between coralline limestones. Because the lithologic character of this blue shale is the same as that of the blue shale just below the wavy layer and because *Fenestella incerta*, sp. nov., *F. vera* Ulrich var. *acuta*, var. nov., *Isotrypa megista*, sp. nov., and *I. anomala*, sp. nov., occur in all of these zones, it is believed that they represent the same depositional environment, although stratigraphically the blue shale zone of Location 14 is not equivalent to the blue shale of Locations 9 and 17.

Central area

There are but few horizons in the central part of the state which contain a fenestellid fauna. Zone 6 of Location 29 and Zone 3 of Location 28 were the only horizons which yielded Fenestellidae in this area and the faunas from these two zones are not the same.

Eastern side

On the eastern side of the state the lowest bed represented in the material studied is the Bell Shale, which occurs in pockets in the limestone of the quarry of the Michigan Limestone and Chemical Co. at Calcite, near Rogers City, Location 31, and as a five-foot bed at the base of the quarry at Rockport, Location 38. See chart, p. 269, for the faunal list of this shale.

The next higher zone, from which two species are identified, is not correlated on the basis of the Fenestellidae because both species have a long vertical range. The stratigraphic position of this zone is based upon the records of the Michigan Limestone and Chemical Co., which were made available by Mr. J. Valentine of that company. These records show this zone to be about fifty-seven feet above the Bell Shale. This zone is exposed at Location 32.

The faunas of the blue shale zones of Locations 40, 41, 46, 47 and 49 each show a considerable number of the same species. It is suggested that the similarity of the faunas from these horizons does not indicate that they are to be correlated as the same geologic formation, but rather that these species are long-lived forms which have a vertical range from the Long Lake Series to the Dock Street Shale. The blue shale of Locations 46 and 47 was described by Grabau [6] as the Dock Street Shale, which he placed at the base of the Upper Traverse or Thunder Bay Series. This horizon lies well above the one at Location 40, which is thought to be a member of the Long Lake Series of Grabau,[7] which he called the upper member of the Lower Traverse Group.

The highest beds exposed are those at Partridge Point, Location 35. These lie at the top of the Thunder Bay Series of Grabau.[8] Although this horizon and the one represented by Zone 8 of Location 40 have four species in common, they are not to be considered as the same horizon, but rather as another example of a recurrence

[6] Grabau, A. W., "Stratigraphy of the Traverse Group of Michigan," *Report of the State Board of Geological Survey of Michigan for the Year 1901*, pp. 163–210.

[7] *Ibid.* [8] *Ibid.*

of similar conditions which permitted these species to return to this area from a parent sea where they lived during the time the intermediate beds were being deposited.

List of Locations

LOCATIONS

9. Quarry of the Charlevoix Rock Products Co. NW. ¼ Sec. 28, T. 34 N., R. 8 W.
10. Charlevoix City quarry, now used as a dump ground. Immediately south of the center of E.–W. road, on the N. line of Sec. 33, T. 34 N., R. 8 W.
14. Quarry of the Petoskey Portland Cement Co.
17. Antrim Lime Co. quarry. SE. ¼ Sec. 1, T. 34 N., R. 6 W.
28. Exposure below dam on Black River one-quarter mile north of Tower, Michigan
29. Abandoned quarry on shore of Black Lake, near the State Park
31. Quarry of the Michigan Limestone and Chemical Co. at Calcite, near Rogers City, Michigan
32. Point on highway M–10, SW. corner of Sec. 35, Rogers Township
35. Exposure at Partridge Point
36. Abandoned quarry about one-quarter mile south of the center of Sec. 29, Alpena Township
38. Quarry at Rockport, Michigan
40. Quarry of the Michigan Alkali Co., at Alpena
41. Spillway below Power House, at the Three (Four) Mile Dam on the Thunder Bay River
46. Shale bank on the south side of the Thunder Bay River one-half mile east of the Three (Four) Mile Dam
47. Shale bank on the south side of the Thunder Bay River at the Seven Mile Dam
49. Abandoned quarry of a cement company near El Cajon Beach

BIBLIOGRAPHY

GRABAU, A. W. Stratigraphy of the Traverse Group of Michigan. Report of the State Board of Geological Survey of Michigan for the Year 1901, pp. 163–210.

HALL, JAMES. On the Mode of Growth and Relations of the Fenestellidae. New York State Geologist, Annual Report 4, pp. 35–46, pls. 1–2 (1885).

—— Descriptions of Fenestellidae of the Hamilton Group of New York. New York State Geologist, Annual Report 6, pp. 41–70, pls. 1–7 (1887).

—— AND SIMPSON, G. B. Natural History of New York. Geological Survey of the State of New York. Paleontology, Vol. 6: Corals and Bryozoa. 1887. Descriptions and figures of species of Fenestellidae from the Lower Helderberg are on pp. 43–66, pls. 17–22; from the Upper Helderberg, on pp. 104–172, pls. 34–54.

KING, WILLIAM. On Some Families and Genera of Corals. Annals and Magazine of Natural History, Sec. Ser., Vol. 3, pp. 388–390 (1849).

—— A Monograph of the Permian Fossils of England. "Bryozoa," pp. 28–49. London, 1850.

LONSDALE, WILLIAM. "Corals," in Sir R. I. Murchison's Silurian System, Part II, pp. 675–694, pls. 15–16. "Bryozoa," pp. 676–680, pl. 15. London, 1839.

McCOY, SIR FREDERICK. A Synopsis of the Characters of the Carboniferous Limestone Fossils of Ireland. "Bryozoa," pp. 194–207, pls. 26–29. Dublin, 1844.

PHILLIPS, JOHN. Figures and Descriptions of the Paleozoic Fossils of Cornwall, Devon, and West Somerset. . . . "Bryozoa," pp. 20–27, pls. 9–13. London, 1841.

POHL, ERWIN R. The Middle Devonian Traverse Group of Rocks in Michigan. A Summary of Existing Knowledge. Proceedings of the United States National Museum, Vol. 76, Art. 14, pp. 1–34, pls. 1–2 (1929).

PROUT, H. A. Third Series of Descriptions of New Species of Bryozoa from the Paleozoic Rocks of the Western States and Territories. Transactions of the Academy of Science of St. Louis, Vol. 1, pp. 443–452, pls. 17–18 (1859).

—— Descriptions of New Species of Bryozoa. Transactions of the Academy of Science of St. Louis, Vol. 2, pp. 410–413 (1866).

SIMPSON, G. B. A Handbook of the Genera of the North American Palaeozoic Bryozoa, with an Introduction upon the Structure of Living Species. New York State Geologist, Annual Report 14, pp. 403–669, pls. 1–25 [1897]. (This report of the State Geologist also forms part of the Forty-eighth Annual Report of the New York State Museum, Albany, 1895 [1897], with the same paging.)

ULRICH, E. O. Paleozoic Bryozoa. Geological Survey of Illinois, Vol. 8: Geology and Paleontology. Text, Part II. Paleontology of Illinois, Section 6, pp. 349–361, 395–397, and 534–614. Plates (in separate volume), pls. 44, 49, 50–52, 54, 57 (1890).

WINCHELL, ALEXANDER. "Appendix" to The Grand Traverse Region, a Report on the Geological and Industrial Resources of the Counties of Antrim, Grand Traverse, Benzie and Leelanaw in the Lower Peninsula of Michigan. Ann Arbor, 1866.

EXPLANATION OF PLATE I

All specimens magnified 12 diameters

PAGE

FIG. 1. *Fenestella eximia* Winchell *problematica*, var. nov.............. 238
Obverse face of frond showing low, rounded carina and deeply depressed dissepiments.
Holotype No. 10505 U.M.
Zone 2, Location 10.

FIG. 2. *Fenestella vera* Ulrich var. *acuta*, var. nov.................... 240
Obverse face of frond showing strongly developed peristomes, and apertures opening laterally.
Holotype No. 10504 U.M.
Shale zone, Location 14.

FIG. 3. *Fenestella vera* Ulrich var. *acuta*, var. nov.................... 240
Reverse face of frond showing granular node at junction of branch and dissepiment.
Holotype No. 10504 U.M.
Shale zone, Location 14.

FIG. 4. *Fenestella idalia* Hall.................................. 239
Obverse face of frond showing rounded branches, with apertures opening laterally.
Plesiotype No. 10482 U.M.
Zone 1, Location 17.

FIG. 5. *Fenestella idalia* Hall.................................. 239
Reverse face of frond showing both straight and sinuous branches and slightly depressed dissepiments.
Plesiotype No. 10482 U.M.
Zone 1, Location 17.

FIG. 6. *Fenestella idalia* Hall.................................. 239
Section cut slightly oblique to plane of frond, showing striae in deeper part of reverse branches.
Plesiotype No. 10482 U.M.
Zone 1, Location 17.

FIG. 7. *Fenestella compacta*, sp. nov.............................. 241
Obverse face of frond. Branches weathered so that they appear to be irregularly nodose.
Holotype No. 10481 U.M.
Shaly limestone, Location 35.

FIG. 8. *Fenestella compacta* sp. nov.............................. 241
Section cut slightly oblique to plane of frond, showing variation in size and shape of fenestrules. Section largely crystallized.
Holotype No. 10481 U.M.
Shaly limestone, Location 35.

PLATE I

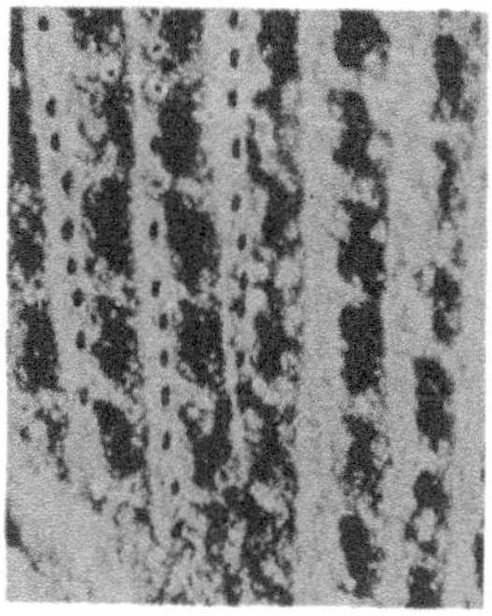

1

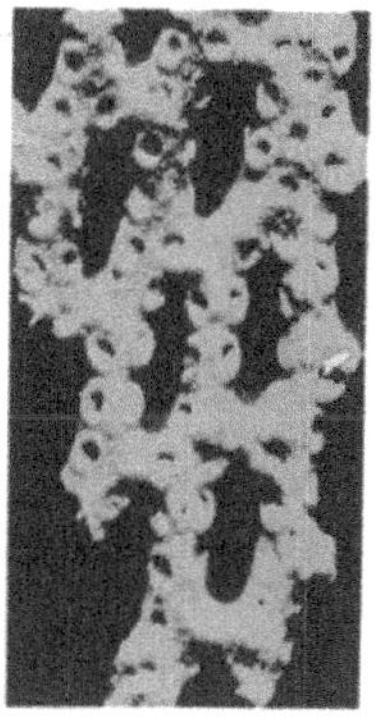

2

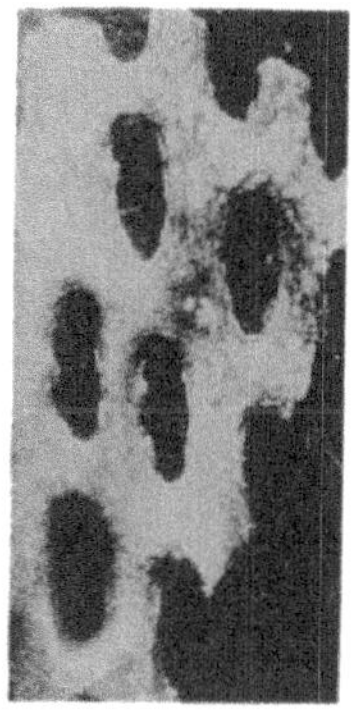

3

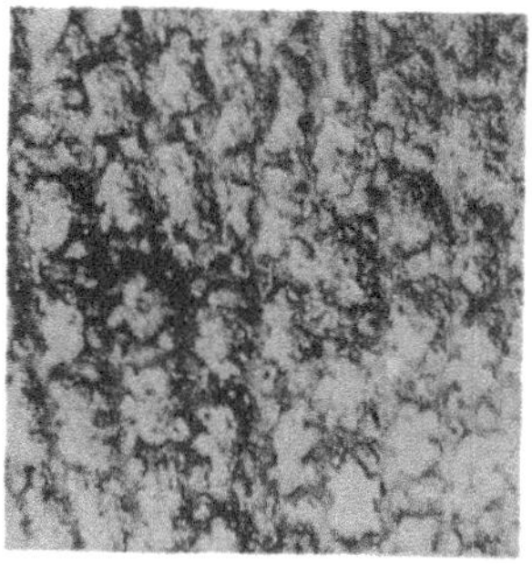

4

6

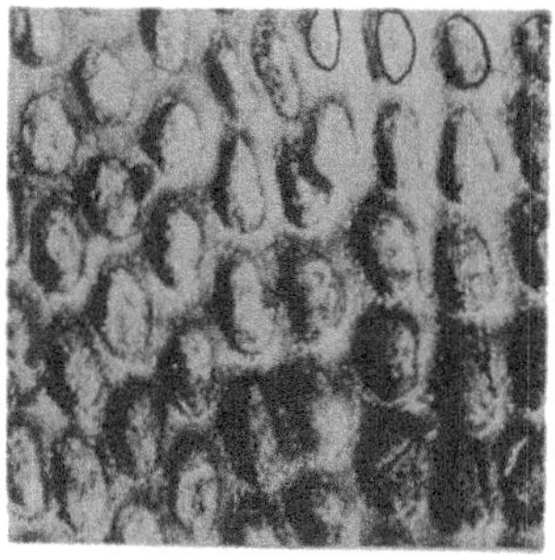

5

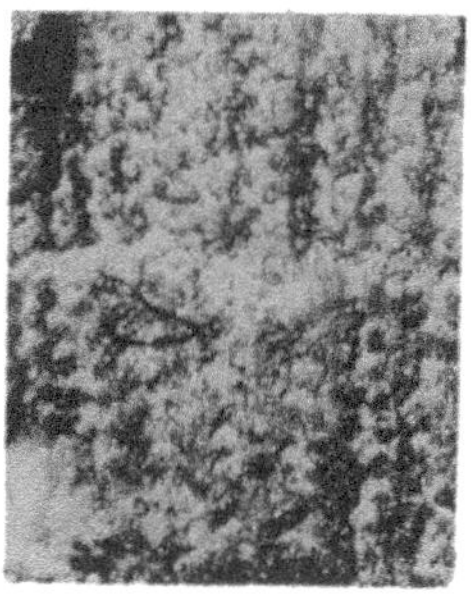

7

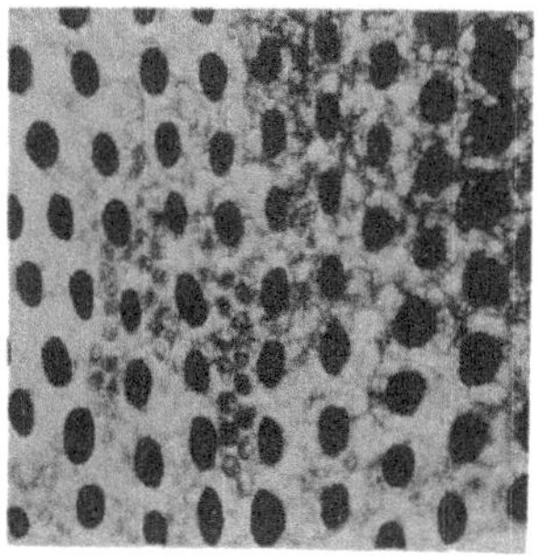

8

EXPLANATION OF PLATE II

All specimens magnified 12 diameters

PAGE

FIG. 1. *Fenestella paridistans*, sp. nov. 242
Obverse face of frond showing low, angular carina with blunt nodes disposed upon its summit.
Holotype No. 10480 U.M.
Zone 1, Location 9.

FIG. 2. *Fenestella paridistans*, sp. nov. 242
Reverse face of frond, showing granulose branches and dissepiments and occasional large granular nodes irregularly disposed upon branches.
Holotype No. 10480 U. M.
Zone 1, Location 9.

FIG. 3. *Fenestella paridistans*, sp. nov. 242
Section cut slightly oblique to plane of frond, showing arrangement of zoœcia in branches and structure of nodes upon obverse carina.
Holotype No. 10480 U.M.
Zone 1, Location 9.

FIG. 4. *Fenestella variifenestrula*, sp. nov. 242
Obverse face of frond, showing rugose appearance of angular branches. Apertures concealed by a thick secondary deposit.
Holotype No. 10479 U.M.
Shaly limestone, Location 35.

FIG. 5. *Fenestella variifenestrula*, sp. nov. 242
Reverse face of frond, showing branches and dissepiments on same plane, and low, broad, granular nodes disposed at junction of branches and dissepiments.
Holotype No. 10479 U.M.
Shaly limestone, Location 35.

FIG. 6. *Fenestella variifenestrula*, sp. nov. 242
Section cut slightly oblique to plane of frond, showing wide variation in size of fenestrules.
Holotype No. 10479 U.M.
Shaly limestone, Location 35.

FIG. 7. *Fenestella nodicula*, sp. nov. 243
Obverse face of frond, showing angular branches and slender, much depressed dissepiments.
Holotype No. 10478 U.M.
Shaly limestone, Location 35.

FIG. 8. *Fenestella nodicula*, sp. nov. 243
Section cut slightly oblique to plane of frond, showing structure of minute nodes disposed upon reverse branches and wide variation in size and shape of fenestrules.
Holotype No. 10478 U.M.
Shaly limestone, Location 35.

FIG. 9. *Fenestella incerta*, sp. nov. 244
Section cut slightly oblique to plane of frond, showing slender carinae on branches and angular, depressed dissepiments with an aperture occasionally disposed upon them.
Holotype No. 10477 U.M.
Zone 1, Location 9.

PLATE II

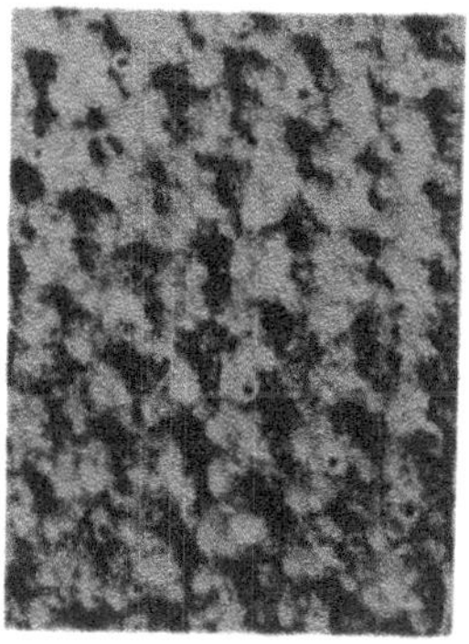

1

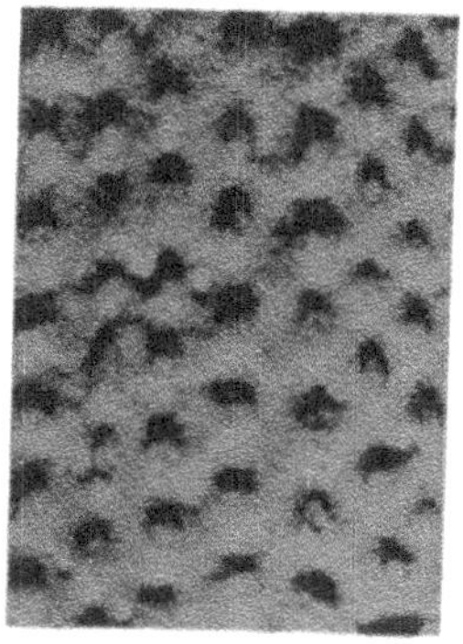

2

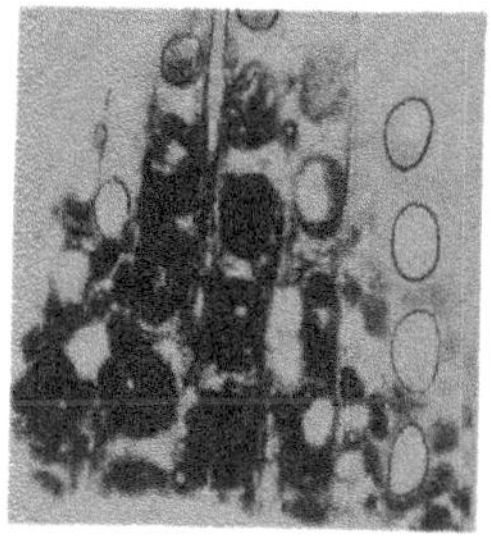

3

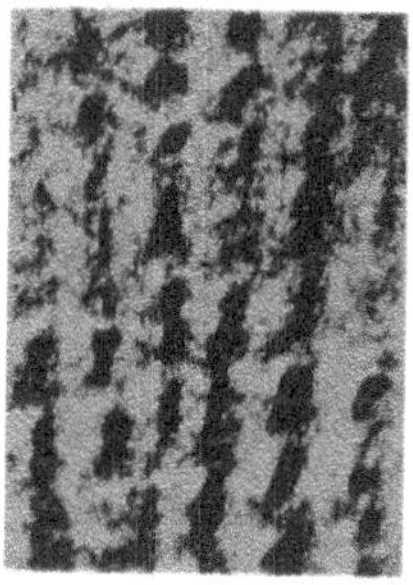

4

5

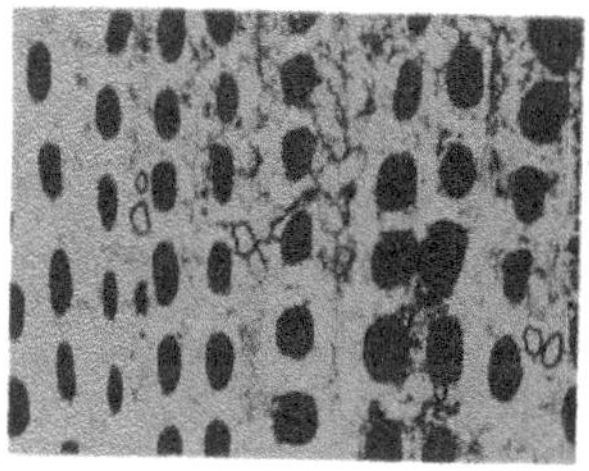

6

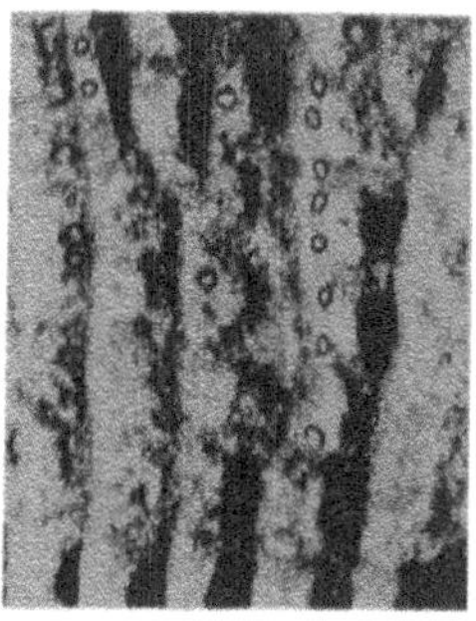

7

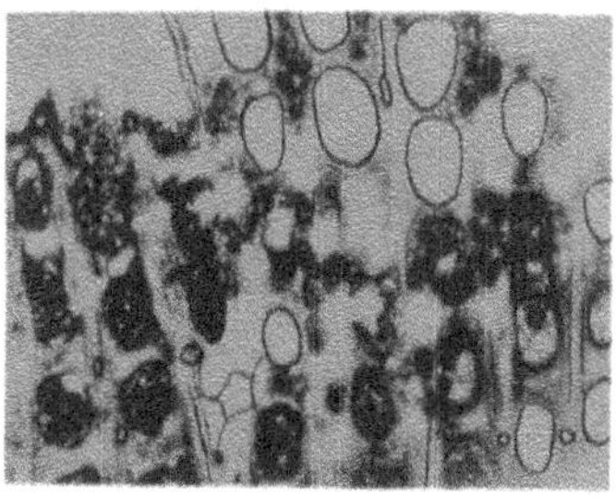

8

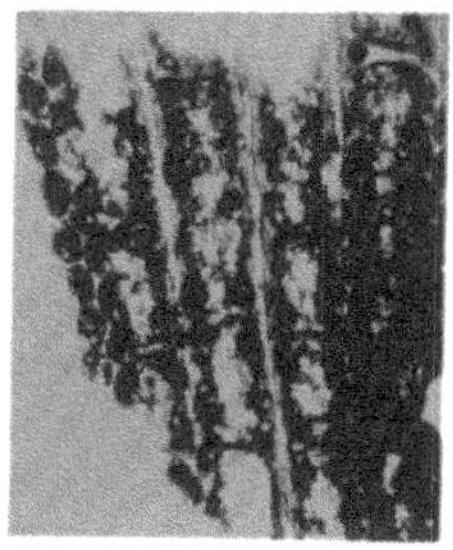

9

EXPLANATION OF PLATE III

All specimens magnified 12 diameters

PAGE

FIG. 1. *Fenestella incerta*, sp. nov. 244
Obverse face of frond, showing slender, rounded carinae with blunt nodes regularly disposed upon them.
Holotype No. 10477 U.M.
Zone 1, Location 9.

FIG. 2. *Fenestella incerta*, sp. nov. 244
Reverse face of frond, showing strong nodes disposed at junction of branches and dissepiments.
Holotype No. 10477 U.M.
Zone 1, Location 9.

FIG. 3. *Fenestella foraminosa*, sp. nov. 245
Obverse face of frond, showing apertures opening directly. Specimen partly covered with a secondary deposit.
Holotype No. 10498 U.M.
Location 36.

FIG. 4. *Fenestella foraminosa*, sp. nov. 245
Section cut slightly oblique to plane of frond. Central portion shows position of apertures with relation to zoœcia.
Holotype No. 10498 U.M.
Location 36.

FIG. 5. *Fenestella incisa*, sp. nov. 246
Obverse face of frond, showing prominent carinae with blunt nodes disposed upon summits.
Holotype No. 10509 U.M.
Top zone, Location 28.

FIG. 6. *Fenestella incisa*, sp. nov. 246
Reverse face of frond, showing minutely granulose branches and dissepiments.
Holotype No. 10509 U.M.
Top zone, Location 28.

FIG. 7. *Fenestella megalopora*, sp. nov. 245
Obverse face of frond, showing apertures opening laterally, closely disposed, and with their peristomes indenting borders of fenestrules.
Holotype No. 10497 U.M.
Top zone, Location 28.

FIG. 8. *Fenestella megalopora*, sp. nov. 245
Reverse face of frond, showing faint carina upon branch, with an occasional granular node.
Holotype No. 10497 U.M.
Top zone, Location 28.

FIG. 9. *Fenestella megalopora*, sp. nov. 245
Section cut slightly oblique to plane of frond, showing arrangement of apertures at bifurcation of a branch, and structure of carinae and elongated nodes on reverse branches.
Holotype No. 10497 U.M.
Top zone, Location 28.

PLATE III

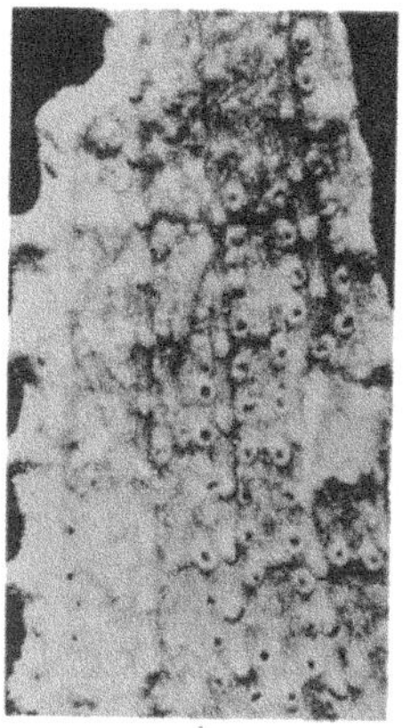

1

2

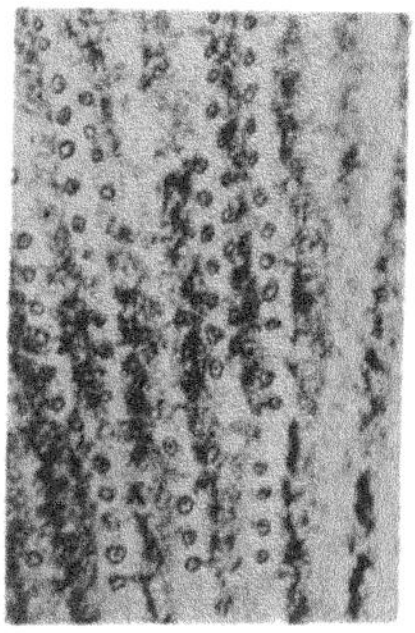

3

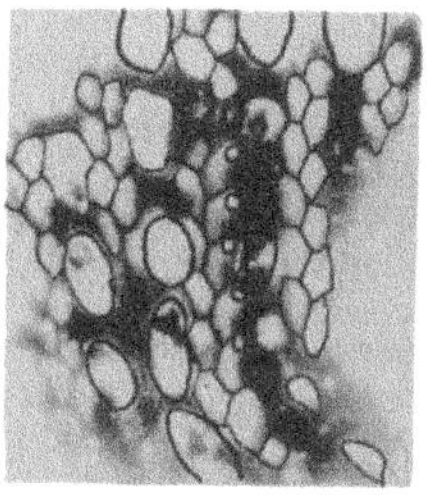

4

5

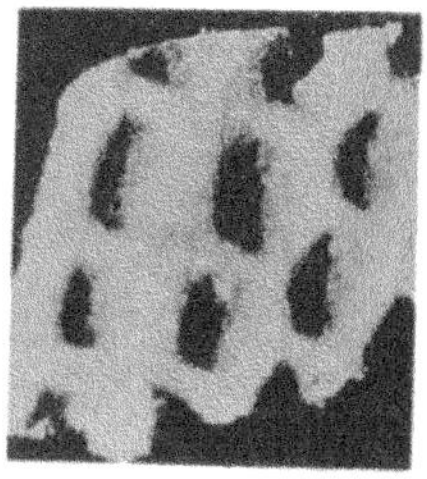

6

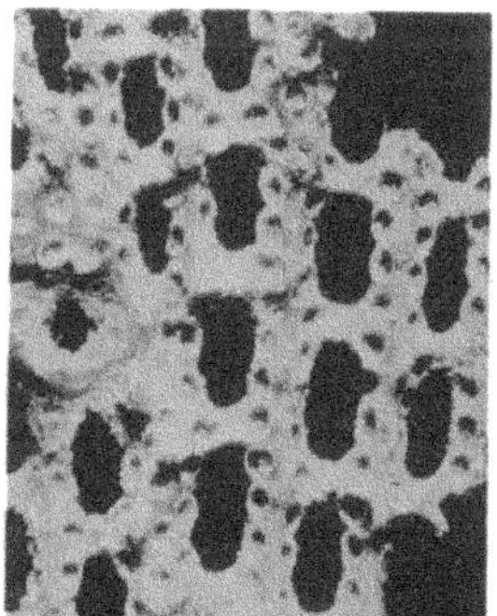

7

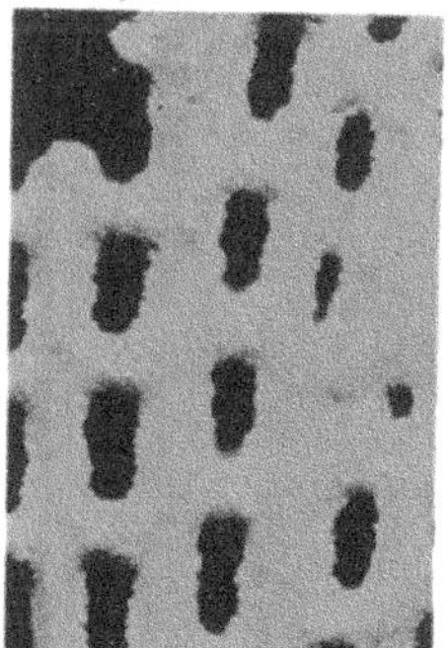

8

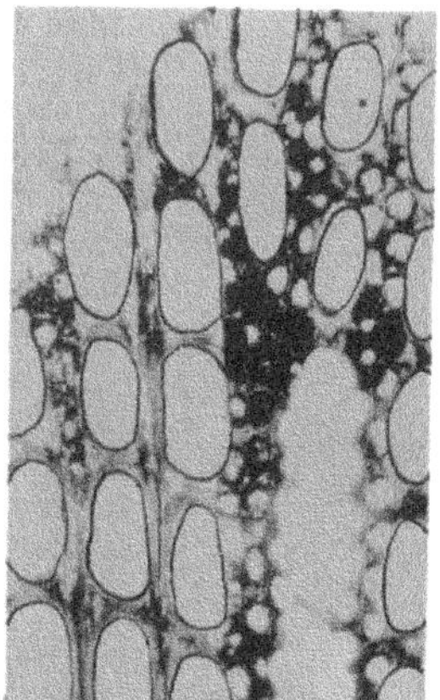

9

EXPLANATION OF PLATE IV

All specimens magnified 12 diameters

PAGE

FIG. 1. *Fenestella longispinosa*, sp. nov. 247
Reverse face of frond, showing dissepiments depressed below plane of branches. Specimen badly worn.
Holotype No. 10476 U.M.
Shaly limestone, Location 35.

FIG. 2. *Hemitrypa variosa*, sp. nov. 247
Obverse face of frond, showing superstructure.
Holotype No. 10475 U.M.
Location 46.

FIG. 3. *Fenestella longispinosa*, sp. nov. 247
Section cut slightly oblique to plane of frond. Left side of figure shows apertures opening directly and minute nodes regularly disposed upon obverse branches. Right side of figure shows structure of reverse branches with occasional node disposed at junction of branches and dissepiments.
Holotype No. 10476 U.M.
Shaly limestone, Location 35.

FIG. 4. *Isotrypa tropozomena*, sp. nov. 248
Reverse face of frond, showing strong, angular dissepiments continuing across branches.
Holotype No. 10474 U.M.
Middle shale zone, Location 14.

FIG. 5. *Hemitrypa variosa*, sp. nov. 247
Section cut oblique to plane of frond, showing structure of branches with carinae supporting superstructure.
Holotype No. 10475 U.M.
Location 46.

FIG. 6. *Isotrypa tropozomena*, sp. nov. 248
Section cut oblique to plane of frond, showing arrangement of apertures and carina on both branches and dissepiments.
Holotype No. 10474 U.M.
Middle shale zone, Location 14.

PLATE IV

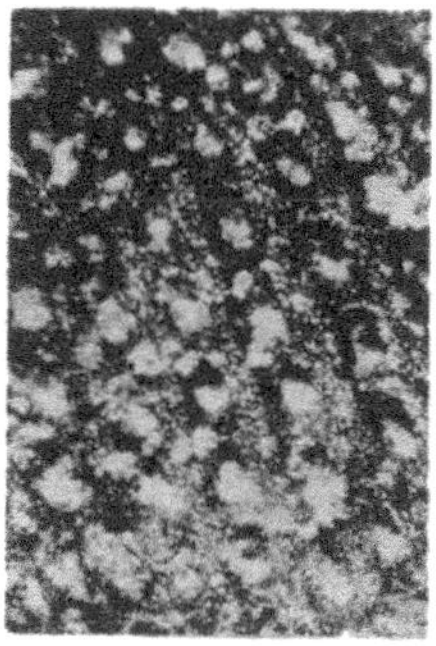

1

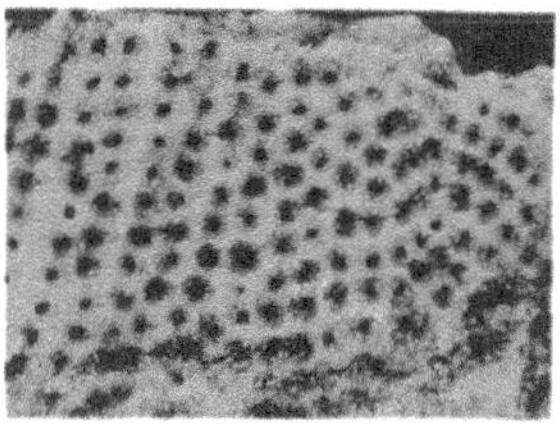

2

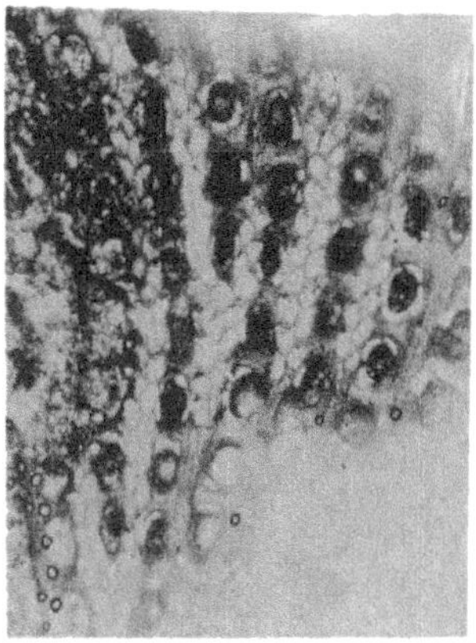

3

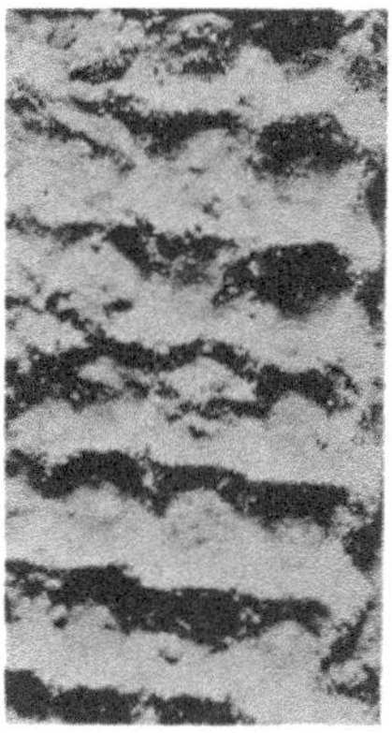

4

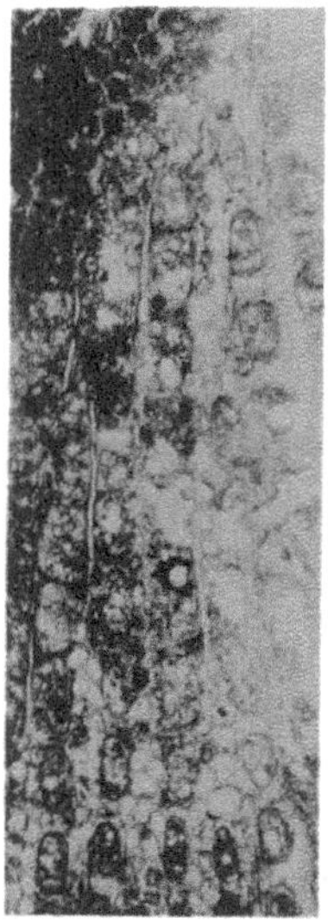

5

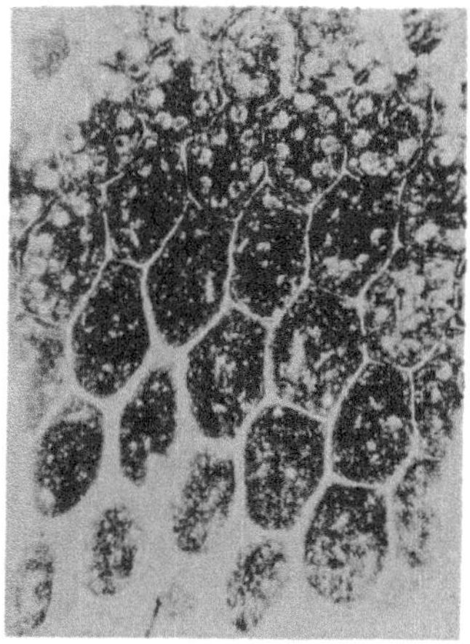

6

EXPLANATION OF PLATE V

All specimens magnified 12 diameters

PAGE

FIG. 1. *Isotrypa megista*, sp. nov. 249
Obverse face of frond, showing angular, granulose superstructure.
Holotype No. 10473 U.M.
Middle shale zone, Location 14.

FIG. 2. *Isotrypa megista*, sp. nov. 249
Reverse face of frond, showing irregular branches with low nodes disposed upon them.
Holotype No. 10473 U.M.
Middle shale zone, Location 14.

FIG. 3. *Isotrypa megista*, sp. nov. 249
Section cut nearly parallel to plane of frond, showing arrangement of zoœcia in zigzag branches, and wide variation in width of dissepiments.
Holotype No. 10473 U.M.
Middle shale zone, Location 14.

FIG. 4. *Isotrypa angulata*, sp. nov. 250
Obverse face of frond, showing superstructure.
Holotype No. 10472 U.M.
Location 47.

FIG. 5. *Isotrypa angulata*, sp. nov. 250
Reverse face of frond, showing branches more sinuous and fenestrules more elongate-oval than those of superstructure.
Holotype No. 10472 U.M.
Location 47.

FIG. 6. *Isotrypa angulata*, sp. nov. 250
Section cut slightly oblique to plane of frond, showing extremely slender carinae supporting superstructure.
Holotype No. 10472 U.M.
Location 47.

FIG. 7. *Isotrypa anomala*, sp. nov. 250
Obverse face of frond, showing faintly carinated superstructure.
Holotype No. 10471 U.M.
Zone 1, Location 9.

FIG. 8. *Isotrypa anomala*, sp. nov. 250
Reverse face of frond, showing irregularity in size of branches and dissepiments occasionally depressed below plane of branches.
Holotype No. 10471 U.M.
Zone 1, Location 9.

FIG. 9. *Isotrypa anomala* var. *sinuosa*, var. nov. 251
Reverse face of frond, showing coarsely granulose branches and dissepiments.
Holotype No. 10470 U.M.
Zone 1, Location 49.

PLATE V

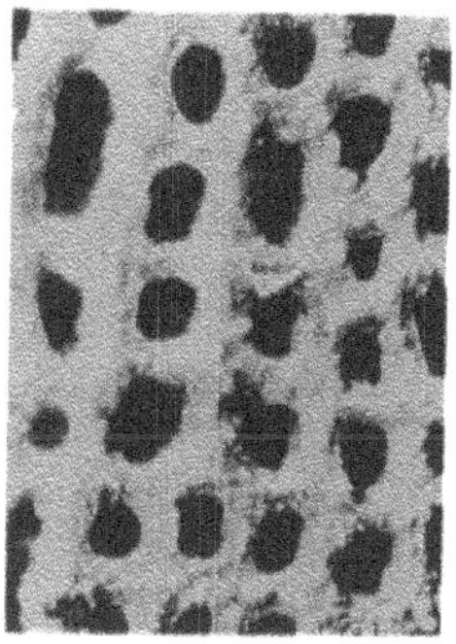

1

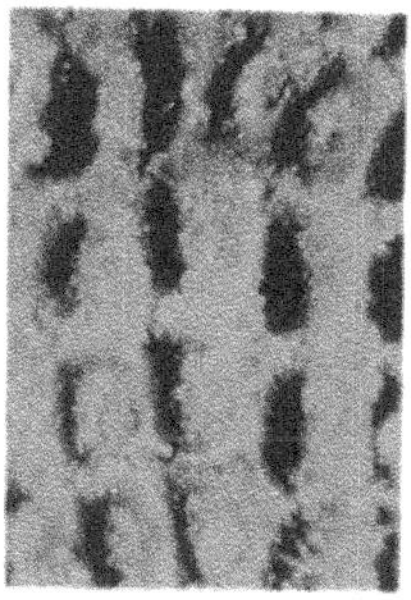

2

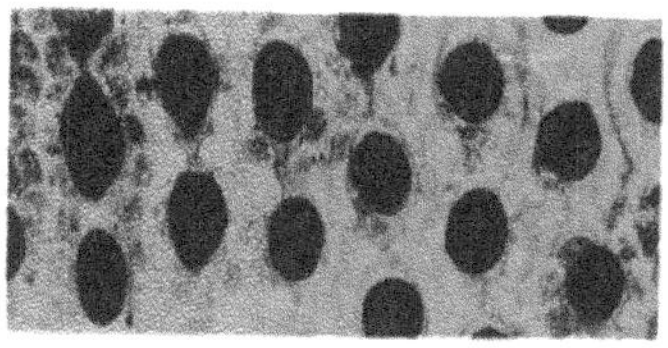

3

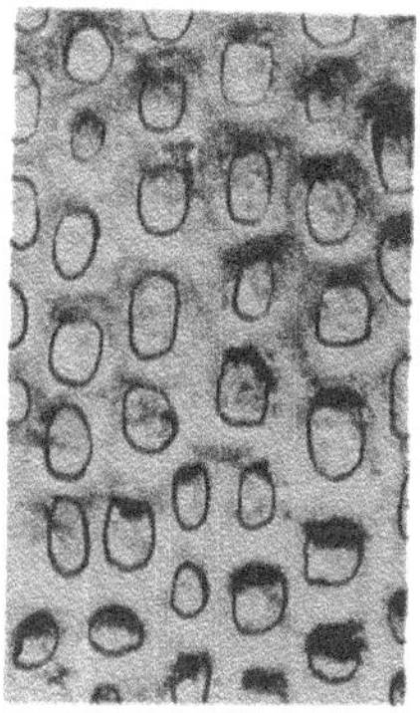

4

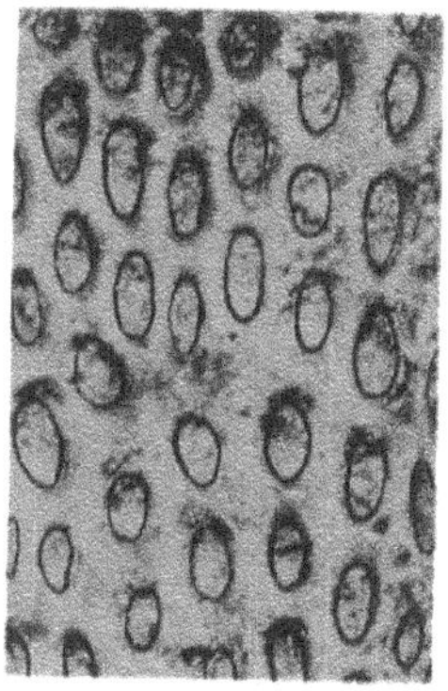

5

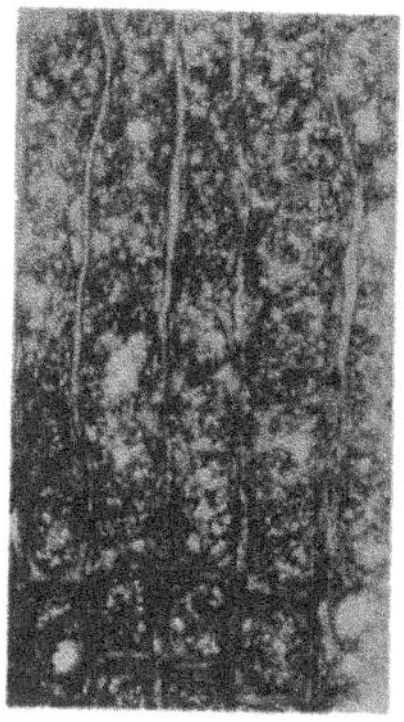

6

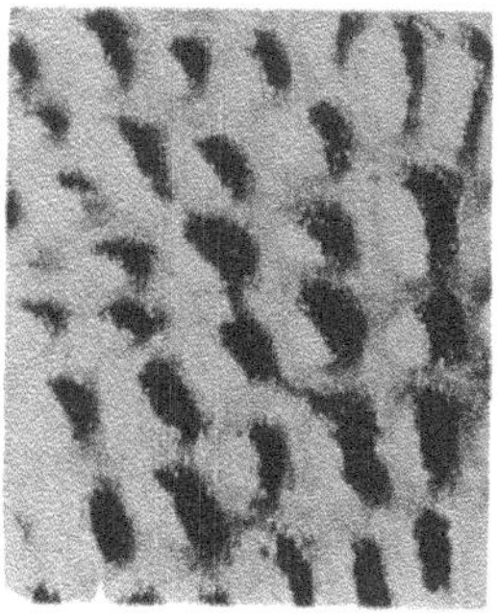

7

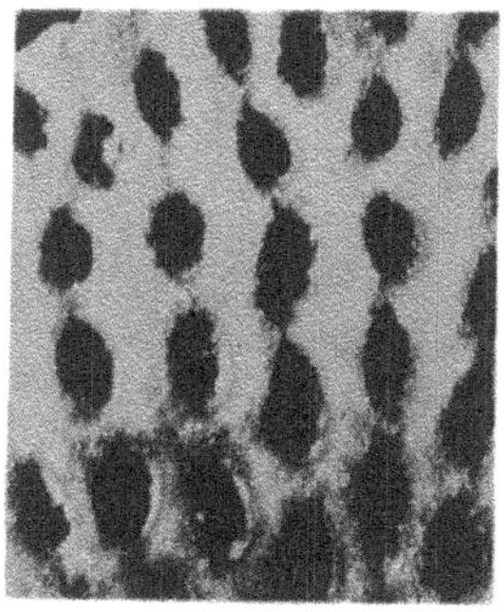

8

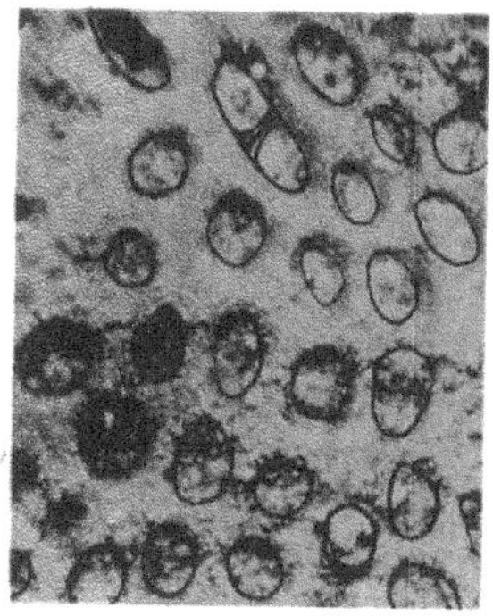

9

EXPLANATION OF PLATE VI

All specimens magnified 12 diameters

PAGE

FIG. 1. *Isotrypa anomala* var. *sinuosa*, var. nov. 251
Obverse face of frond, showing sharply carinated superstructure.
Holotype No. 10470 U.M.
Zone 1, Location 49.

FIG. 2. *Isotrypa anomala* var. *sinuosa*, var. nov. 251
Section cut oblique to plane of frond, showing sinuous to zigzag reverse branches, arrangement of zoœcia in branches,, and thin carinae supporting superstructure.
Holotype No. 10470 U.M.
Zone 1, Location 49.

FIG. 3. *Isotrypa rara*, sp. nov. 251
Obverse face of frond, showing faint striae in places on superstructure.
Holotype No. 10469 U.M.
Zone 8, Location 40.

FIG. 4. *Isotrypa rara*, sp. nov. 251
Section cut oblique to plane of frond, showing reverse branches and extremely thin carinae supporting superstructure.
Holotype No. 10469 U.M.
Zone 8, Location 40.

FIG. 5. *Isotrypa ovata*, sp. nov. 252
Obverse face of frond, showing superstructure.
Holotype No. 10468 U.M.
Zone 1, Location 17.

FIG. 6. *Isotrypa ovata*, sp. nov. 252
Reverse face of frond, showing irregularly disposed, granular nodes and very strong dissepiments.
Holotype No. 10468 U.M.
Zone 1, Location 17.

FIG. 7. *Isotrypa ovata*, sp. nov. 252
Section cut slightly oblique to plane of frond, showing arrangement and structure of nodes on reverse branches.
Holotype No. 10468 U.M.
Zone 1, Location 17.

PLATE VI

1

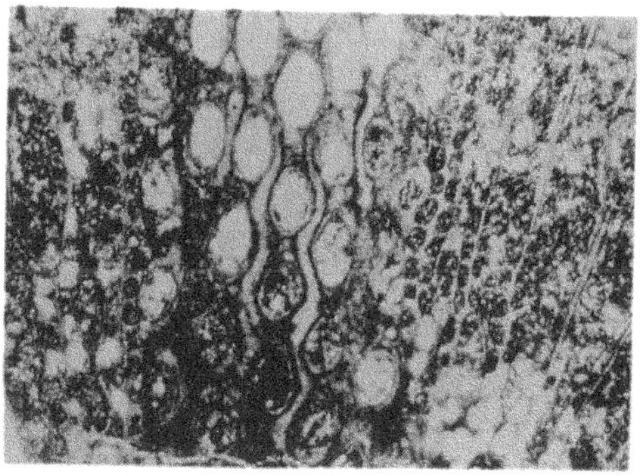

2

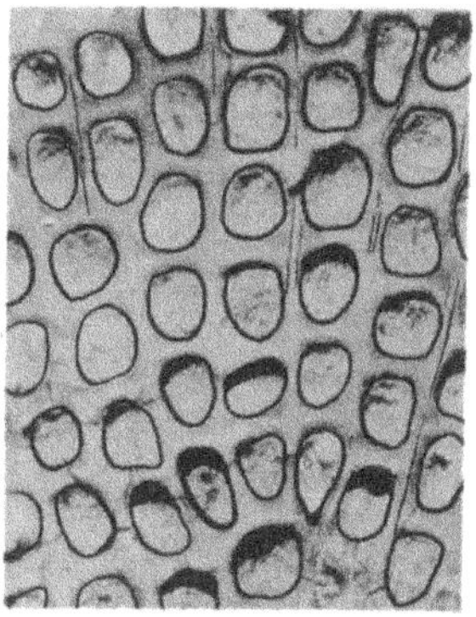

3

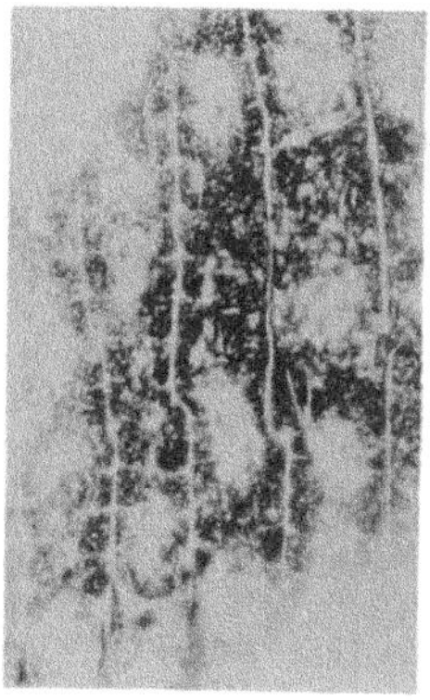

4

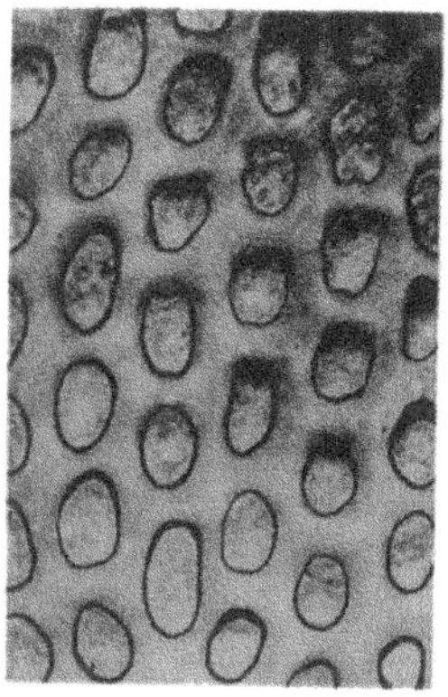

5

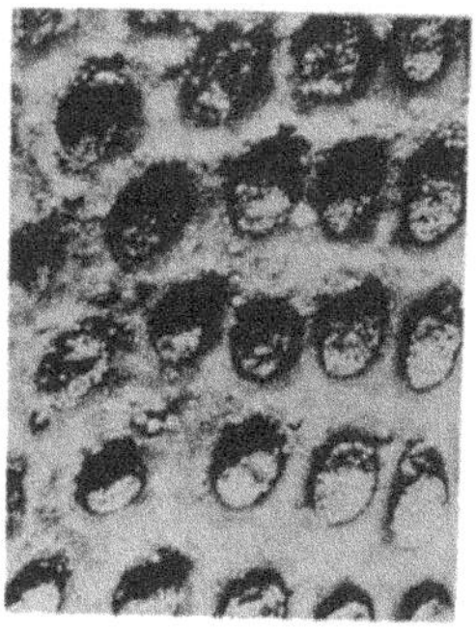

6

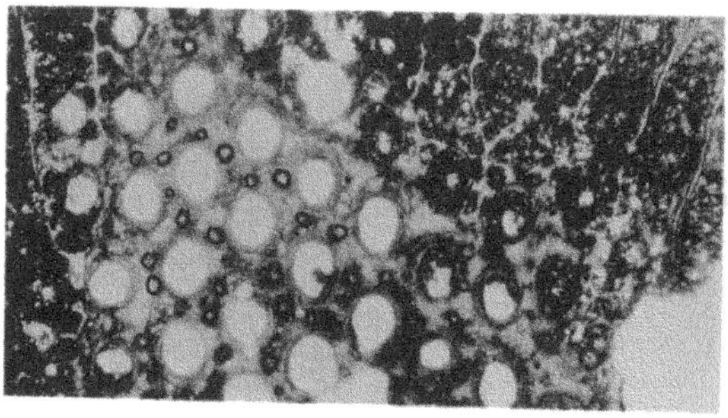

7

EXPLANATION OF PLATE VII

All specimens magnified 12 diameters

PAGE

FIG. 1. *Isotrypa hexagona*, sp. nov. 252
Reverse face of frond, showing sinuous carina on branches, and dissepiments stronger than branches.
Holotype No. 10467 U.M.
Zone 1, Location 17.

FIG. 2. *Isotrypa hexagona*, sp. nov. 252
Section cut oblique to plane of frond, showing superstructure, very thin carinae on obverse branches, and arrangement of zoœcia in branches.
Holotype No. 10467 U.M.
Zone 1, Location 17.

FIG. 3. *Isotrypa vibrata*, sp. nov. 253
Obverse face of frond, showing coarsely granulose appearance and regularity of superstructure.
Holotype No. 10466 U.M.
Zone 1, Location 17.

FIG. 4. *Isotrypa vibrata*, sp. nov. 253
Section cut oblique to plane of frond, showing structure of superstructure and anastomosed reverse branches.
Holotype No. 10466 U.M.
Zone 1, Location 17.

FIG. 5. *Isotrypa vibrata*, sp. nov. 253
Reverse face of frond, showing anastomosed parts of branches stronger than plane of normal branches and raised above them.
Holotype No. 10466 U.M.
Zone 1, Location 17.

FIG. 6. *Isotrypa oxytropis*, sp. nov. 254
Obverse face of frond, showing sharply carinated superstructure.
Holotype No. 10465 U.M.
Bell Shale, Location 31.

FIG. 7. *Isotrypa oxytropis*, sp. nov. 254
Reverse face of frond, showing large pores upon anastomosed parts of branches. Raised margins of pores give to branches a nodose appearance on first examination.
Holotype No. 10465 U.M.
Bell Shale, Location 31.

PLATE VII

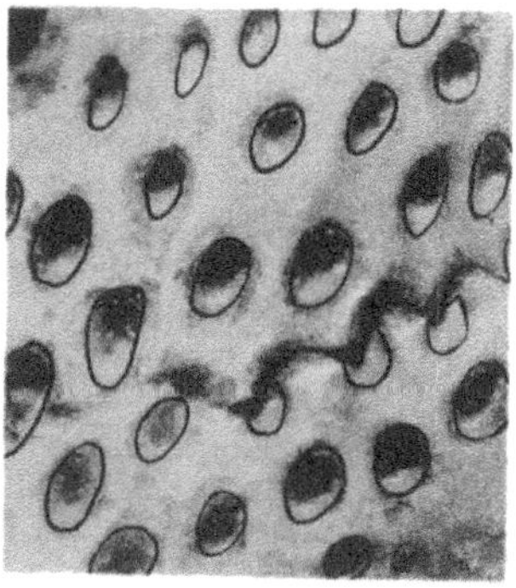

1

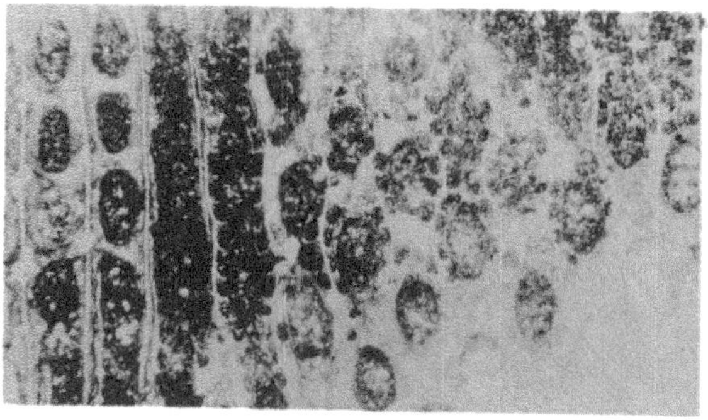

2

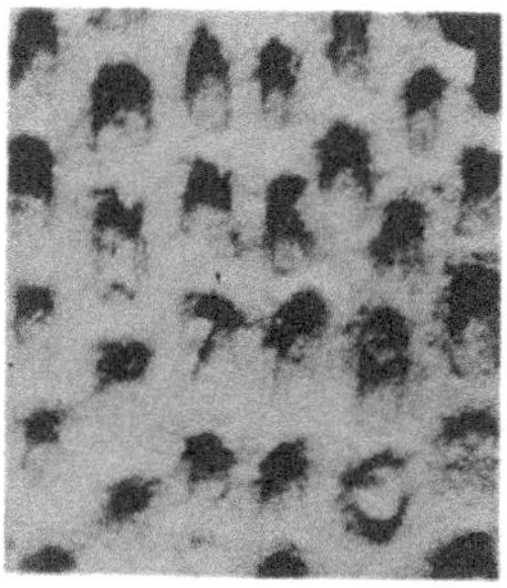

3

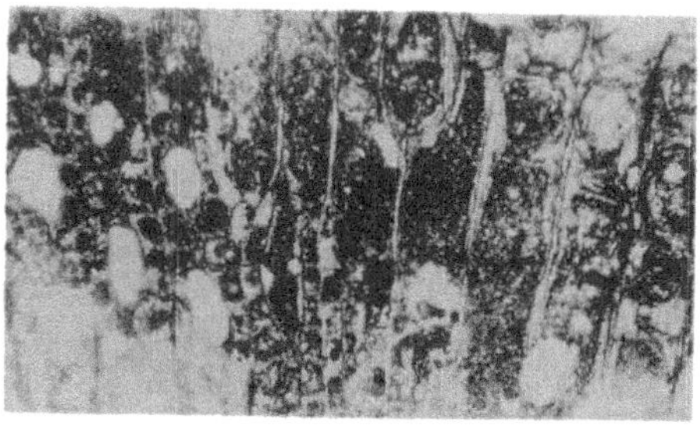

4

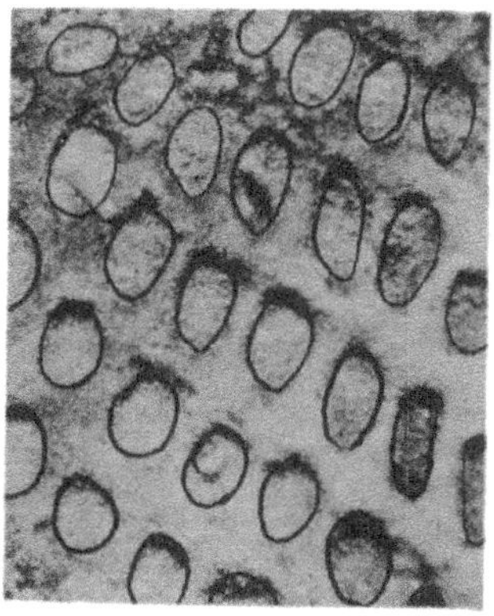

5

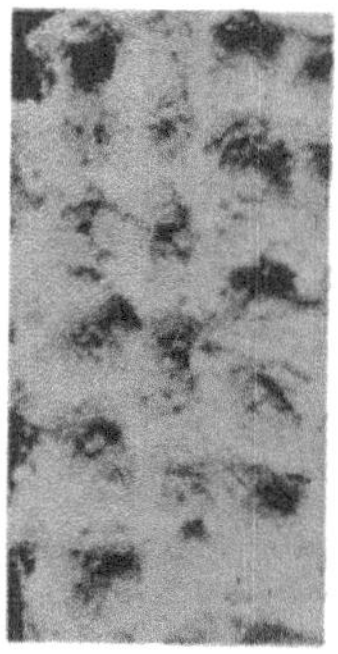

6

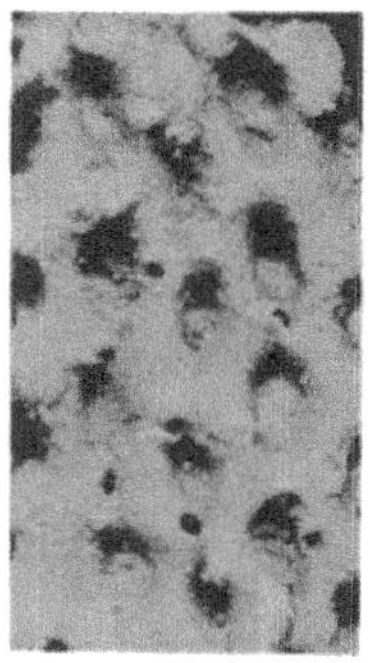

7

EXPLANATION OF PLATE VIII

All specimens magnified 12 diameters

PAGE

FIG. 1. *Isotrypa gigantica*, sp. nov. 254
Obverse face of frond, showing zigzag appearance of expanded summits of carinae forming superstructure.
Holotype No. 10500 U.M.
Zone 8, Location 40.

FIG. 2. *Isotrypa gigantica*, sp. nov. 254
Reverse face of frond, showing granulose appearance of branches and dissepiments.
Holotype No. 10500 U.M.
Zone 8, Location 40.

FIG. 3. *Isotrypa isopeda*, sp. nov. 255
Obverse face of frond, showing irregular appearance of carinae on surface of superstructure.
Holotype No. 10464 U.M.
Location 47.

FIG. 4. *Isotrypa isopeda*, sp. nov. 255
Section cut nearly parallel to plane of frond, showing superstructure.
Holotype No. 10464 U.M.
Location 47.

FIG. 5. *Isotrypa isopeda*, sp. nov. 255
Section cut oblique to plane of frond, showing carinae below superstructure.
Holotype No. 10464 U.M.
Location 47.

FIG. 6. *Phyllopora aequirotundata*, sp. nov. 256
Reverse face of frond, showing large, subcircular fenestrules and anastomosed branches.
Holotype No. 10499 U.M.
Zone 1, Location 17.

FIG. 7. *Phyllopora aequirotundata*, sp. nov. 256
Section cut oblique to plane of frond, showing minutely granulose appearance of reverse branches and arrangement of apertures upon obverse branches.
Holotype No. 10499 U.M.
Zone 1, Location 17.

PLATE VIII

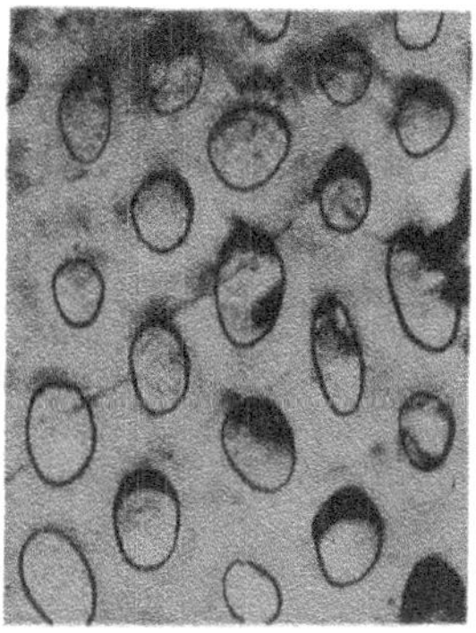

1

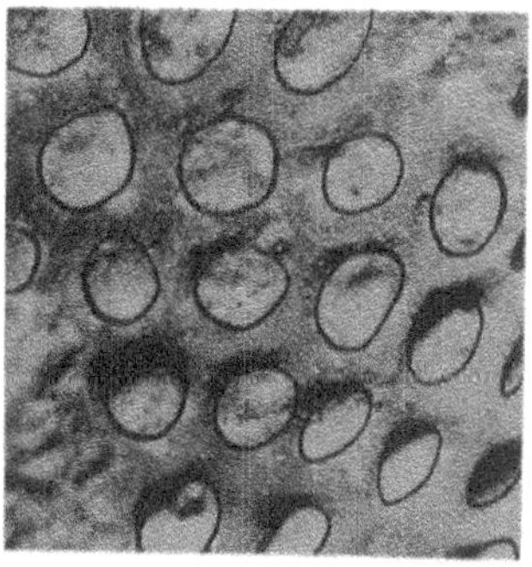

2

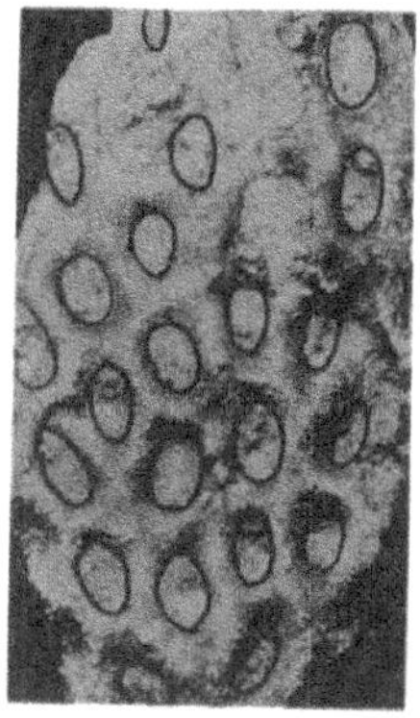

3

4

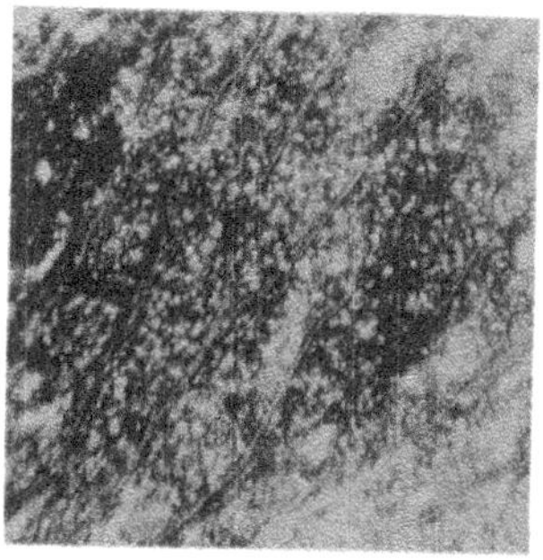

5

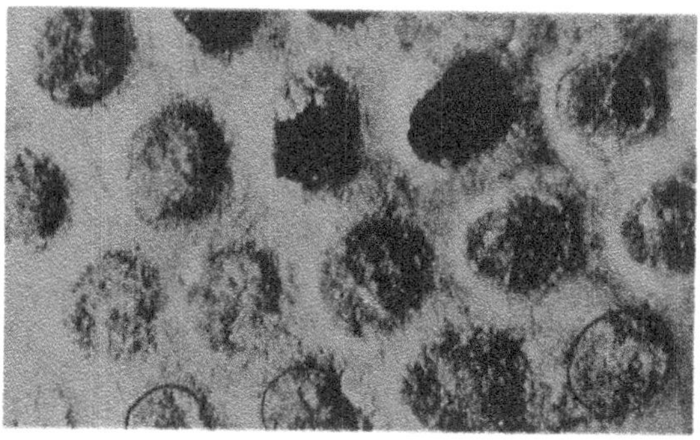

6

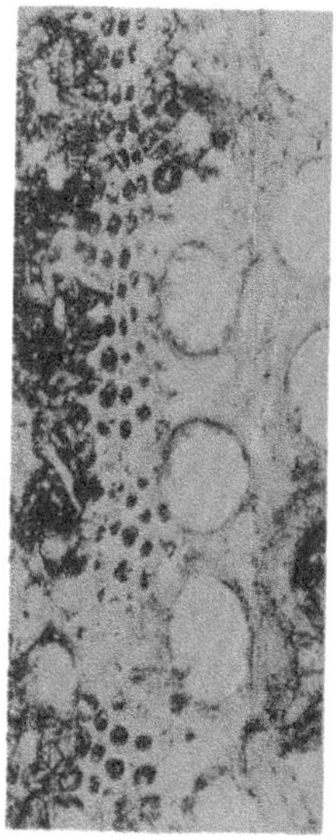

7

EXPLANATION OF PLATE IX

All specimens magnified 12 diameters

PAGE

FIG. 1. *Polypora ambiplana*, sp. nov. 256
Reverse face of frond, showing rounded branches with granular nodes occasionally disposed at junction of branches and dissepiments.
Holotype No. 10463 U.M.
Zone 1, Location 49.

FIG. 2. *Polypora ambiplana*, sp. nov. 256
Section cut slightly oblique to plane of frond, showing arrangement of apertures upon branches and regularity of size and shape of fenestrules on obverse face.
Holotype No. 10463 U.M.
Zone 1, Location 49.

FIG. 3. *Polypora finitima*, sp. nov. 257
Obverse and reverse face of frond. Surface of obverse face removed, showing arrangement of zooecia in branches. Reverse face shows gently rounded branches and regularity of size and shape of fenestrules.
Holotype No. 10506 U.M.
Bell Shale, Location 31.

FIG. 4. *Polypora amorpha*, sp. nov. 257
Reverse face of frond, showing both dissepiments and anastomosed portions of branches and variation in shape of fenestrules from circular to oval to subquadrangular.
Holotype No. 10460 U.M.
Zone 6, Location 29.

FIG. 5. *Polypora amorpha*, sp. nov. 257
Section cut nearly parallel to plane of frond, showing two dissepiments with apertures disposed upon them and distribution of ranges of apertures upon branches.
Holotype No. 10460 U.M.
Zone 6, Location 29.

FIG. 6. *Polypora magnifica*, sp. nov. 258
Reverse face of frond, showing branches and dissepiments faintly carinated and minutely granulose.
Holotype No. 10495 U.M.
Zone 1, Location 49.

PLATE IX

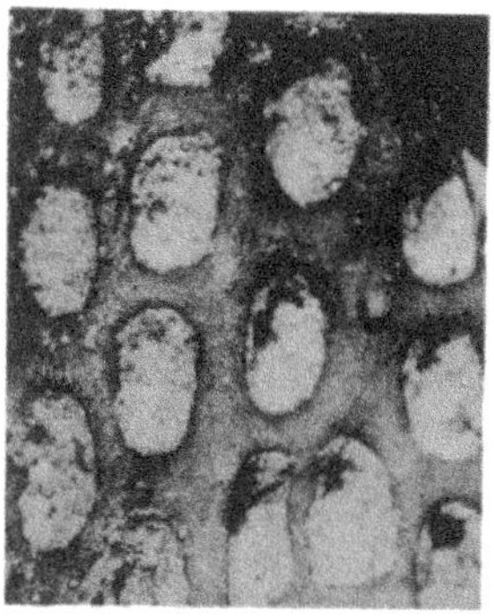

1

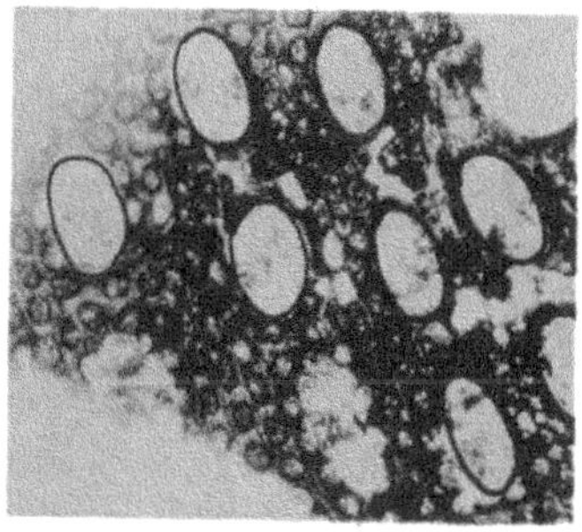

2

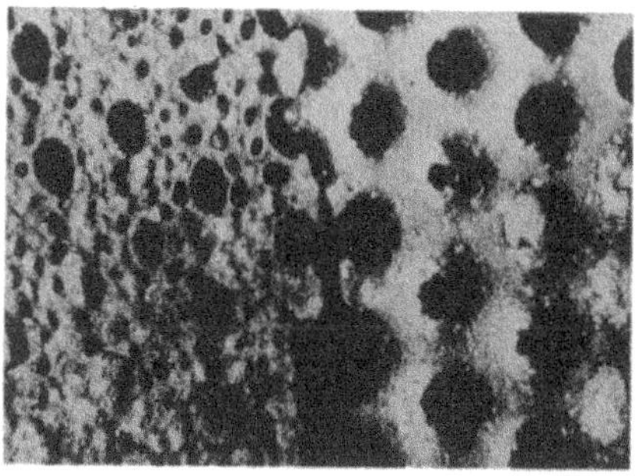

3

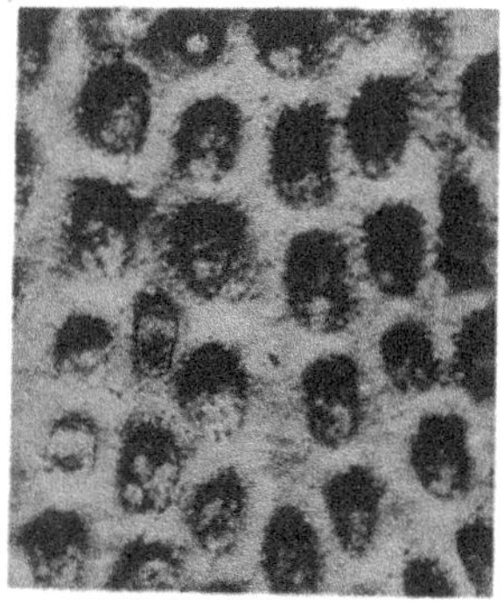

4

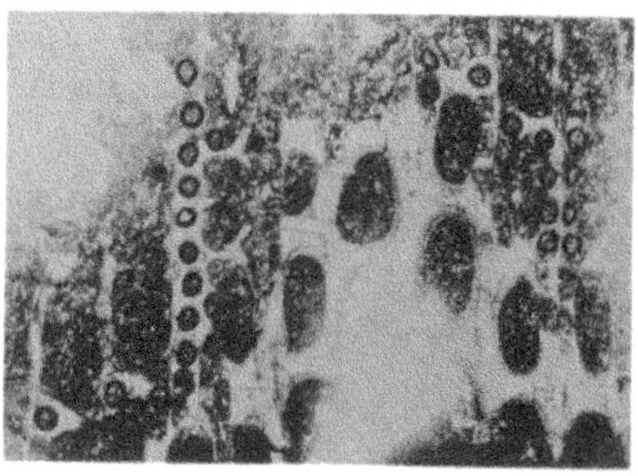

5

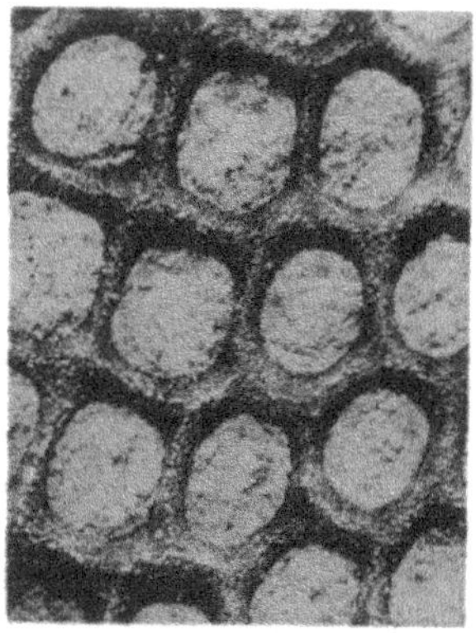

6

EXPLANATION OF PLATE X

All specimens magnified 12 diameters

PAGE

FIG. 1. *Polypora magnifica*, sp. nov. 258
Section cut oblique to plane of frond, showing arrangement of apertures on branches and structure of granules and carinae on reverse branches and dissepiments.
Holotype No. 10495 U.M.
Zone 1, Location 49.

FIG. 2. *Polypora brevissima*, sp. nov. 259
Reverse face of frond, showing great variation in width of branches and strong dissepiments.
Holotype No. 10494 U.M.
Zone 1, Location 49.

FIG. 3. *Polypora brevissima*, sp. nov. 259
Section cut nearly parallel to plane of frond, showing broad, flatly rounded and much depressed dissepiments.
Holotype No. 10494 U.M.
Zone 1, Location 49.

FIG. 4. *Polypora pyramidata*, sp. nov. 259
Reverse face of frond, showing angular branches and dissepiments.
Holotype No. 10493 U.M.
Zone 1, Location 49.

FIG. 5. *Polypora pyramidata*, sp. nov. 259
Section cut slightly oblique to plane of frond, showing granulose structure of reverse branches.
Holotype No. 10493 U.M.
Zone 1, Location 49.

FIG. 6. *Polypora indentata*, sp. nov. 260
Reverse face of frond, showing slender dissepiments broadly expanded at junction with branches, and striated appearance of worn parts of branches.
Holotype No. 10492 U.M.
Zone 6, Location 29.

FIG. 7. *Polypora indentata*, sp. nov. 260
Section cut slightly oblique to plane of frond, showing angular, depressed dissepiments on obverse face.
Holotype No. 10492 U.M.
Zone 6, Location 29.

FIG. 8. *Polypora uniplana*, sp. nov. 260
Reverse face of frond, showing coarsely striated appearance of branches. Specimen largely embedded in matrix.
Holotype No. 10491 U.M.
Middle shale zone, Location 14.

FIG. 9. *Polypora uniplana*, sp. nov. 260
Section cut slightly oblique to plane of frond, showing regularity of size and shape of fenestrules.
Holotype No. 10491 U.M.
Middle shale zone, Location 14.

PLATE X

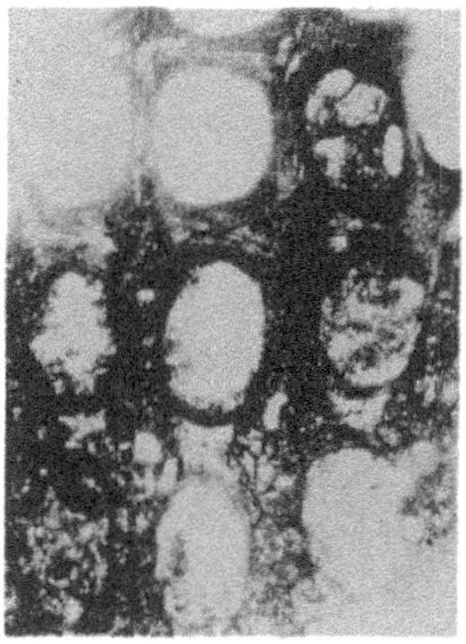

1

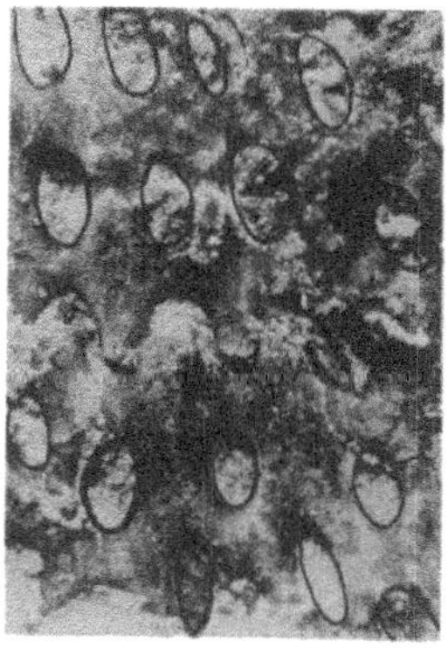

2

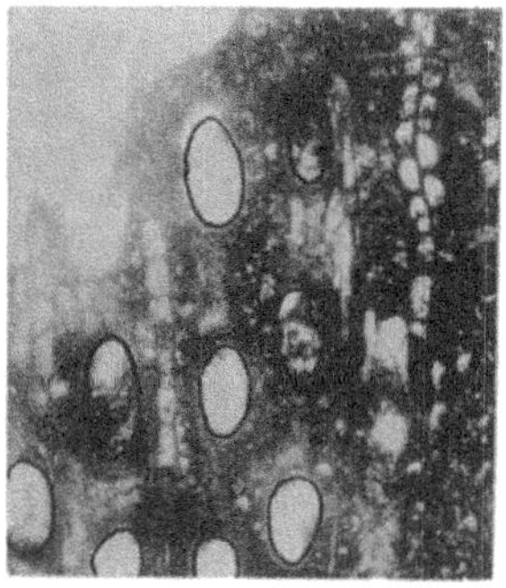

3

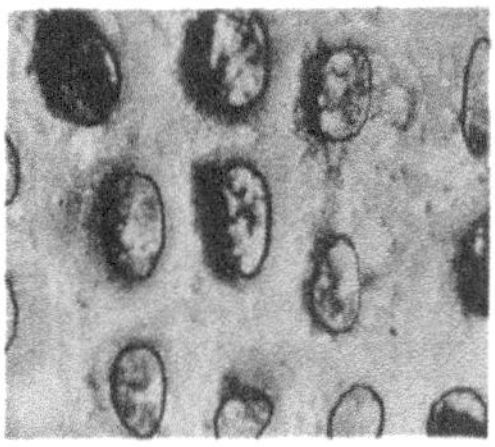

4

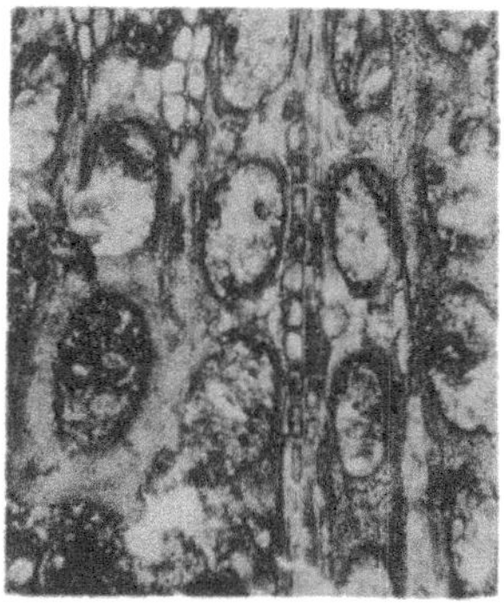

5

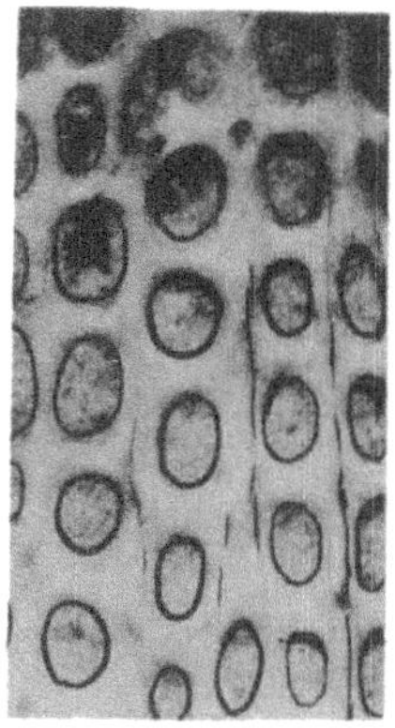

6

7

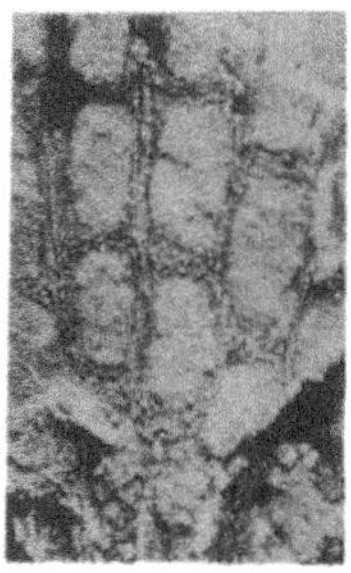

8

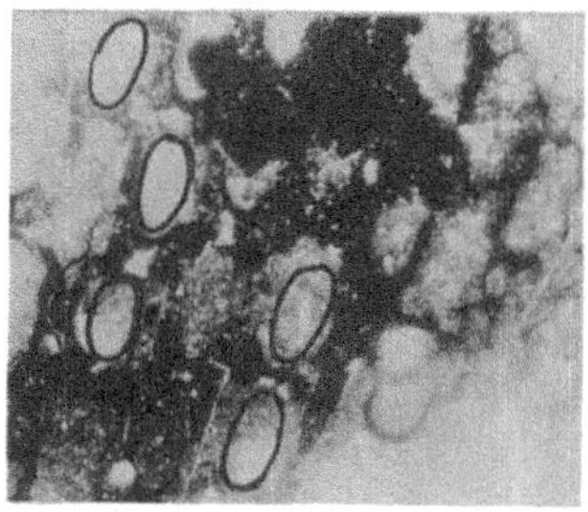

9

EXPLANATION OF PLATE XI

All specimens magnified 12 diameters

PAGE

FIG. 1. *Polypora minuta*, sp. nov. 261
Obverse face of frond, showing peristomes of lateral ranges of apertures indenting borders of fenestrules.
Holotype No. 10490 U.M.
Bell Shale, Location 31.

FIG. 2. *Polypora minuta*, sp. nov. 261
Reverse face of frond, showing rounded branches and dissepiments and variation in shape of fenestrules.
Holotype No. 10490 U.M.
Bell Shale, Location 31.

FIG. 3. *Polypora minuta*, sp. nov. 261
Section cut oblique to plane of frond, showing minutely granulose structure of reverse branches and dissepiments depressed below plane of obverse branches.
Holotype No. 10490 U.M.
Bell Shale, Location 31.

FIG. 4. *Polypora allelomorpha*, sp. nov. 261
Obverse face of frond, showing disposal of apertures upon branches and anastomosis of branches.
Holotype No. 10507 U.M.
Bell Shale, Location 31.

FIG. 5. *Polypora allelomorpha*, sp. nov. 261
Reverse face of frond, showing sharply rounded branches and absence of dissepiments owing to anastomosis of branches.
Holotype No. 10507 U.M.
Bell Shale, Location 31.

FIG. 6. *Polypora labellosa*, sp. nov. 262
Obverse face of frond, showing sharp peristomes.
Holotype No. 10462 U.M.
Bell Shale, Location 31.

FIG. 7. *Polypora labellosa*, sp. nov. 262
Reverse face of frond, showing minute, granulose nodes disposed upon branches and dissepiments, and regular arrangement of fenestrules.
Holotype No. 10462 U.M.
Bell Shale, Location 31.

FIG. 8. *Polypora tenuosa*, sp. nov. 263
Obverse face of frond, showing apertures widely disposed, both longitudinally and transversely on branches.
Holotype No. 10461 U.M.
Zone 1, Location 49.

FIG. 9. *Polypora tenuosa*, sp. nov. 263
Reverse face of frond, showing dissepiments on same plane as branches, and small size of branches and fenestrules.
Holotype No. 10461 U.M.
Zone 1, Location 49.

PLATE XI

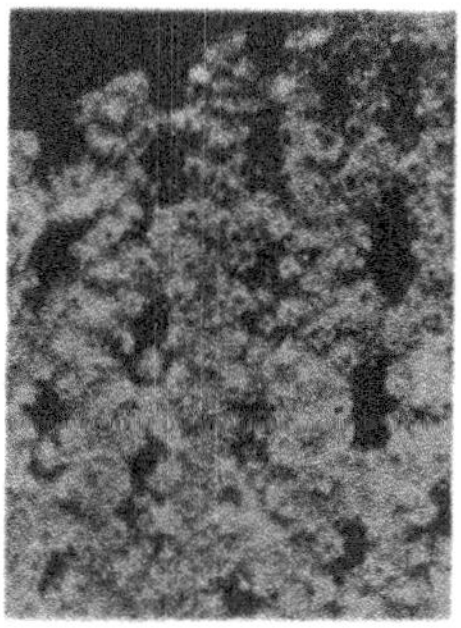

1

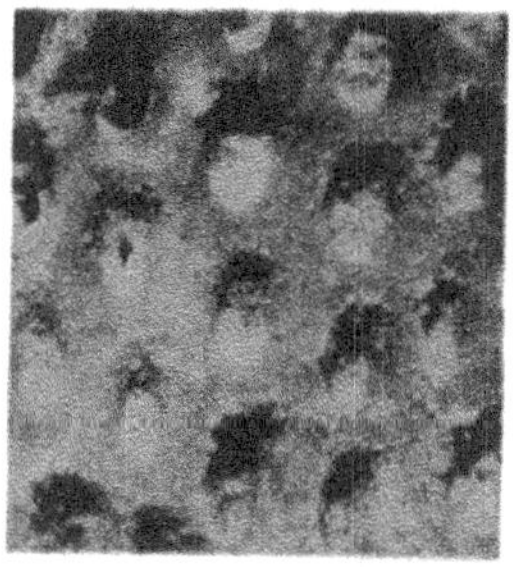

2

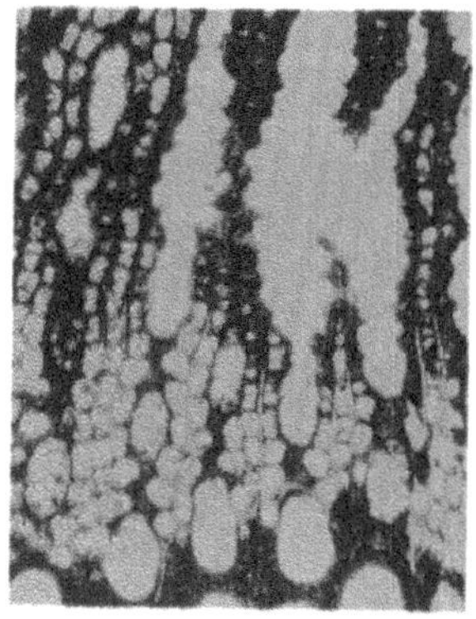

3

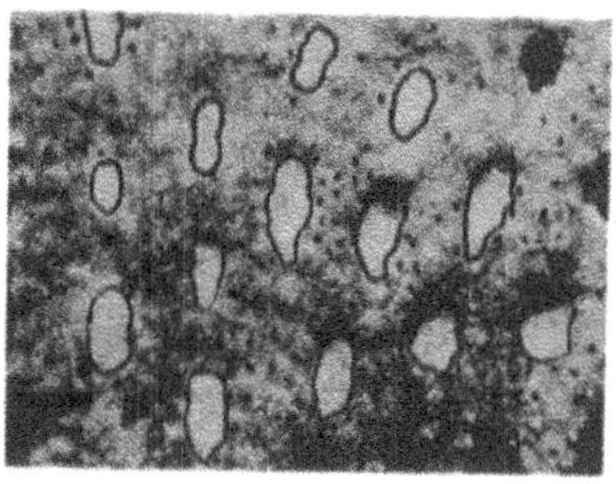

4

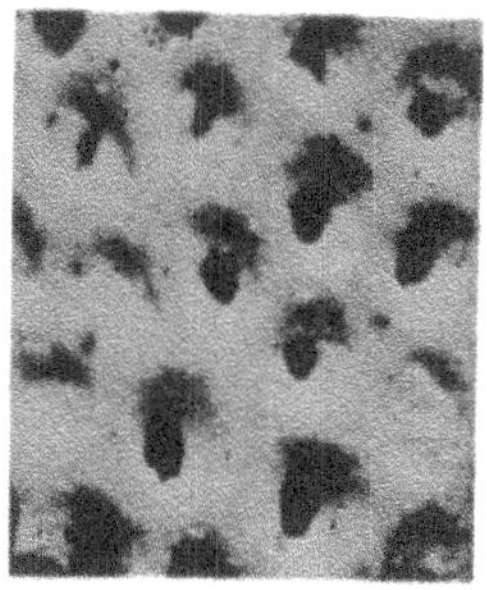

5

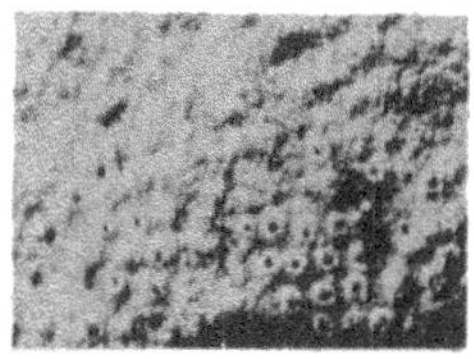

6

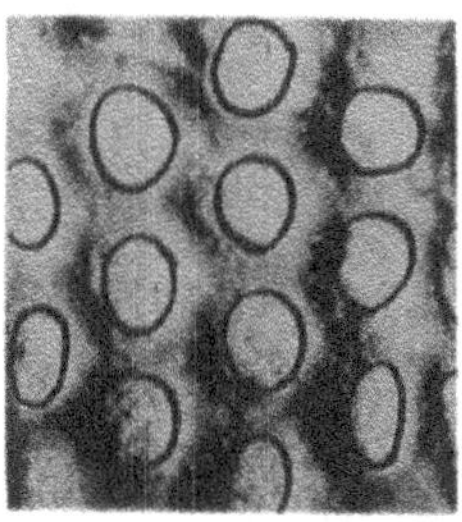

7

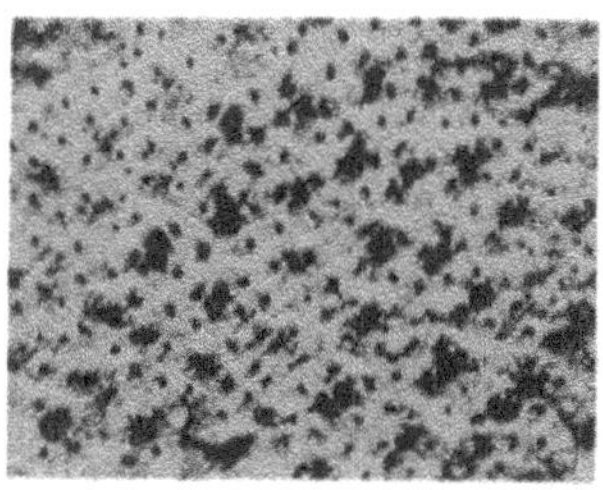

8

9

EXPLANATION OF PLATE XII

All specimens magnified 12 diameters

PAGE

FIG. 1. *Polypora exemplaria*, sp. nov. 263
Obverse face of frond, showing angularity of branches and absence of peristomes.
Holotype No. 10489 U.M.
Zone 1, Location 17.

FIG. 2. *Polypora exemplaria*, sp. nov. 263
Reverse face of frond, showing relationship of rounded branches and dissepiments.
Holotype No. 10489 U.M.
Zone 1, Location 17.

FIG. 3. *Polypora exemplaria*, sp. nov. 263
Section cut oblique to plane of frond, showing regularity in shape of fenestrules and in width of dissepiments.
Holotype No. 10489 U.M.
Zone 1, Location 17.

FIG. 4. *Polypora muricula*, sp. nov. 264
Obverse face of frond, showing distribution of apertures and irregularly disposed minute nodes.
Holotype No. 10488 U.M.
Zone 1, Location 17.

FIG. 5. *Polypora muricula*, sp. nov. 264
Section cut slightly oblique to plane of frond, showing granulose structure of reverse branches and subcircular fenestrules.
Holotype No. 10488 U.M.
Zone 1, Location 17.

FIG. 6. *Polypora muricula*, sp. nov. 264
Section cut nearly parallel to plane of frond, showing minute nodes disposed between ranges of apertures.
Holotype No. 10488 U.M.
Zone 1, Location 17.

FIG. 7. *Polypora modesta*, sp. nov. 264
Obverse face of frond, showing slender dissepiments depressed below plane of branches and peristomes with anterior edge more strongly developed.
Holotype No. 10496 U.M.
Shaly limestone, Location 35.

FIG. 8. *Polypora modesta*, sp. nov. 264
Section cut slightly oblique to plane of frond, showing carinae on reverse branches and dissepiments.
Holotype No. 10496 U.M.
Shaly limestone, Location 35.

FIG. 9. *Ptiloporina jugosa*, sp. nov. 265
Reverse face of frond, showing strong primary branch with secondary branches diverging laterally from it.
Holotype No. 10487 U.M.
Middle shale zone, Location 14.

PLATE XII

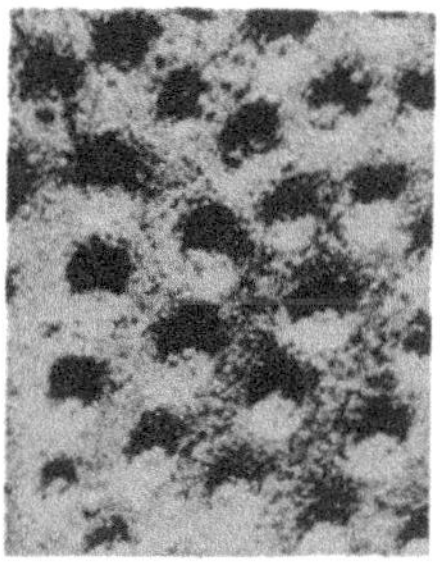
2

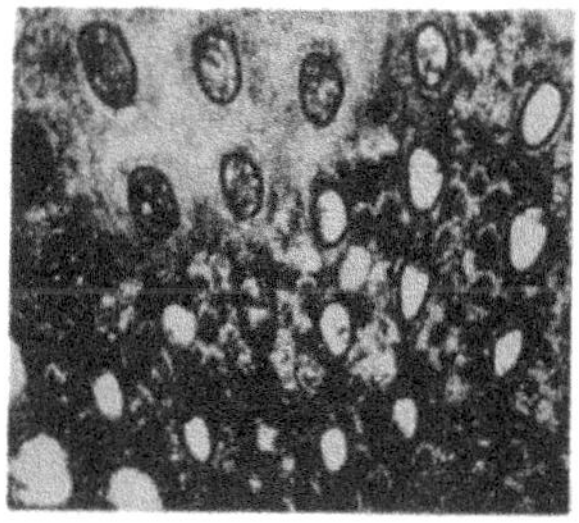
3

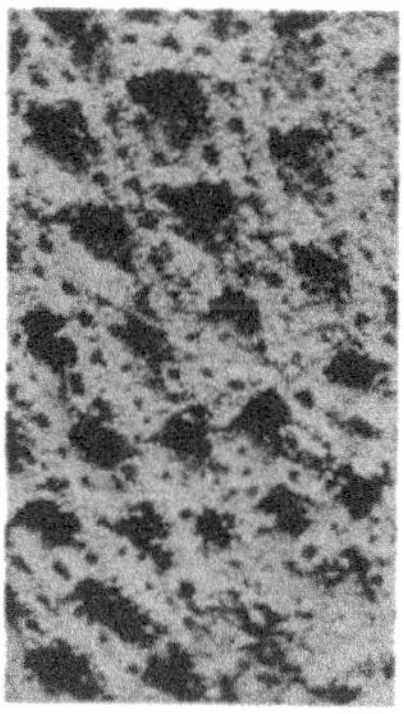
1

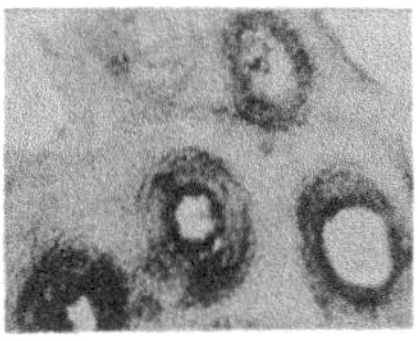
5

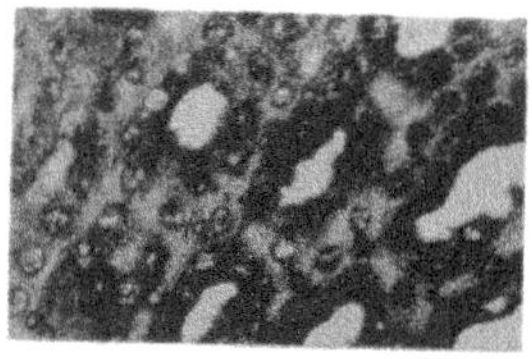
6

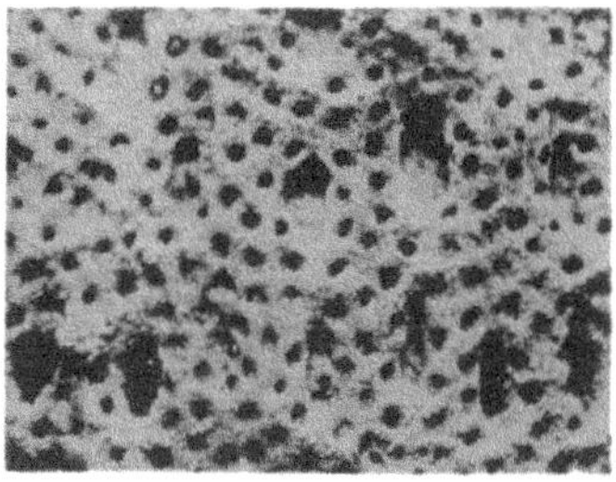
4

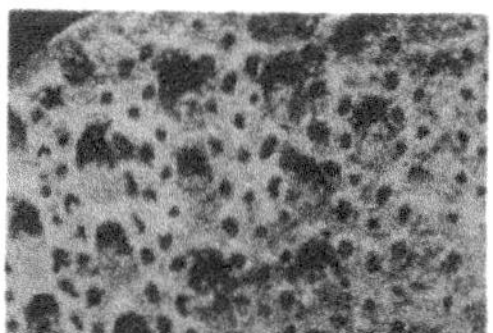
7

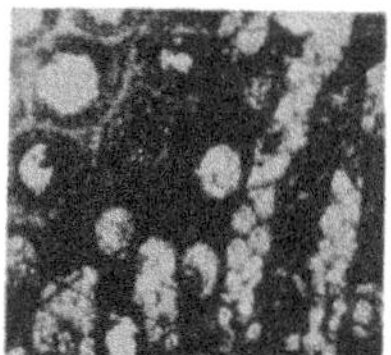
8

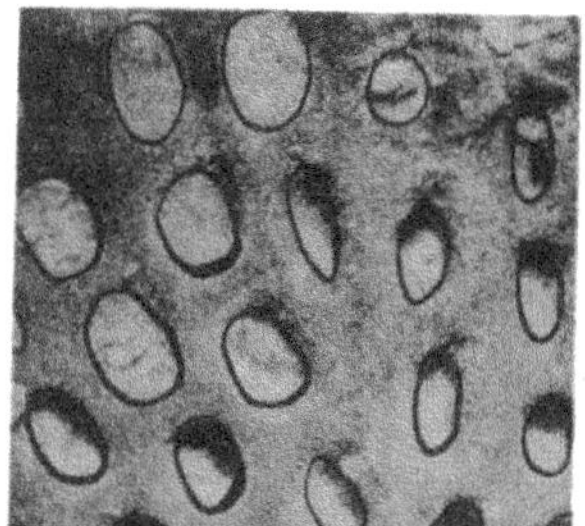
9

EXPLANATION OF PLATE XIII

All specimens magnified 12 diameters

PAGE

FIG. 1. *Semicosinium rhombicum* Ulrich var. *obliquum*, var. nov....... 266
Obverse face of frond, showing large accessory pores disposed upon a few dissepiments.
Holotype No. 10486 U.M.
Zone 1, Location 49.

FIG. 2. *Ptiloporina jugosa*, sp. nov.................................. 265
Section cut oblique to plane of branch, showing arrangement of apertures upon branches and sharp conical nodes regularly disposed upon summits of obverse branches.
Holotype No. 10487 U.M.
Middle shale zone, Location 14.

FIG. 3. *Semicosinium rhombicum* Ulrich var. *obliquum*, var. nov........ 266
Reverse face of frond, showing straight branches, comparatively strong dissepiments, and regularity in size and shape of fenestrules.
Holotype No. 10486 U.M.
Zone 1, Location 49.

FIG. 4. *Semicosinium thyene* Hall var. *michiganense*, var. nov......... 266
Reverse face of frond, showing faintly carinated branches, with granulose nodes disposed upon them.
Holotype No. 10485 U.M.
Location 46.

FIG. 5. *Semicosinium rhombicum* Ulrich var. *obliquum*, var. nov....... 266
Section cut oblique to plane of frond, showing anastomosis of branches and regularity in size of oval fenestrules on obverse face.
Holotype No. 10486 U.M.
Zone 1, Location 49.

FIG. 6. *Semicosinium thyene* Hall var. *michiganense*, var. nov......... 266
Section cut oblique to plane of frond, showing slightly expanded summit of carina, apertures opening directly, and regularity of size and shape of reverse fenestrules.
Holotype No. 10485 U.M.
Location 46.

PLATE XIII

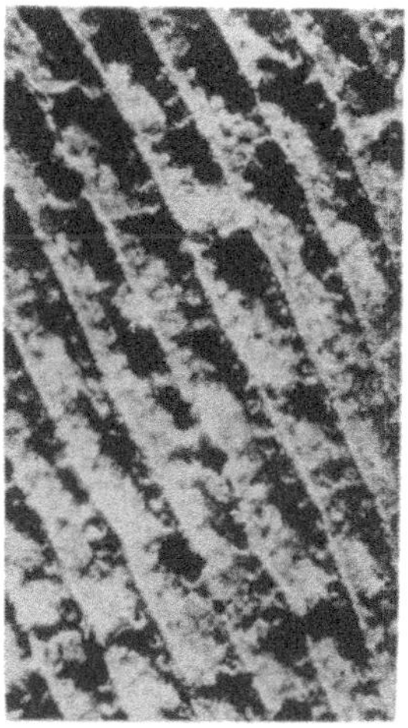

1

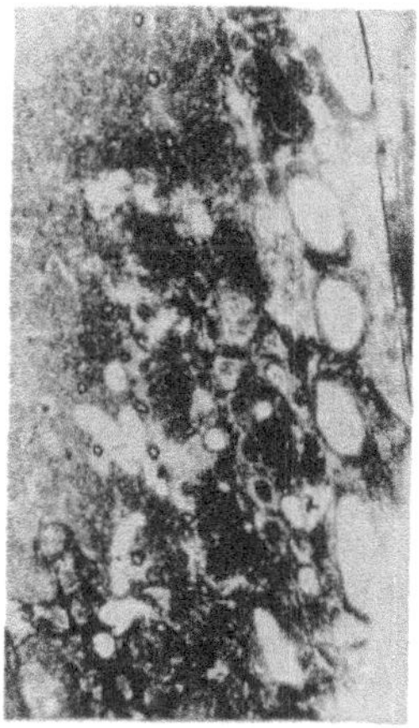

2

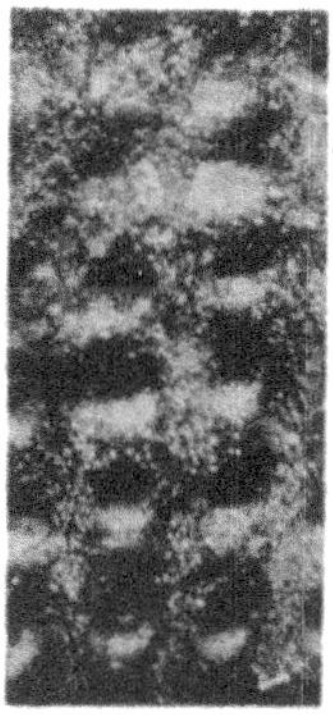

3

4

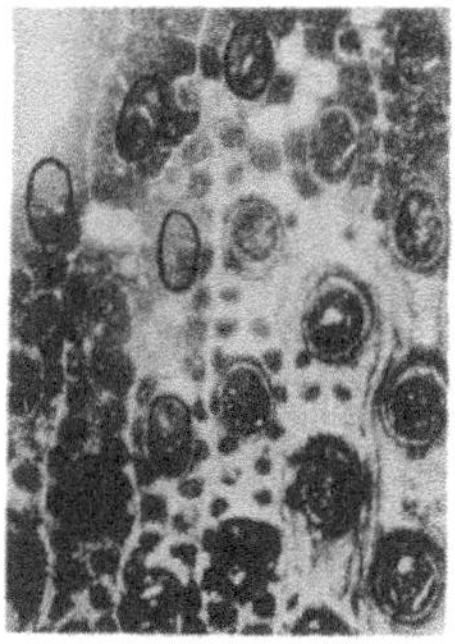

5

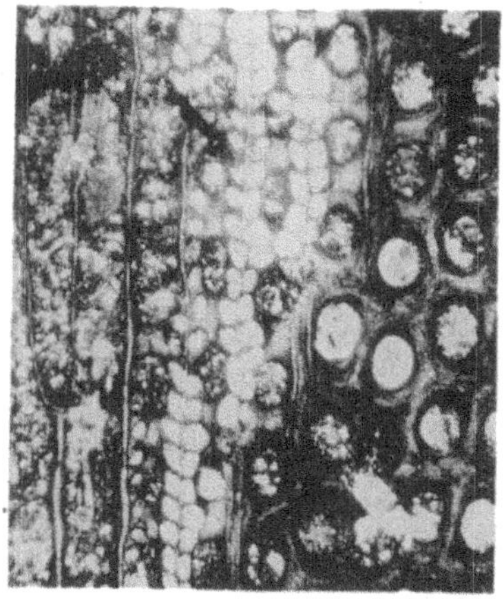

6

EXPLANATION OF PLATE XIV

All specimens magnified 12 diameters

PAGE

FIG. 1. *Semicosinium approximatum*, sp. nov. 267
Obverse face of frond, showing low, thin, sinuous carinae separating apertures, which open directly, and small regularly oval fenestrules.
Holotype No. 10484 U.M.
Shaly limestone, Location 35.

FIG. 2. *Semicosinium approximatum*, sp. nov. 267
Section cut slightly oblique to plane of frond, showing faintly carinated reverse branches and circular fenestrules.
Holotype No. 10484 U.M.
Shaly limestone, Location 35.

FIG. 3. *Semicosinium flexuosum*, sp. nov. 268
Reverse face of frond, showing rounded, zigzag branches and variation in size of fenestrules in younger and older portions of frond.
Holotype No. 10483 U.M.
Zone 1, Location 17.

FIG. 4. *Semicosinium flexuosum*, sp. nov. 268
Section cut slightly oblique to plane of frond, showing flexuous carinae with slightly expanded summits.
Holotype No. 10483 U.M.
Zone 1, Location 17.

PLATE XIV

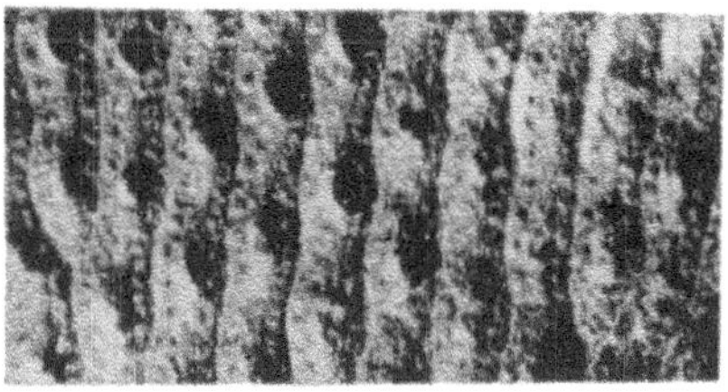

1

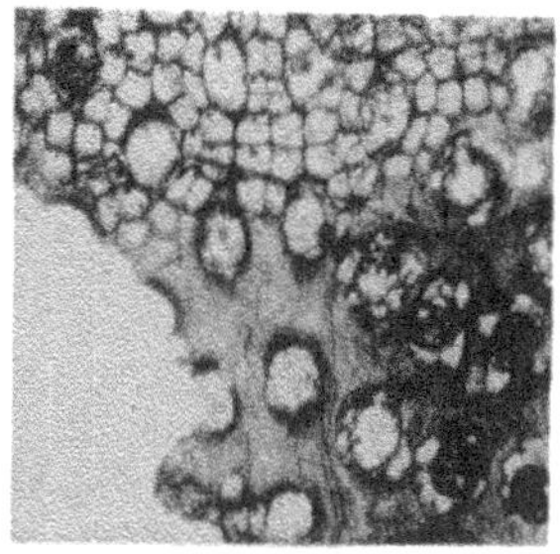

2

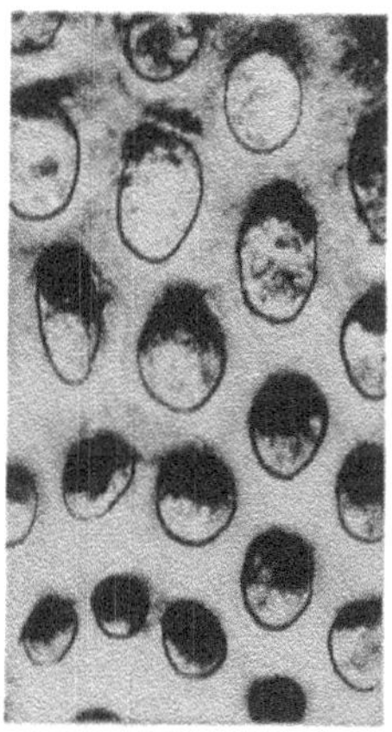

3

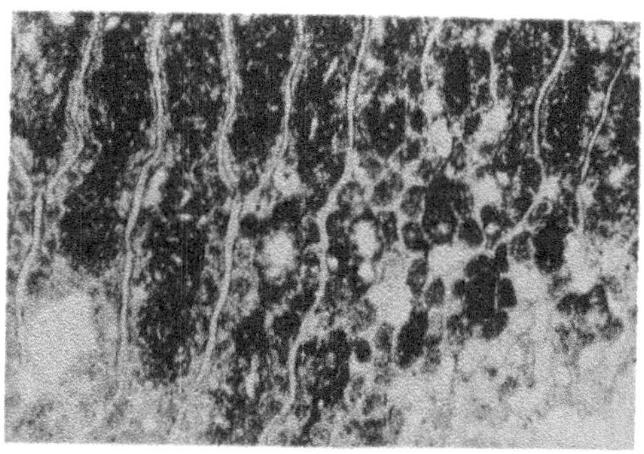

4

VOL. VII. ATHENIAN LEKYTHOI WITH OUTLINE DRAWING IN MATT COLOR ON A WHITE GROUND, AND AN APPENDIX: ADDITIONAL LEKYTHOI WITH OUTLINE DRAWING IN GLAZE VARNISH ON A WHITE GROUND. By Arthur Fairbanks. With 41 plates. Pp. x + 275. $3.50. Postage extra.

VOL. VIII. THE OLD TESTAMENT MANUSCRIPTS IN THE FREER COLLECTION. By Henry A. Sanders. With 9 plates showing pages of the Manuscripts in facsimile. Pp. viii + 357. $3.50. Postage extra.

Parts Sold Separately in Paper Covers:

Part I. THE WASHINGTON MANUSCRIPT OF DEUTERONOMY AND JOSHUA. With 3 folding plates. Pp. vi + 104. $1.25. Postage extra.

Part II. THE WASHINGTON MANUSCRIPT OF THE PSALMS. With 1 single plate and 5 folding plates. Pp. viii + 105–349. $2.00. Postage extra.

VOL. IX. THE NEW TESTAMENT MANUSCRIPTS IN THE FREER COLLECTION. By Henry A. Sanders. With 8 plates showing pages of the Manuscripts in facsimile. Pp. x + 323. $3.50. Postage extra.

Parts Sold Separately in Paper Covers:

Part I. THE WASHINGTON MANUSCRIPT OF THE FOUR GOSPELS. With 5 plates. Pp. vii + 247. $2.00. Postage extra.

Part II. THE WASHINGTON MANUSCRIPT OF THE EPISTLES OF PAUL. With 3 plates. Pp. ix + 251–315. $1.25. Postage extra.

VOL. X. THE COPTIC MANUSCRIPTS IN THE FREER COLLECTION. By William H. Worrell. With 12 plates. Pp. xxvi + 396. $4.75. Postage extra.

Parts Sold Separately in Paper Covers:

Part I. THE COPTIC PSALTER. The Coptic Text in the Sahidic Dialect, with an Introduction, and with 6 plates showing pages of the Manuscript and Fragments in Facsimile. Pp. xxvi + 112. $2.00. Postage extra.

Part II. A HOMILY ON THE ARCHANGEL GABRIEL BY CELESTINUS, BISHOP OF ROME, AND A HOMILY ON THE VIRGIN BY THEOPHILUS, ARCHBISHOP OF ALEXANDRIA, FROM MANUSCRIPT FRAGMENTS IN THE FREER COLLECTION AND THE BRITISH MUSEUM. The Coptic Text with an Introduction and Translation, and with 6 plates showing pages of the Manuscripts in facsimile. Pp. 113–396. $2.50. Postage extra.

VOL. XI. CONTRIBUTIONS TO THE HISTORY OF SCIENCE. By Louis C. Karpinski and John G. Winter. With 11 plates. Pp. xi + 283. $3.50. Postage extra.

Parts Sold Separately:

Part I. ROBERT OF CHESTER'S LATIN TRANSLATION OF THE ALGEBRA OF AL-KHOWARIZMI. With an Introduction, Critical Notes, and an English Version. By Louis C. Karpinski, University of Michigan. With 4 plates showing pages of manuscripts in facsimile, and 25 diagrams in the text. Pp. vii + 164. $2.00. Postage extra.

Part II. THE PRODROMUS OF NICOLAUS STENO'S LATIN DISSERTATION CONCERNING A SOLID BODY ENCLOSED BY PROCESS OF NATURE

WITHIN A SOLID. Translated into English by John G. Winter, University of Michigan, with a Foreword by Professor William H. Hobbs With 7 plates. Pp. vii + 169–283. $1.30. Postage extra.

VOL. XII. STUDIES IN EAST CHRISTIAN AND ROMAN ART. By Charles R. Morey and Walter Dennison. With 67 plates (10 colored) and 91 illustrations in the text. Pp. xiii + 175. *(Out of print.)*

Parts Sold Separately:

Part I. EAST CHRISTIAN PAINTINGS IN THE FREER COLLECTION. By Charles R. Morey. With 13 plates (10 colored) and 34 illustrations in the text. Pp. xiii + 86. *(Out of print.)*

Part II. A GOLD TREASURE OF THE LATE ROMAN PERIOD. By Walter Dennison. With 54 plates and 57 illustrations in the text. Pp. 89–175. Bound in cloth. $2.50. Postage extra.

VOL. XIII. FRAGMENTS FROM THE CAIRO GENIZAH IN THE FREER COLLECTION. By Richard Gottheil, Columbia University, and William H. Worrell, University of Michigan. Text, with Translation and an Introduction. With 52 plates showing the different styles of writing in facsimile. Pp. xxxi + 273. Bound in cloth. $4.00. Postage extra.

VOL. XIV. TWO STUDIES IN LATER ROMAN AND BYZANTINE ADMINISTRATION. By Arthur E. R. Boak and James E. Dunlap, University of Michigan. Pp. x + 324. Bound in cloth. $2.25. Postage extra.

Parts Sold Separately in Paper Covers:

Part I. THE MASTER OF THE OFFICES IN THE LATER ROMAN AND BYZANTINE EMPIRES. By Arthur E. R. Boak. Pp. x + 160. $1.00. Postage extra.

Part II. THE OFFICE OF THE GRAND CHAMBERLAIN IN THE LATER ROMAN AND BYZANTINE EMPIRES. By James E. Dunlap. Pp. 164–324. $1.00. Postage extra.

VOL. XV. GREEK THEMES IN MODERN MUSICAL SETTINGS. By Albert A. Stanley, University of Michigan. With 10 plates. Pp. xxii + 385. *(Out of print.)*

Parts Sold Separately in Paper Covers:

Part I. INCIDENTAL MUSIC TO PERCY MACKAYE'S DRAMA OF SAPPHO AND PHAON. Pp. 1–68. $0.90 net.

Part II. MUSIC TO THE ALCESTIS OF EURIPIDES WITH ENGLISH TEXT. Pp. 71–120. $0.80 net.

Part III. MUSIC FOR THE IPHIGENIA AMONG THE TAURIANS BY EURIPIDES WITH GREEK TEXT. Pp. 123–214. $0.75 net.

Part IV. TWO FRAGMENTS OF ANCIENT GREEK MUSIC. Pp. 217–225. $0.30 net.

Part V. MUSIC TO CANTICA OF THE MENAECHMI OF PLAUTUS. Pp. 229–263. $0.60 net.

Part VI. ATTIS: A SYMPHONIC POEM. Pp. 265–384. $1.00 net.

VOL. XVI. NICOMACHUS OF GERASA: INTRODUCTION TO ARITHMETIC. Translated into English by Martin Luther D'Ooge, with Studies in

Greek Arithmetic by Frank Egleston Robbins and Louis C. Karpinski. Pp. vii + 318. (*Out of print.*)

VOLS. XVII–XX. ROYAL CORRESPONDENCE OF THE ASSYRIAN EMPIRE. Translated into English, with a transliteration of the text and a Commentary. By Leroy Waterman, University of Michigan.

VOL. XVII. Translation and Transliteration. Pp. x + 490. $4.50. Postage extra.

VOL. XVIII. Translation and Transliteration. Pp. iv + 524. $4.50. Postage extra.

VOL. XIX. Commentary. Pp. x + 377. $4.00. Postage extra.

VOL. XX. Supplement and Indexes. (*In preparation.*)

VOL. XXI. THE MINOR PROPHETS IN THE FREER COLLECTION AND THE BERLIN FRAGMENT OF GENESIS. By Henry A. Sanders and Carl Schmidt. With 7 plates. Pp. xiii + 436. $3.50. Postage extra.

VOL. XXII. A PAPYRUS CODEX OF THE SHEPHERD OF HERMAS. By Campbell Bonner, University of Michigan. (*In press.*)

VOL. XXIII. THE COMPLETE COMMENTARY OF OECUMENIUS ON THE APOCALYPSE: Now printed for the first time from Manuscripts at Messina, Rome, Salonika and Athos. By H. C. Hoskier. Pp. viii + 260. $4.00. Postage extra.

VOL. XXIV. ZENON PAPYRI IN THE UNIVERSITY OF MICHIGAN COLLECTION. By C. C. Edgar. Pp. xiv + 211. With 6 plates. $3.50. Postage extra.

VOL. XXV. KARANIS: TOPOGRAPHICAL AND ARCHITECTURAL REPORT OF EXCAVATIONS DURING THE SEASONS 1924–28. By A. E. R. Boak and E. Peterson. Pp. viii + 69. With 42 plates, 19 plans, and 1 map. $2.00. Postage extra.

VOL. XXVI. COPTIC SOUNDS. By William H. Worrell, University of Michigan.

Part I. THE MAIN CURRENTS OF THEIR HISTORY. Pp. x + 59. Bound in paper. $1.50. Postage extra.

Part II. (*In preparation.*)

VOL. XXVII. ATHENIAN FINANCIAL DOCUMENTS OF THE FIFTH CENTURY. By B. D. Meritt, University of Michigan. Pp. xiv + 192. $3.50. Postage extra.

FACSIMILES OF MANUSCRIPTS

Size, 40.5 × 35 cm.

FACSIMILE OF THE WASHINGTON MANUSCRIPT OF DEUTERONOMY AND JOSHUA IN THE FREER COLLECTION. With an Introduction by Henry A. Sanders. Pp. x; 201 heliotype plates. The University of Michigan. Ann Arbor, Michigan, 1910.

Limited edition, distributed only to Libraries, under certain conditions. A list of Libraries containing this Facsimile is printed in *University of Michigan Studies, Humanistic Series*, Volume VIII, pp. 351–353.

Size, 34 × 26 cm.

FACSIMILE OF THE WASHINGTON MANUSCRIPT OF THE FOUR GOSPELS IN THE FREER COLLECTION. With an Introduction by Henry A. Sanders. Pp. x;

372 heliotype plates and 2 colored plates. The University of Michigan. Ann Arbor, Michigan, 1912.

Limited edition, distributed only to Libraries, under certain conditions. A list of Libraries containing this Facsimile is printed in *University of Michigan Studies, Humanistic Series*, Volume IX, pp 317–320.

Size, 30.5 × 40.6 cm.

FACSIMILE OF THE WASHINGTON MANUSCRIPT OF THE MINOR PROPHETS IN THE FREER COLLECTION AND THE BERLIN FRAGMENT OF GENESIS. With an Introduction by Henry A. Sanders. With 130 plates. The University of Michigan. Ann Arbor, Michigan, 1927.

Limited edition, distributed only to Libraries, under certain conditions. A list of Libraries containing this Facsimile is printed in *University of Michigan Studies, Humanistic Series*, Volume XXI, pp. 431–434.

SCIENTIFIC SERIES

Size, 28 × 18.5 cm. 4°. Bound in Cloth.

VOL. I. THE CIRCULATION AND SLEEP. By John F. Shepard, University of Michigan. Pp. ix + 83, with an Atlas of 63 plates, bound separately. Text and Atlas, $2.50. Postage extra.

VOL. II. STUDIES ON DIVERGENT SERIES AND SUMMABILITY. By Walter B. Ford, University of Michigan. Pp. xi + 194. $2.50. Postage extra.

Size, 16 × 23.6 cm.

VOL. III. THE GEOLOGY OF THE NETHERLANDS EAST INDIES. By H. A. Brouwer. With 18 plates and 17 text figures. Pp. xii + 160. $3.00. Postage extra.

VOL. IV. THE GLACIAL ANTICYCLONES: THE POLES OF THE ATMOSPHERIC CIRCULATION. By William Herbert Hobbs. With 3 plates and 53 figures. Pp. xxiv + 198. $2.75. Postage extra.

VOLS. V–VIII. REPORTS OF THE GREENLAND EXPEDITIONS OF THE UNIVERSITY OF MICHIGAN (1926–31). W. H. Hobbs, Director.

VOL. V. Aërology, Expeditions of 1926 and 1927–29. With 23 plates and 30 text figures. Pp. x + 262. $6.00. Postage extra.

VOL. VI. Aërology, Expeditions of 1930–31. (*In preparation.*)

VOL. VII. Meteorology. (*In preparation.*)

VOL. VIII. Geology, Glaciology, Botany, etc. (*In preparation.*)

MEMOIRS OF THE UNIVERSITY OF MICHIGAN MUSEUMS

Size, 26 × 17 cm. 4°. Bound in Cloth.

VOL. I. THE WHIP SNAKES AND RACERS: GENERA MASTICOPHIS AND COLUBER. By A. I. Ortenburger, University of Oklahoma. With 36 plates and 64 text figures. Pp. xviii + 247. $6.00. Postage extra.

VOL. II. DESCRIPTION OF THE SKULL OF A NEW FORM OF PHYTOSAUR, WITH NOTES ON THE CHARACTERS OF DESCRIBED NORTH AMERICAN PHYTOSAURS. By E. C. Case, University of Michigan. With 7 plates and 24 text figures. Pp. vi + 56. $2.00. Postage extra.

Orders should be addressed to The Librarian, University of Michigan, Ann Arbor, Michigan. Postage extra.

UNIVERSITY OF MICHIGAN PUBLICATIONS

HUMANISTIC PAPERS

General Editor: EUGENE S. McCARTNEY

Size, 22.7 × 15.2 cm. 8°. Bound in Cloth.

The Life and Work of George Sylvester Morris. A Chapter in the History of American Thought in the Nineteenth Century. By Robert M. Wenley. Pp. xv + 332. $1.50. Postage extra.

Latin and Greek in American Education, with Symposia on the Value of Humanistic Studies, Revised Edition. Edited by Francis W. Kelsey. Pp. xiii + 360. $2.00. Postage extra.

The Menaechmi of Plautus. The Latin Text, with a Translation by Joseph H. Drake, University of Michigan. Pp. xi + 130. Paper covers. $0.60. Postage extra.

LANGUAGE AND LITERATURE

Vol. I. Studies in Shakespeare, Milton and Donne. By Members of the English Department of the University of Michigan. Pp. viii + 232. $2.50. Postage extra.

Vol. II. Elizabethan Proverb Lore in Lyly's 'Euphues' and in Pettie's 'Petite Pallace,' with Parallels from Shakespeare. By Morris P. Tilley. Pp. x + 461. $3.50. Postage extra.

Vol. III. The Social Mode of Restoration Comedy. By Kathleen M. Lynch. Pp. x + 242. $2.50. Postage extra.

Vol. IV. Stuart Politics in Chapman's 'Tragedy of Chabot.' By Norma D. Solve. Pp. x + 176. Cloth. $2.50. Postage extra.

Vol. V. El Libro del Cauallero Zifar: Part I, Text. By C. P. Wagner, University of Michigan. Pp. xviii + 532, with 9 plates. Cloth. $5.00. Postage extra.

Vol. VI. El Libro del Cauallero Zifar: Part II, Commentary. By C. P. Wagner. (*In preparation.*)

Vol. VII. Strindberg's Dramatic Expressionism. By C. E. W. L. Dahlström. Pp. xii + 242. Cloth. $2.50. Postage extra.

Vol. VIII. Essays and Studies in English and Comparative Literature. By Members of the English Department of the University of Michigan. Pp. viii + 231. Cloth. $2.50. Postage extra.

Vol. IX. Toward the Understanding of Shelley. By Bennett Weaver, University of Michigan. Pp. xii + 258. Cloth. $2.50. Postage extra.

HISTORY AND POLITICAL SCIENCE

(*The first three volumes of this series were published as "Historical Studies," under the direction of the Department of History. Volumes IV and V were published without numbers.*)

Vol. I. A History of the President's Cabinet. By Mary Louise Hinsdale. Pp. ix + 355. (*Out of print.*)

Orders should be addressed to The Librarian, University of Michigan, Ann Arbor, Michigan. Postage extra.

Vol. II. English Rule in Gascony, 1199–1259, with Special Reference to the Towns. By Frank Burr Marsh. Pp. xi + 178. Cloth. $1.25. Postage extra.

Vol. III. The Color Line in Ohio: A History of Race Prejudice in a Typical Northern State. By Frank Uriah Quillan. Pp. xvi + 178. Cloth. $1.50. Postage extra.

Vol. IV. The Senate and Treaties, 1789–1817. The Development of the Treaty-Making Functions of the United States Senate during Their Formative Period. By Ralston Hayden, University of Michigan. Pp. xvi + 237. Cloth. $1.50. Postage extra.

Vol. V. William Plumer's Memorandum of Proceedings in the United States Senate, 1803–1807. Edited by Everett Somerville Brown, University of Michigan. Pp. xi + 673. Cloth. $3.50. Postage extra.

Vol. VI. The Grain Supply of England during the Napoleonic Period. By W. F. Galpin, Syracuse University. Pp. xi + 305. Cloth. $3.00. Postage extra.

Vol. VII. Eighteenth Century Documents relating to the Royal Forests, the Sheriffs and Smuggling: Selected from the Shelburne Manuscripts in the William L. Clements Library. By Arthur Lyon Cross, University of Michigan. With 4 plates. Pp. xviii + 328. $3.00. Postage extra.

Vol. VIII. The Low Countries and the Hundred Years' War, 1326–1347. By Henry S. Lucas, University of Chicago. Pp. xviii + 696. Cloth. $4.00. Postage extra.

Vol. IX. The Anglo-French Treaty of Commerce of 1860 and the Progress of the Industrial Revolution in France. By A. L. Dunham, University of Michigan. Pp. xiv + 409. Cloth. $3.00. Postage extra.

Vol. X. The Youth of Erasmus. By A. Hyma. Pp. xii + 350. With 8 plates and 2 maps. $3.00. Postage extra.

CONTRIBUTIONS FROM THE MUSEUM OF PALEONTOLOGY

Vol. I. The Stratigraphy and Fauna of the Hackberry Stage of the Upper Devonian. By Carroll Lane Fenton and Mildred Adams Fenton. With 45 plates, 9 text figures and 1 map. Pp. xi + 260. Cloth. $2.75. Postage extra.

Vol. II. Consisting of 14 papers. With 41 plates, 39 text figures and 1 map. Pp. ix + 240. Cloth. $3.00. Postage extra.

Parts Sold Separately in Paper Covers:

No. 1. A Possible Explanation of Fenestration in the Primitive Reptilian Skull, with Notes on the Temporal Region of the Genus Dimetrodon, by E. C. Case. Pp. 1–12, with five illustrations. $0.30.

No. 2. Occurrence of the Collingwood Formation in Michigan, by R. Ruedemann and G. M. Ehlers. Pp. 13–18. $0.15.

No. 3. Silurian Cephalopods of Northern Michigan, by Aug. F. Foerste, Pp. 19–86, with 17 plates. $1.00.

No. 4. A Specimen of *Stylemys nebrascensis* Leidy, with the Skull Preserved, by E. C. Case. Pages 87–91, with 7 text figures. Price, $0.20.

Orders should be addressed to The Librarian, University of Michigan, Ann Arbor, Michigan. Postage extra.

No. 9. Arthrodiran Remains from the Devonian of Michigan, by E. C. Case. Pages 163–181, with 5 plates and 13 text figures. Price, $0.35.

No. 10. Life Models of the Heads of Two Types of Phytosaurs, by E. C. Case. Pages 183–185, with 1 plate. Price, $0.20.

No. 11. Description of a New Species of *Buettneria*, with a Discussion of the Brain Case, by E. C. Case. Pages 187–206, with 3 plates and 11 text figures. Price, $0.35.

No. 12. On *Callixylon Newberryi* (Dawson) Elkins et Wieland, by Chester A. Arnold. Pages 207–232, with 7 plates and 9 text figures. Price, $0.50.

No. 13. A Description and Stratigraphic Correlation of the Fenestellidae from the Devonian of Michigan, by Charles F. Deiss. Pages 233–275, with 14 plates. Price, $0.60.

VOLUME IV

No. 1. A Collection of Stegocephalians from Scurry County, Texas, by E. C. Case. Pages 1–56, with 7 plates and 45 text figures. Price, $0.60.

No. 2. A Perfectly Preserved Segment of the Armor of a Phytosaur, with Associated Vertebrae, by E. C. Case. Pages 57–80, with 8 plates and 6 text figures. Price, $0.30.

No. 3. On the Caudal Region of *Coelophysis Sp.* and on Some New or Little Known Forms from the Upper Triassic of Western Texas, by E. C. Case. Pages 81–91, with 11 text figures. Price, $0.20.

No. 4. *Cylindrophyllum panicum* (Winchell) and *Cylindrophyllum hindshawi*, Sp. Nov., Tetracoralla from the Traverse Group of Michigan, by G. M. Ehlers and T. E. White. Pages 93–100, with 5 plates. Price, $0.20.

No. 5. *Gypidula petoskeyensis*, Sp. Nov., a New Brachiopod from the Traverse Group of Michigan, by R. W. Imlay. Pages 101–103, with 1 plate. Price, $0.15.

ARCHAEOLOGICAL REPORT

PRELIMINARY REPORT UPON THE EXCAVATIONS AT TEL UMAR, IRAQ, CONDUCTED BY THE UNIVERSITY OF MICHIGAN AND THE TOLEDO MUSEUM OF ART. Leroy Waterman, Director. With 13 plates and 7 text figures. Pp. x + 62. $1.50. Bound in paper.

UNIVERSITY OF MICHIGAN COLLECTIONS

CATALOGUE OF THE STEARNS COLLECTION OF MUSICAL INSTRUMENTS (Second edition). By Albert A. Stanley. With 40 plates. Pp. 276. $4.00.

PAPERS OF THE MICHIGAN ACADEMY OF SCIENCE, ARTS AND LETTERS

(Containing Papers submitted at Annual Meetings)

Editors: EUGENE S. McCARTNEY AND PETER OKKELBERG

Size, 24.2 × 16.5 cm. 8°. Bound in Cloth.

VOL. I (1921). With 38 plates, 1 text figure and 5 maps. Pp. xi + 424. $2.00. Postage extra.

VOL. II (1922). With 11 plates and 7 text figures. Pp. xi + 226. $2.00. Postage extra. Bound in paper, $1.50. Postage extra.

Orders should be addressed to The Librarian, University of Michigan, Ann Arbor, Michigan. Postage extra.

Printed and bound by CPI Group (UK) Ltd, Croydon, CR0 4YY

16/09/2026

14963492-0003